D1015133

Concrete Structures Reference Guide

The McGraw-Hill Engineering Reference Guide Series

This series makes available to professionals and students a wide variety of engineering information and data available in McGraw-Hill's library of highly acclaimed books and publications. The books in the series are drawn directly from this vast resource of titles. Each one is either a condensation of a single title or a collection of sections culled from several titles. The Project Editors responsible for the books in the series are highly respected professionals in the engineering areas covered. Each Editor selected only the most relevant and current information available in the McGraw-Hill library, adding further details and commentary where necessary.

Church · EXCAVATION PLANNING REFERENCE GUIDE

Gaylord and Gaylord · CONCRETE STRUCTURES REFERENCE GUIDE

Hicks · BUILDING SYSTEMS REFERENCE GUIDE

Hicks · CIVIL ENGINEERING CALCULATIONS REFERENCE GUIDE

Hicks · MACHINE DESIGN CALCULATIONS REFERENCE GUIDE

Hicks · PLUMBING DESIGN AND INSTALLATION REFERENCE GUIDE

Hicks · POWER GENERATION CALCULATIONS REFERENCE GUIDE

Hicks · POWER PLANT EVALUATION AND DESIGN REFERENCE GUIDE

Higgins · PRACTICAL CONSTRUCTION EQUIPMENT MAINTENANCE REFERENCE GUIDE

Johnson and Jasik · ANTENNA APPLICATIONS REFERENCE GUIDE

Markus and Weston · ESSENTIAL CIRCUITS REFERENCE GUIDE

Merritt · CIVIL ENGINEERING REFERENCE GUIDE

Ross · HIGHWAY DESIGN REFERENCE GUIDE

Rothbart · MECHANICAL ENGINEERING ESSENTIALS REFERENCE GUIDE

Woodson · HUMAN FACTORS REFERENCE GUIDE FOR ELECTRONICS AND COMPUTER PROFESSIONALS

Woodson · HUMAN FACTORS REFERENCE GUIDE FOR PROCESS PLANTS

Concrete Structures Reference Guide

Edited by

Edwin H. Gaylord, Jr.
Professor of Civil Engineering, Emeritus
University of Illinois, Urbana

Charles N. Gaylord (Deceased)
Professor of Civil Engineering, Emeritus
University of Virginia

Jeremy Robinson
Project Editor

McGraw-Hill Book Company
New York St. Louis San Francisco Auckland
Bogotá Hamburg London Madrid Mexico
Milan Montreal New Delhi Panama
Paris São Paulo Singapore
Sydney Tokyo Toronto

Library of Congress Cataloging-in-Publication Data

Concrete structures reference guide.

(The McGraw-Hill engineering reference guide series)
"The material in this volume has been published previously in Structural engineering handbook, 2nd ed., edited by Edwin H. Gaylord, Jr., and Charles N. Gaylord."
—Verso t.p.
Includes index.
1. Concrete construction—Handbooks, manuals, etc.
I. Gaylord, Edwin Henry. II. Gaylord, Charles N.
III. Robinson, Jeremy, date. IV. Gaylord, Edwin Henry. Structural engineering handbook. V. Series.
TA681.C755 1988 624.1'834 88-12708
ISBN 0-07-023067-6

1234567890 DOC/DOC 8921098

ISBN 0-07-023067-6

Printed and bound by R. R. Donnelley & Sons Company.

Information contained in this work has been obtained by McGraw-Hill, Inc. from sources belived to be reliable. However, neither McGraw-Hill nor its authors guarantees the accuracy or completeness of any information published herein and neither McGraw-Hill nor its authors shall be responsible for any errors, omissions or damages arising out of use of this information. This work is published with the understanding that McGraw-Hill and its authors are supplying information but are not attempting to render engineering or other professional services. If such services are required, the assistance of an appropriate professional should be sought.

Contents

Index follows the Appendix.

Contributors

David P. Billington *Professor of Civil Engineering, Princeton University (Thin-Shell Concrete Structures)*

Walter L. Dickey *Consulting Civil and Structural Engineer, Los Angeles, Cal. (Masonry Construction)*

Phil M. Ferguson *Professor Emeritus of Civil Engineering, University of Texas, Austin (Design of Reinforced-Concrete Structural Members)*

German Gurfinkel *Professor of Civil Engineering, University of Illinois, Urbana (Reinforced-Concrete Bunkers and Silos)*

T. Y. Lin *Professor Emeritus of Civil Engineering, University of California, Berkeley (Design of Prestressed Concrete Structural Members)*

Raymond C. Reese *Consulting Engineer, Toledo, Ohio (Design of Reinforced-Concrete Structural Members)*

Francis A. Vitolo *Former President, Corbetta Construction Company, Inc., White Plains, N.Y. (Concrete Construction Methods)*

Paul Zia *Professor of Civil Engineering, North Carolina State University (Design of Prestressed Concrete Structural Members)*

Preface

The second edition of Gaylord and Gaylord's *Structural Engineering Handbook* (McGraw-Hill, 1979), on which this book is based, provides engineers, architects, and students of civil engineering and architecture with an authoritative reference work on structural engineering by assembling in one volume a concise, up-to-date treatment of the planning, design, and construction of a variety of engineered structures.

The purpose of this abridged version is more modest: It is intended to present the fundamental concepts of the design of concrete structures in a concise manner. Each chapter in this Reference Guide has been taken in full from the *Structural Engineering Handbook*. For a more thorough treatment, however, the reader is urged to consult the *Handbook*.

The six sections have been written by eight authors chosen for their eminence and wide experience in specific areas of analysis, design, and construction. They have presented their material in ready-to-use form wherever possible. To this end, derivations of formulas are omitted in all but a few instances and many worked-out examples are given. Background information, descriptive matter, and explanatory material have been condensed and omitted. Because each section treats a subject which is broad enough to fill a book in itself, the authors have had to select that material which in their judgment is likely to be most useful to the greatest number of users. However, sources of additional material are noted for most of the topics which could not be treated in sufficient detail.

Edwin H. Gaylord, Jr.

Concrete Structures Reference Guide

Section 1

Design of Reinforced-Concrete Structural Members

RAYMOND C. REESE
Consulting Engineer, Toledo, Ohio

PHIL M. FERGUSON
Professor Emeritus of Civil Engineering, University of Texas, Austin

1. Concrete Concrete has a compressive strength (in standard 6- × 12-in. cylinders at 28 days) of about 3000 to 4000 psi, with a possible range from 2500 or less to 10,000 or 12,000 psi or more if special methods are used.

The tensile strength of concrete is roughly 10 percent of its compressive strength or, perhaps more precisely, $5\sqrt{f_c'}$, where f_c' is the 28-day cylinder compressive strength. For both direct tensile stress and diagonal tensile stresses of considerable magnitude that are induced by shears, steel reinforcement is ordinarily used to replace or supplement the strength of the concrete.

2. Reinforcement Steel reinforcement in the United States consists largely of bars, almost always deformed, in a dozen sizes, standardized by the American Society for Testing and Materials and the U.S. Department of Commerce (Table 1). These bars come in a variety of grades and strengths, the higher strengths usually being accompanied by a lesser ductility and vice versa. Properties of bars are covered by ASTM specifications, which are revised as need arises. Table 2 summarizes the main properties of the commonly available types.

Reinforcing bars are usually cut and bent in the detailed form at a fabricator's yard, tagged and delivered, ready for placement. In certain areas, the bars are delivered cut to length and are bent at the job site. Sometimes they are delivered in stock lengths of appropriate sizes and are cut and bent at the job site. When bars are prefabricated, the various shapes are designated by type numbers (Fig. 1).

Welded-wire fabric consists of sheets or rolls of mesh made by welding each intersection of crossing layers of the same or different gage wires at equal spacings each way (square mesh) or different spacings (rectangular mesh). Sheets or rolls may have a maximum width of 12 or 13 ft (depending upon the width of welding equipment available) by whatever length can be handled.

Welded-wire fabric is designated by two numbers and two letter-number combinations, such as 6 × 8–W8.0 × W4.0, where the first number gives the spacing in inches of the longitudinal wires and the second number gives the spacing of the transverse wires in inches. The first letter-number combination gives the type and area of the longitudinal wire; the second combination, the information on the transverse wire. The Wire Reinforc-

TABLE 1 Deformed Bar Designation Numbers, Nominal Weights, Nominal Dimensions, and Deformation Requirements

		Nominal dimensions*				Deformation requirements, in.		
Bar designation No.†	Nominal weight, lb/ft	Diameter, in.	Cross-sectional area, in.²	Perimeter, in.	Max OD,‡ in.	Max average spacing	Min average height	Max gap (chord of 12½% of nominal perimeter)
3§	0.376	0.375	0.11	1.178	7/16	0.262	0.015	0.143
4	0.668	0.500	0.20	1.571	9/16	0.350	0.020	0.191
5	1.043	0.625	0.31	1.963	11/16	0.437	0.028	0.239
6	1.502	0.750	0.44	2.356	7/8	0.525	0.038	0.286
7	2.044	0.875	0.60	2.749	1	0.612	0.044	0.334
8	2.670	1.000	0.79	3.142	1⅛	0.700	0.050	0.383
9	3.400	1.128	1.00	3.544	1¼	0.790	0.056	0.431
10	4.303	1.270	1.27	3.990	1 7/16	0.889	0.064	0.487
11	5.313	1.410	1.56	4.430	1⅝	0.987	0.071	0.540
14	7.65	1.693	2.25	5.32	1 15/16	1.185	0.085	0.648
18	13.60	2.257	4.00	7.09	2½	1.58	0.102	0.864

The above weights were adopted as standards by the CRSI in 1934. They have been approved through the U.S. Department of Commerce Simplified Practice Recommendation R26.

Sizes 14 and 18 are large bars generally not carried in regular stock. They are available by arrangement with the supplier.

*The nominal dimensions of a deformed bar are equivalent to those of a plain round bar having the same weight per foot as the deformed bar.

†Bar numbers are based on the number of eighths of an inch in the nominal diameter of the bars.

‡The maximum outside diameter including deformations may be important as when punching holes in structural-steel members to accommodate bars, or in fitting couplings, or in nesting or bundling bars. Exact dimensions vary among manufacturers. Tabulated values allow for deformations, longitudinal ribs, and out of round.

§While No. 3 bars are used for stirrups, column ties, etc., it is suggested that slab bars, in general, be No. 4 or larger.

ing Institute (WRI) established the letter-number designation which relates to the cross-sectional area of the wire. The letter W designates smooth wire and the letter D describes deformed wire. The number following the letters W or D is the cross-sectional area of the wire in hundredths of a square inch. A W8.0 wire is a smooth wire with a cross-sectional area of 0.08 in.²; a W4.0 wire has a cross-sectional area of 0.04 in.² There are three widely used styles of fabric, namely, 4, 6, and 10 gage, with the corresponding W number (for smooth fabric) of W4.0, W2.9, and W1.4. Commonly available styles of welded-wire fabric and their properties are given in Table 3.

Because fabric may be made of drawn wire, and even high-strength or deformed wire, and has welded cross wires, such fabric is permitted certain higher stresses in code specifications.

Some experimenting has been done with glass fibers, plastic threads, chopped wire, and similar reinforcements.

3. Specifications, Codes, and Standards The ASTM maintains specifications for the quality of all the items incorporated into reinforced concrete, for methods of mixing, and detailed methods for sampling, testing, and approving both the individual items and the finished concrete.

The American Concrete Institute publishes Building Code Requirements for Reinforced Concrete (ACI 318) and a large collection of guides and manuals, including the Detailing Manual (ACI 315) and the Manual of Inspection, SP-2.

The Portland Cement Association has aided, supplemented, and greatly extended the work of these groups. The Concrete Reinforcing Steel Institute has played a parallel role in the field of reinforcing steel.

4. Strength Design and Working-Stress Design ACI318 is written from the viewpoint of *strength design* but permits *working-stress* design.

In strength design loads are multiplied by *load factors* and member strengths by *capacity reduction factors*. Load factors depend on the type and combination of loads and are greater than 1, except for dead load when it is combined with wind load or earthquake load. Capacity reduction factors depend on the relative importance of the member in the structure, the expected quality control in construction, and the precision with which strength calculations represent particular responses (shear, moment, etc.).

TABLE 2 Physical Requirements for ASTM Deformed Reinforcing Bars

Type of steel and ASTM specification	Size Nos. inclusive	Grade	Yield[a] min, psi	Tensile[b] strength min, psi	Elongation in 8 in. min., %	Cold-bend test[c] 180°	
Billet steel A615[d]	3–11[e]	40	40,000[f]	70,000	Note *g*	Nos. 3, 4, 5:	$d = 4t$
						Nos. 6–11:	$d = 5t$
	3–18	60	60,000	90,000	Note *h*	Nos. 3, 4, 5:	$d = 4t$
						Nos. 6, 7, 8:	$d = 6t$
						Nos. 9, 10, 11:	$d = 8t$
Rail steel A616[d]	3–11	50	50,000[f]	80,000	Note *i*	Note *m*	
	3–11	60	60,000	90,000	Note *j*	Note *m*	
Axle steel A617[d]	3–11[e]	40	40,000[f]	70,000	Note *k*	Same as A615	
	3–11	60	60,000	90,000	Note *l*	Nos. 3–5:	$d = 4t$
						No. 6:	$d = 5t$
						Nos. 7–8:	$d = 6t$
						Nos. 9–11:	$d = 8t$

[a]Yield point for A615 Grade 40. Yield strength for all others.

[b]Tensile strength determined on a full-sized specimen except that Nos. 11, 14, and 18 may be turned down as follows: 8-in. test specimen, ¾ in. min diameter; 2-in. test specimen, 0.500 in. The ACI Building Code accepts only full-sized specimens.

[c]d = diameter of pin around which specimen is bent; t = nominal diameter of specimen. This test not required for Nos. 14 and 18 unless ordered in accordance with supplemental requirements of A615.

[d]Weldability not a part of the specification.

[e]Bar Nos. 7 to 11 may not be readily available in Grade 40.

[f]Plain rounds up to and including 1¼ in. diameter, in coils or cut lengths, shall be furnished under this specification in Grade 40 or Grade 60. (Weight for plain rounds smaller than ⅜ in. diameter shall be computed on the basis of the size in ASTM A510.) For bending properties, test provisions of nearest nominal diameter deformed bar size shall apply. Those requirements for deformations and marking shall not be applicable.

[g]Bar No. 3, 11 percent; 4 to 6, 12 percent; 7, 11 percent; 8, 10 percent; 9, 9 percent; 10, 8 percent; 11, 7 percent.

[h]Bar Nos. 3 to 6, 9 percent; 7 to 8, 8 percent; 9 to 18, 7 percent. (Elongation in 2 in., min, bar Nos. 11, 14, 18, 9 percent.)

[i]Bar No. 3, 6 percent; 4 to 6, 7 percent; 7, 6 percent; 8 to 11, 4.5 percent.

[j]Bar Nos. 3 to 6, 6 percent; 7, 5 percent; 8 to 11, 4.5 percent.

[k]Bar No. 3, 11 percent; 4 to 6, 12 percent; 7, 11 percent; 8, 10 percent; 9, 9 percent; 10, 8 percent; 11, 7 percent.

[l]Bar Nos. 3 to 7, 8 percent; 8 to 11, 7 percent.

[m]Bend tests not required on bars fabricated by the producer.

In working-stress design, load factors are unity and the resulting stresses in the concrete and the reinforcement are low enough to assure an adequate factor of safety and an approximately linear stress-strain relationship in both steel and concrete.

5. ACI Load and Reduction Factors According to ACI 318-77 the required strength U provided to resist dead load D and live load L must be at least equal to

$$U = 1.4D + 1.7L \tag{1a}$$

If a wind load W must be considered, the following combinations of D, L, and W should be investigated:

$$U = 0.75(1.4D + 1.7L + 1.7W) \tag{1b}$$
$$U = 0.9D + 1.3W \tag{1c}$$

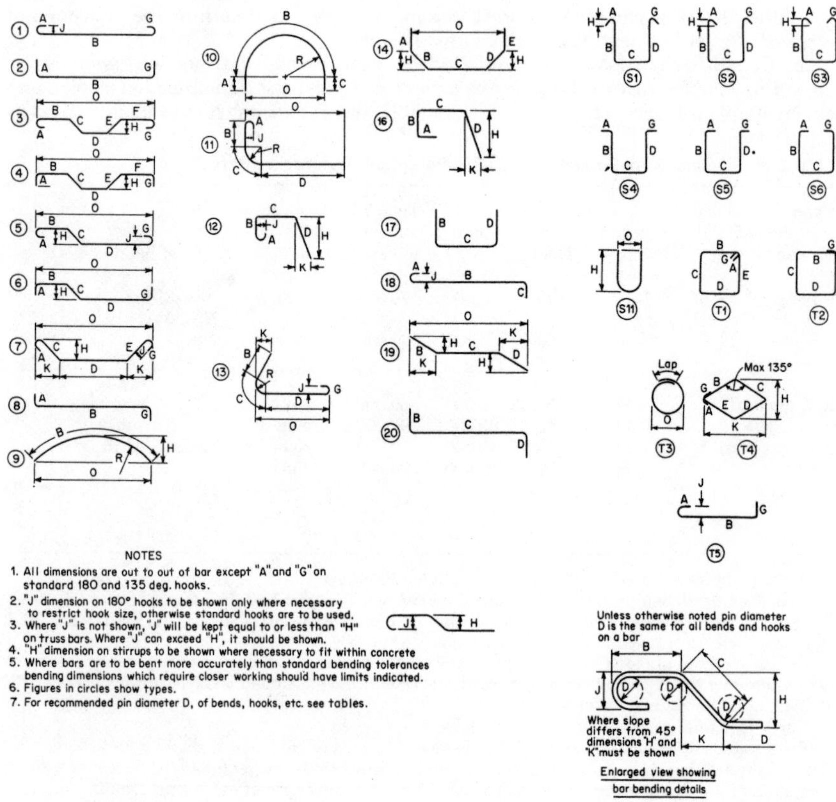

NOTES
1. All dimensions are out to out of bar except "A" and "G" on standard 180 and 135 deg. hooks.
2. "J" dimension on 180° hooks to be shown only where necessary to restrict hook size, otherwise standard hooks are to be used.
3. Where "J" is not shown, "J" will be kept equal to or less than "H" on truss bars. Where "J" can exceed "H", it should be shown.
4. "H" dimension on stirrups to be shown where necessary to fit within concrete
5. Where bars are to be bent more accurately than standard bending tolerances bending dimensions which require closer working should have limits indicated.
6. Figures in circles show types.
7. For recommended pin diameter D, of bends, hooks, etc. see tables.
Lap
Max 135°
Unless otherwise noted pin diameter D is the same for all bends and hooks on a bar
Where slope differs from 45° dimensions "H" and "K" must be shown
Enlarged view showing bar bending details

Fig. 1a Typical bar bends. (*From Ref. 4.*)

Equation (1*b*) must be checked for two cases: (1) L equal to its full value, and (2) L equal zero. In any case, the strength of the member or structure must not be less than required by Eq. (1*a*).

Resistance to earthquake loads E can be investigated by substituting $1.1E$ for W in Eqs. (1*b*) and (1*c*).

If lateral earth pressure H must be considered,

$$U = 1.4D + 1.7L + 1.7H \tag{1d}$$

If D or L reduce the effect of H, their coefficients must be taken as 0.90 for D and zero for L.

For lateral pressures F from liquids, the value $1.4F$ should be substituted for $1.7H$ in Eq. (1*d*). The vertical pressure of the liquid is considered as dead load.

Any impact effects are to be included with the live load L. When structural effects of differential settlement, creep, shrinkage, or temperature change may be significant,

$$U = 0.75(1.4D + 1.4T + 1.7L) \tag{1e}$$

but not less than

$$U = 1.4(D + T) \tag{1f}$$

Capacity reduction factors ϕ are given in Table 4.

6. Precision Calculations need only be made to the following recommended degrees of precision:

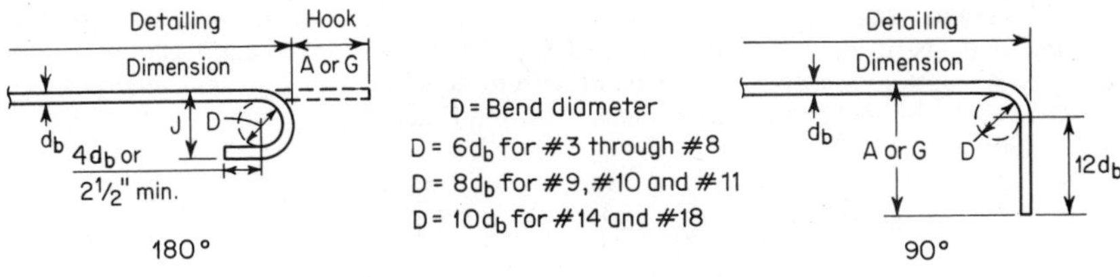

Bar size	Dimensions of standard 180 deg hooks, all grades			Dimensions of standard 90 deg hooks, all grades	
	A or G	J	D	A or G	D
#3	5"	3"	2 1/4"	6"	2 1/4"
#4	6	4	3	8	3
#5	7	5	3 3/4	10	3 3/4
#6	8	6	4 1/2	1'-0"	4 1/2
#7	10	7	5 1/4	1-2	5 1/4
#8	11	8	6	1-4	6
#9	1'-3"	11 1/4	9	1-7	9
#10	1-5	1'-0 3/4	10 1/4	1-10	10 1/4
#11	1-7	1-2 1/4	11 1/4	2-0	11 1/4
#14	2-2	1-8 1/2	17	2-7	17
#18	2-11	2-3	22 3/4	3-5	22 3/4

Note: When available depth is limited, #3 through #11 Grade 40 bars having 180-deg hooks may be bent with $D=5d_b$, and correspondingly smaller A and J dimensions.

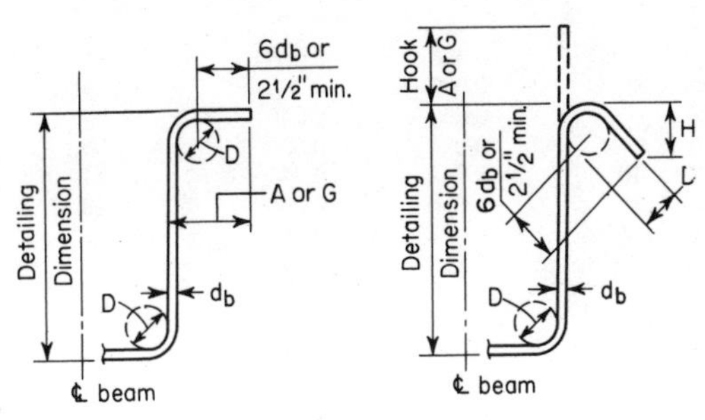

Stirrup hooks
(tie bends similar)

Stirrup and tie hook dimensions (in.)
Grades 40-50-60 ksi

Bar size	D	90° Hook	135° Hook	
		Hook A or G	Hook A or G	H approx.
#3	1 1/2	4	4	2 1/2
#4	2	4 1/2	4 1/2	3
#5	2 1/2	6	5 1/2	3 1/4

Note: 135-deg column tie hook may not be bent less than diameter of column vertical bar enclosed in hook.

Hooks and bends of welded wire fabric

Inside diameter of bends in welded wire fabric, plain or deformed, for stirrups and ties shall be at least four wire diameters for wire larger than D6 or W6 and two wire diameters for all other wires. Bends with inside diameter of less than eight wire diameters shall not be less than four wire diameters from nearest welded intersection.

Fig. 1b Standard hook details. (*From Ref. 4.*)

Loads to nearest 1 psf, 10 plf, 100-lb concentration
Span lengths to about 0.01 ft
Total loads and reactions to 0.1 kip
Moments to nearest 0.1 kip-in.
Individual bar areas to 0.01 in.^{2}
Concrete sizes to ½ in.
Bar spacings to ½ in. (supports are crimped at 1-in. intervals)
Effective depth of beam to 0.1 in.

7. Rectangular Beams Various shapes of compression block have been suggested for flexural analysis in the postelastic stage. ACI 10.2.6 permits any shape that predicts the ultimate strength in reasonable agreement with tests, while 10.2.7 permits an equivalent rectangle. For the rectangular stress block shown in Fig. 2 the following relationships can be established:

$$M_u = \phi f_y A_s \left(d - \frac{a}{2}\right) = \phi f_y \rho b d^2 \left(1 - 0.59 \frac{\rho f_y}{}\right) \tag{2}$$

where $a = \dfrac{f_y A_s}{0.85 f'_c b}$

Then, with $q = \rho f_y / f'_c$,

$$k_u = \frac{M_u}{\phi b d^2} = q f'_c (1 - 0.59q) \tag{3}$$

For balanced reinforcement

$$\rho_b = \frac{0.85 \beta_1 f'_c}{f_y} \frac{87{,}000}{87{,}000 + f_y} \tag{4}$$

where $\beta_1 = 0.85$ for $f'_c \leq 4000$ psi and decreases by 0.05 for each 1000 psi above 4,000. To

TABLE 3 Common Stock Styles of Welded-Wire Fabric

Style designation		Steel area, sq in./ft		Weight,
New designation (by W number)	Old designation (by steel wire gage)	Longit.	Trans.	approx. lb/100 sq ft
Rolls:				
6 × 6–W1.4 × W1.4	6 × 6–10 × 10	0.028	0.028	21
6 × 6–W2.0 × W2.0	6 × 6–8 × 8*	0.040	0.040	29
6 × 6–W2.9 × W2.9	6 × 6–6 × 6	0.058	0.058	42
6 × 6–W4.0 × W4.0	6 × 6–4 × 4	0.080	0.080	58
4 × 4–W1.4 × W1.4	4 × 4–10 × 10	0.042	0.042	31
4 × 4–W2.0 × W2.0	4 × 4–8 × 8*	0.060	0.060	43
4 × 4–W2.9 × W2.9	4 × 4–6 × 6	0.087	0.087	62
4 × 4–W4.0 × 4.0	4 × 4–4 × 4	0.120	0.120	85
Sheets:				
6 × 6–W2.9 × W2.9	6 × 6–6 × 6	0.058	0.058	42
6 × 6–W4.0 × W4.0	6 × 6–4 × 4	0.080	0.080	58
6 × 6–W5.5 × W5.5	6 × 6–2 × 2†	0.110	0.110	80
4 × 4–W4.0 × W4.0	4 × 4–4 × 4	0.120	0.120	85

*Exact W number size for 8 gage is W2.1.
†Exact W number size for 2 gage is W5.4.

TABLE 4 ACI Capacity Reduction Factors ϕ

Type of stress	ϕ
Axial tension, and bending with or without axial tension	0.90
Axial compression with or without bending:	
Members with spiral reinforcement	0.75*
Other reinforced members	0.70*
Shear and torsion	0.85
Bearing on concrete	0.70
Bending in plain concrete	0.65

*If $f_y \leq 60{,}000$ psi, these values may be increased linearly to 0.90 as P_u decreases from $0.10 f'_c A_g$ to zero for sections with symmetrical reinforcement and $(h\text{-}d'\text{-}d_s)/h \geq 0.70$. For sections with small axial compression not satisfying this requirement, these values may be increased linearly to 0.90 as P_u decreases from $0.10 f'_c A_g$ or P_b, whichever is smaller, to zero.

guarantee failure by yielding of the reinforcement, the reinforcement ratio ρ must not exceed $0.75\ \rho_b$.

The control of computations for flexure lies in k_u. Values are tabulated in Table 5, which starts with $\rho = 200/f_y$ in accordance with ACI 10.5.1 to ensure an underreinforced member and ends with $\rho = 0.75\ \rho_b$ in accordance with ACI 10.3.2.

Table 6 gives areas of groups of bars and Table 7 areas of bars in slabs.

The depth of a continuous beam ordinarily can be estimated at 1 in./ft of span. For the fairly heavy load and for the noncontinuous beam, this may have to be increased somewhat. Shear can usually be provided for by web reinforcement.

Example 1 Design a single-span, simply supported, rectangular reinforced-concrete beam 8 in. wide carrying a block wall between two shafts in a reinforced-concrete building whose floor-to-floor height is 10 ft 0 in. (Fig. 3). $f'_c = 4000$, $f_y = 60{,}000$ psi.

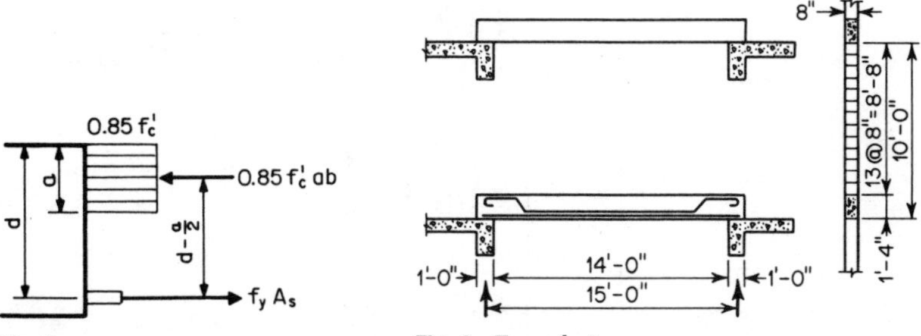

Fig. 2

Fig. 3 Example 1.

SOLUTION. Making the conservative assumptions of no arching of the wall, a wall weight of 51 psf, and a probable beam depth of 14 in. (i.e., 1 in. of depth per foot of span and, in this case, about two block courses), compute:

LOADS

8.83 ft of wall at 51 psf		= 450
1.17 ft of concrete at 100 psf		= 117
10.0 ft	Total	= 567 plf

According to Art. 5, this service load, all being dead, is to be multiplied by a load factor of 1.4; so

$$W = 15.0 \text{ ft at } 567 \times 1.4 = 11{,}910 \text{ lb}$$

and the simple-beam moment $WL/8 = M = 11{,}910 \times 15.0 \times 12/8 = 267{,}980$ in.-lb.

A fairly reasonable design results if the steel ratio is kept at about 1 percent, for which $k_u = M_u/\phi bd^2$ would be 547 (Table 5); so one can calculate

$$d = \sqrt{M_u/k_u\phi b} = \sqrt{267{,}980/(547 \times 0.9 \times 8)} = 8.23 \text{ in.}$$

Hence, try a beam depth of 12 in. with $d = 10$ in.; then $k_u = M_u/\phi bd^2 = 267{,}980/(0.9 \times 8 \times 10 \times 10) = 372$.

From Table 5, $\rho = 0.0066$,

$$A_s = \rho bd = 0.0066 \times 8 \times 10 = 0.53 \text{ sq. in.}$$

Use 2 No. 5 = 0.62 sq in. Table 8 gives minimum beam widths to accommodate various groupings of equal-sized bars. One bar is trussed up at each end to help resist diagonal tension.

8. Continuity ACI 318 requires that frames be analyzed by methods based upon elastic behavior, though it does permit a redistribution of from 10 to 20 percent of the negative moment at those supports where $(\rho - \rho') \leqq 0.5\rho_b$. Code 8.6 defines the permitted percentage as $20\,[1 - (\rho - \rho')/\rho_b]$ and requires that such redistributed moments be taken into account in computing positive moments within the spans. Any further application of plastic distribution or limit design is not recognized, except in slab systems.

Inherent in continuity is the fact that moments usually change sign within the span; reinforcement *must be* provided at *every* point where it is likely to be needed under any reasonable load pattern or load intensity. If the outline of the concrete changes, as from a T-section at midspan to a rectangular one at the support, or from a shallow depth at the

TABLE 5 Flexural-Strength Coefficients k_u for Rectangular Sections*

	f'_c = 3000 psi			4000 psi			5000 psi					
	f_y	40 ksi	60		40	60		40	60	Any $f'_c \leqq 5000$†		
$\omega = \rho f_y / f'_c$	k_u	ρ	ρ	k_u	ρ	ρ	k_u	ρ	ρ	c/d†	a/d	z/d
0.020	59	0.0015	0.0010	79	0.0020	0.0013	99	0.0025	0.0017	0.028	0.024	0.988
0.030	88	0.0023	0.0015	118	0.0030	0.0020	147	0.0038	0.0025‡	0.042	0.035	0.982
0.040	117	0.0030	0.0020	156	0.0040	0.0027‡	195	0.0050	0.0033	0.056	0.047	0.976
0.050	146	0.0038	0.0025	194	0.0050	0.0033	243	0.0063	0.0042	0.069	0.059	0.971
0.060	174	0.0045	0.0030	232	0.0060	0.0040	289	0.0075	0.0050	0.083	0.071	0.965
0.070	201	0.0053	0.0035	268	0.0070	0.0047	336	0.0088	0.0058	0.097	0.083	0.959
0.080	229	0.0060	0.0040	305	0.0080	0.0053	381	0.0100	0.0067	0.111	0.094	0.953
0.090	256	0.0068	0.0045	341	0.0090	0.0060	426	0.0113	0.0075	0.125	0.106	0.947
0.100	282	0.0075	0.0050	376	0.0100	0.0067	470	0.0125	0.0083	0.139	0.118	0.941
0.110	309	0.0083	0.0055	411	0.0110	0.0073	514	0.0138	0.0092	0.153	0.130	0.935
0.120	335	0.0090	0.0060	446	0.0120	0.0080	558	0.0150	0.0100	0.167	0.142	0.929
0.130	360	0.0098	0.0065	480	0.0130	0.0087	600	0.0163	0.0108	0.180	0.153	0.923
0.140	385	0.0105	0.0070	514	0.0140	0.0093	642	0.0175	0.0117	0.194	0.165	0.917
0.150	410	0.0113	0.0075	547	0.0150	0.0100	684	0.0188	0.0125	0.208	0.177	0.912
0.160	435	0.0120	0.0080	580	0.0160	0.0107	724	0.0200	0.0133	0.222	0.189	0.906
0.170	459	0.0128	0.0085	612	0.0170	0.0113	765	0.0213	0.0142	0.236	0.201	0.900
0.180	483	0.0135	0.0090	644	0.0180	0.0120	804	0.0225	0.0150	0.250	0.212	0.894
0.190	506	0.0143	0.0095	675	0.0190	0.0127	844	0.0238	0.0158	0.264	0.224	0.888
0.200	529	0.0150	0.0100	706	0.0200	0.0133	882	0.0250	0.0167	0.278	0.236	0.882

0.210	552	0.0158	0.0105	736	0.0210	0.0140	920	0.0263	0.0175	0.292	0.248	0.876
0.220	574	0.0165	0.0110	766	0.0220	0.0147	957	0.0275	0.0183	0.305	0.260	0.870
0.230	596	0.0173	0.0115	795	0.0230	0.0153	994	0.0288	0.0192	0.319	0.271	0.864
0.240	618	0.0180	0.0120	824	0.0240	0.0160	1030	0.0300	0.0200	0.333	0.283	0.858
0.250	639	0.0188	0.0125	853	0.0250	0.0167	1066	0.0313	0.0208	0.347	0.295	0.853
0.260	660	0.0195	0.0130	880	0.0260	0.0173	1101	0.0325	0.0217	0.361	0.307	0.847
0.270	681	0.0203	0.0135	908	0.0270	0.0180	1135	0.0338	0.0225	0.375	0.319	0.841
0.280	701	0.0210	0.0140	935	0.0280	0.0187	1169	0.0350	0.0233	0.389	0.330	0.835
0.290	721	0.0218	0.0145	962	0.0290	0.0193	1202	0.0363	0.0242	0.403	0.342	0.829
0.300	741	0.0225	0.0150	988	0.0300	0.0200	1234	0.0375	0.0250	0.416	0.354	0.823
0.310	760	0.0233	0.0155	1013	0.0310	0.0207	1267	0.0388		0.430	0.366	0.817
0.320	779	0.0240	0.0160	1038	0.0320	0.0213	1298	0.0400		0.444	0.378	0.811
0.330	797	0.0248		1063	0.0330		1329	0.0413		0.458	0.389	0.805
0.340	815	0.0255		1087	0.0340		1359	0.0425	>0.75ρ_b	0.472	0.401	0.799
0.350	833	0.0263	>0.75ρ_b	1111	0.0350	>0.75ρ_b	1389			0.486	0.413	0.794
0.360	851	0.0270		1134	0.0360		1418	>0.75ρ_b		0.500	0.425	0.788
0.370	868	0.0278		1157	0.0370		1446			0.514	0.437	0.782

*Adapted from Ref. 10.

†For f'_c = 5000 psi, the given c/d must be multiplied by 0.85/0.80.

‡Above these lines $\rho < 200/f_y$.

TABLE 6 Areas of Groups of Bars, in.²

Bar No.	Number of bars 1	2	3	4	5	6	7	8	9	10	11	12
4	0.20	0.40	0.60	0.80	1.00	1.20	1.40	1.60	1.80	2.00	2.20	2.40
5	0.31	0.62	0.93	1.24	1.55	1.86	2.17	2.48	2.79	3.10	3.41	3.72
6	0.44	0.88	1.32	1.76	2.20	2.64	3.08	3.52	3.96	4.40	4.84	5.28
7	0.60	1.20	1.80	2.40	3.00	3.60	4.20	4.80	5.40	6.00	6.60	7.20
8	0.79	1.58	2.37	3.16	3.95	4.74	5.53	6.32	7.11	7.90	8.69	9.48
9	1.00	2.00	3.00	4.00	5.00	6.00	7.00	8.00	9.00	10.0	11.0	12.0
10	1.27	2.54	3.81	5.08	6.35	7.62	8.89	10.2	11.4	12.7	14.0	15.2
11	1.56	3.12	4.68	6.24	7.80	9.36	10.9	12.5	14.0	15.6	17.2	18.7
14	2.25	4.50	6.75	9.00	11.2	13.5	15.7	18.0	20.2	22.5	24.7	27.0
18	4.00	8.00	12.0	16.0	20.0	24.0	28.0	32.0	36.0	40.0	44.0	48.0

TABLE 7 Areas of Bars in Slabs, in.²/ft

Bar No.	Spacing, in. 3	3½	4	4½	5	5½		7	8	9	10	11	12
4	0.80	0.69	0.60	0.53	0.48	0.44	0.40	0.34	0.30	0.27	0.24	0.22	0.20
5	1.24	1.06	0.93	0.83	0.74	0.68	0.62	0.53	0.46	0.41	0.37	0.34	0.31
6	1.76	1.51	1.32	1.17	1.06	0.96	0.88	0.75	0.66	0.59	0.53	0.48	0.44
7	2.40	2.06	1.80	1.60	1.44	1.31	1.20	1.03	0.90	0.80	0.72	0.65	0.60
8	3.16	2.71	2.37	2.11	1.90	1.72	1.58	1.35	1.18	1.05	0.95	0.86	0.79
9	4.00	3.43	3.00	2.67	2.40	2.18	2.00	1.71	1.50	1.33	1.20	1.09	1.00
10	5.08	4.35	3.81	3.39	3.05	2.77	2.54	2.18	1.90	1.69	1.52	1.38	1.27
11	6.24	5.35	4.68	4.16	3.74	3.40	3.12	2.67	2.34	2.08	1.87	1.70	1.56

TABLE 8 Minimum Beam Widths, ACI Code

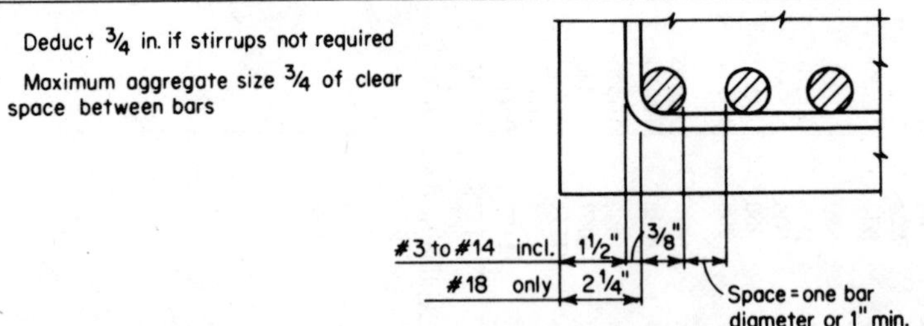

Size of bars	No. of bars in single layer of reinforcement 2	3	4	5	6	7	8	Add for each added bar
No. 4	5¾	7¼	8¾	10¼	11¾	13¼	14¾	1½
No. 5	6	7¾	9¼	11	12½	14¼	15¾	1⅝
No. 6	6¼	8	9¾	11½	13¼	15	16¾	1¾
No. 7	6½	8½	10¼	12¼	14	16	17¾	1⅞
No. 8	6¾	8¾	10¾	12¾	14¾	16¾	18¾	2
No. 9	7¼	9½	11¾	14	16¼	18½	20¾	2¼
No. 10	7¾	10¼	12¾	15¼	17¾	20¼	23	2⅝
No. 11	8	11	13¾	16½	19½	22¼	25	2⅞
No. 14	9	12¼	15¾	19	22½	25¾	29¼	3⅜
No. 18	10½	15	19½	24	28½	33		

support to a much deeper one at midspan, a check on the compressive capacity may also be necessary.

Example 2 Design for flexure a rectangular, continuous, reinforced-concrete beam of three equal 20-ft spans, with ends simply supported, carrying 2 klf live and 1 klf dead load (Fig. 4*a*), f'_c = 4000 psi, f_y = 60,000 psi. The load factors are 1.4 for dead load and 1.7 for live load (Art. 5).

SOLUTION. The three-moment equation gives $M_1 l_1 + 2M_2 (l_1 + l_2) + M_3 l_2 = -w_1 l_1^3/4 \quad -w_2 l_2^3/4$

CASE I. Maximum positive moment at center of interior span. Since $M_1 = 0$,

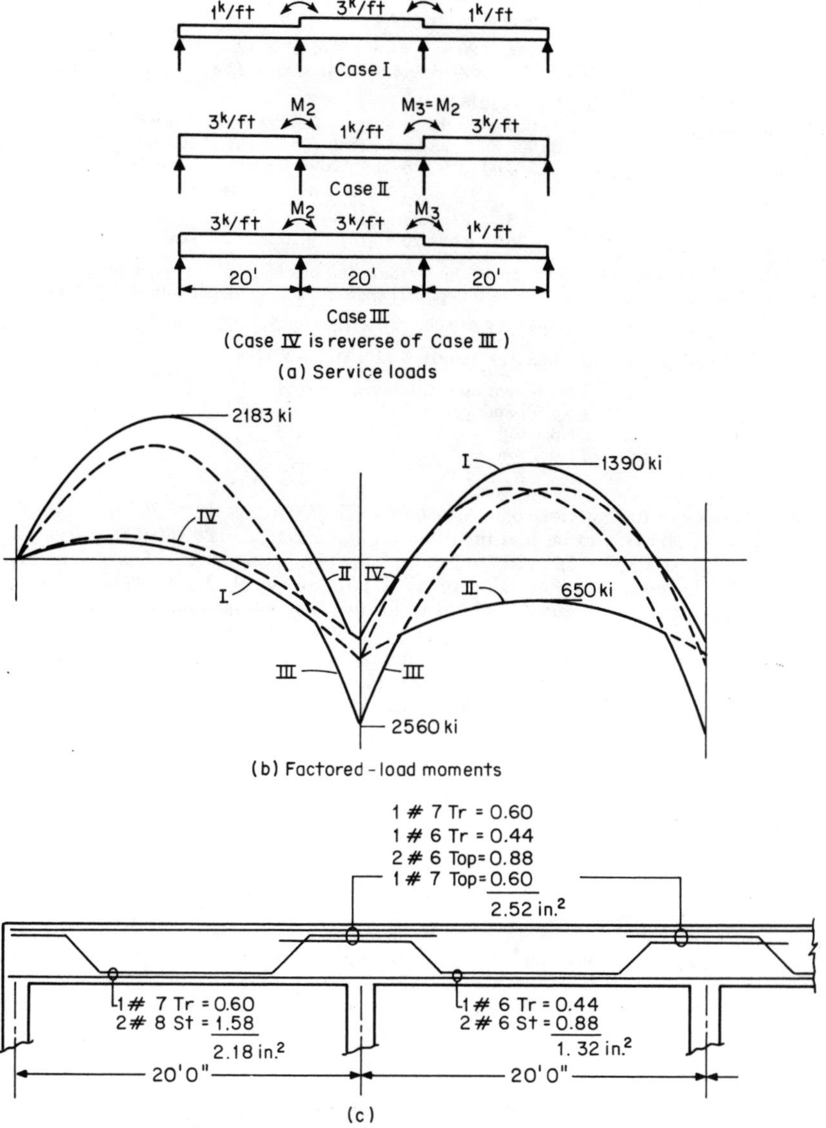

Fig. 4 Example 2.

$$2M_2(20+20)+20M_2=-\frac{1.4\times 1\times 20^3}{4}-\frac{(1.4+1.7\times 2)20^3}{4}$$

$$M_2=-\frac{(1.4+4.8)\times 20^3\times 12}{4(2\times 40+20)}=-1490\text{ kip-in.}$$

$$M=4.8\times 20^2\times 12/8-1490=+1390\text{ kip-in. at midspan}$$

CASE II. Minimum moment at center of interior span.

$$M_2=-\frac{(4.8+1.4)\times 20^3\times 12}{4(2\times 40+20)}=-1490\text{ kip-in.}$$

$$M=1.4\times 20^2\times 12/8-1490=-650\text{ kip-in. at midspan}$$

CASE III. Maximum negative moment at interior support.

$$2M_2\times 40+M_3\times 20=-(4.8+4.8)\times 20^3\times 12/4$$
$$M_2\times 20+2M_3\times 40=-(4.8+1.4)\times 20^3\times 12/4$$

from which $M_2=-2560$ kip-in., $M_3=-1216$ kip-in.

The moments of resistance must encompass the envelopes of all the moment curves (Fig. 4*b*). The maximum negative moment is 2,560,000 in.-lb, and from Table 5, with a reinforcement ratio of 0.010, $M_u/\phi bd^2$ is 547. Therefore, $bd^2=2{,}560{,}000/(0.9\times 547)=5200$, which can be met by the following values for b and d:

b	8	10	12	14	16
d	25.5	22.8	20.8	19.3	18.0

Using the 12 × 20.8 beam size and adding ⅜ in. for the radius of a longitudinal bar, ⅜ in. for a stirrup leg, and 1½ in. of cover gives 23.05 in. total depth. Then, using a 23-in. depth with $d=20.7$ in.,

$$k_u=M_u/\phi bd^2=2{,}560{,}000/(0.9\times 12\times 20.7^2)=553$$

for which $\rho=0.0101$ (Table 5). Then $A_s=0.0101\times 12\times 20.7=2.51$ in.²

1 No. 6 bent up in this beam	= 0.44
1 No. 7 bent up from end span	= 0.60
2 No. 6 top	= 0.88
1 No. 7 top	= 0.60
	2.52 in.²

At the center of the interior span, $M_u/\phi bd^2=650{,}000/(0.9\times 12\times 20.7^2)=140$, for which Table 5 shows ρ to be less than the minimum $200/f_y=200/60{,}000=0.00333$. Therefore, $A_s=0.00333\times 12\times 20.7=0.83$ in.² Use two No. 6 = 0.88 in.², which gives $\rho=0.00354$. The corresponding value of k_u is 207, for which $M_u=958{,}000$ kip-in. Therefore, the two No. 6 bars are adequate for the negative moment over the central portion of the interior span of length l_0 given by $(1490-650)(l_0/20)^2=958-650$, which gives $l_0=12.1$ ft. The No. 7 top bar at each interior support must extend $(20-12.1)/2=4$ ft into the interior span, plus the necessary development length.

For maximum positive moment in the interior span, $M_u/\phi bd^2=1{,}390{,}000/(0.9\times 12\times 20.7^2)=300$; so $\rho=0.0052$ and $A_s=0.0052\times 12\times 20.7=1.29$ in.² Use one No. 6 truss bar and two No. 6 straight bars = 1.32 in.²

For maximum positive moment in the end span, $M_u/\phi bd^2=2{,}183{,}000/(0.9\times 12\times 20.7^2)=472$; so $\rho=0.0085$ and $A_s=0.0085\times 12\times 20.7=2.11$ in.² Use one No. 7 truss bar and two No. 6 straight bars = 2.18 in.²

9. Doubly Reinforced Beams Since strength design assumes a high intensity of compression distributed over a relatively small area, compression reinforcement is seldom necessary for resisting stress. Compression steel is one of the most effective ways to reduce deflection and is much more likely to be supplied for that purpose. It is then the designer's option to include such compression steel in computations for load-carrying capacity.

The ultimate resisting moment is given by

$$M_u=\phi\left[(A_s-A_s')f_y\left(d-\frac{a}{2}\right)+A_s'f_y(d-d')\right]\tag{5}$$

where $a=(A_s-A_s')f_y/0.85f_c'b$ and the use of f_y with A_s' ignores the concrete displaced by A_s'.

The equation is based on the assumption that both tension steel and compression steel are at yield stress when the beam fails. To guarantee this condition requires that

$$\rho - \rho' \geq \frac{0.85\beta_1 f'_c}{f_y} \frac{d'}{d} \frac{87{,}000}{87{,}000 - f_y} \tag{6}$$

In order to assure failure by yielding of the tension reinforcement, rather than by crushing of the concrete, the steel ratio must also satisfy the requirement $\rho - \rho' \leqq 0.75\rho_b$, where ρ_b is given by Eq. (4).

Example 3 A beam required to develop an ultimate moment of 7000 in.-kips is limited to 12 × 24 in., $f'_c = 4000$, $f_y = 60{,}000$ psi. Determine the reinforcement. Assume two rows of tensile reinforcement with $d = 20$ in.

SOLUTION. Using $\rho = \rho_{max}$, the capacity of a singly reinforced section is determined from Eq. (4) with $\beta_1 = 0.85$:

$$\rho_b = 0.85 \times 0.85 \frac{4}{60} \times \frac{87}{87 + 60} = 0.0285$$

$$\rho_{max} = 0.75 \times 0.0285 = 0.02$$

$$A_s = 0.02 \times 12 \times 20 = 4.80 \text{ in.}^2$$

$$a = \frac{4.80 \times 60}{0.85 \times 4 \times 12} = 7.05 \text{ in.} \qquad d - \frac{a}{2} = 16.48$$

$$M_u = 0.90 \times 4.80 \times 60 \times 16.48 = 4270 < 7000 \text{ kip-in.}$$

Assuming one layer of compressive reinforcement, the moment arm of the steel couple will be about 24 − 4.0 − 2.5 = 17.5 in. Therefore, to gain the 2730 kip-in. required, add

$$A_s = A'_s = \frac{2730}{0.9 \times 60 \times 17.5} = 2.89 \text{ in.}^2$$

Try 3 No. 9 = 3.00 in.^{2} for A'_s. Then $A_s = 4.80 + 2.89 = 7.69$ in.^{2}

Try 3 No. 11 + 3 No. 9 = 7.68 in.^{2} For the arrangement shown in Fig. 5, $d = 20.53$ in. and $d' = 2.5$ in.

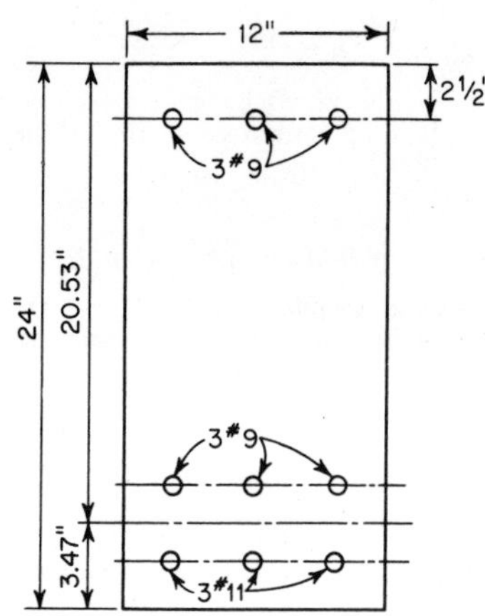

Fig. 5 Example 3.

Then

$$\rho - \rho' = \frac{7.68 - 3.00}{12 \times 20.53} = 0.0190$$

which satisfies the requirements

$$\rho - \rho' \leqq 0.75\rho_b \leqq 0.75 \times 0.0285 = 0.021$$

$$\rho - \rho' \geq 0.85 \times 0.85 \times \frac{4}{60} \times \frac{2.5}{20.53} \times \frac{87}{87 - 60} = 0.0189$$

10. Tee Beam Figure 6*a* shows an isolated tee beam such as might be used for a crane girder or similar freestanding beam. At *b*, the crosshatched area shows the theoretical outline of a tee beam made up of a portion of a monolithic beam-and-slab floor system, while *c* is taken from a concrete-joist floor system. Since the compression is not uniformly distributed across a tee of extreme width, the maximum symmetrical flange for assumed uniformly distributed stress is the smallest of (1) one-quarter of the beam span, (2) a projection of eight slab thicknesses on each side of the stem, and (3) one-half the distance to the next beam on either side.

In one-sided beams, the limits are a projection beyond the stem of $L/12$ or of $6t$, or one-half the distance to the next beam.

Tee beams are of three varieties. If the neutral axis is in the flange or at its bottom, the

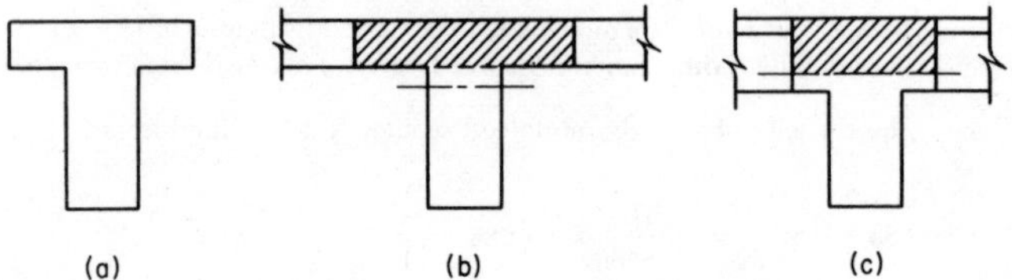

Fig. 6 Tee beam.

beam can be designed as a rectangular cross section of width equal to the width of the flange. If the neutral axis is in the stem and the flange is not too thin (say, $t \geq 0.15d$), the resistance of the small piece of stem between bottom of flange and neutral axis may be neglected. If the neutral axis is in the stem and the flange is thin, compression in the stem should be considered.

Rectangular beams can be reinforced with 2 percent or more of tension steel and still be below 75 percent of balanced reinforcement. As a result, tee beams are not often necessary with ultimate-strength design. However, formulas are available.

If the flange thickness exceeds $1.18qd/\beta_1$, the tee beam is, in effect, rectangular. If it is less,

$$M_u = \phi \left[(A_s - A_{sf}) f_y \left(d - \frac{a}{2} \right) + A_{sf} f_y (d - 0.5t) \right] \tag{7}$$

where $A_{sf} = 0.85(b - b')\,tf'_c/f_y$ = steel area needed to develop compressive strength of overhanging flanges

$a = (A_s - A_{sf})f_y/0.85f'_c b$

The quantity $\rho_w - \rho_f$ must not exceed $0.75\rho_b$, where $\rho_w = A_s/b'd$ and $\rho_f = A_{sf}/b'd$.

Example 4 Redesign the beam of Example 3 as a tee with a flange 4 in. thick.

SOLUTION. The flange must replace strength of A'_s = 2.89 in.² as found in Example 3. A flange 4 in. thick would have equal compressive strength if

$$b - b' = \frac{2.89 \times 60}{0.85 \times 4 \times 4} = 12.8 \text{ in.}$$

$$a = \frac{60(7.75 - 2.89)}{0.85 \times 4 \times 12} = 7.15 \text{ in.}$$

This is slightly on the safe side, since the centroid of the flange lies 0.5 in. higher than A'_s. Check M_u provided by the web and flange (Fig. 7):

$$M_u = 0.90 \times 0.85 \times 4[12 \times 7.15(20.53 - 3.58) + 2 \times 7 \times 4 \times 18.53] = 7625 \text{ kip-in.} > 7000$$

Example 5 With a 3-in. slab and beams spanning 20 ft at 6 ft on centers, design an interior tee beam for a live load of 3 klf and a dead load of 1 klf. $f'_c = 4000$, $f_y = 60{,}000$ psi, maximum depth of beam 20 in. (Fig. 8*a*).

Assume a maximum positive bending moment of $wL^2/16$ and a maximum negative moment of $wL^2/11$ as suggested in ACI 318-77. Sketch the moment envelopes (Fig. 8*b*).

POSITIVE MOMENT

$$M_{LL} = 3.0 \times 1.7 \times 20^2 \times 12/16 = 1530$$
$$M_{DL} = 1.0 \times 1.4 \times 20^2 \times 12/16 = \underline{420}$$
$$M_{LL+DL} = 1950 \text{ kip-in.}$$

NEGATIVE MOMENT

$$M_{LL} = 3.0 \times 1.7 \times 20^2 \times 12/11 = -2225$$
$$M_{DL} = 1.0 \times 1.4 \times 20^2 \times 12/11 = \underline{-610}$$
$$= -2835 \text{ kip-in.}$$

The flange width is governed by ¼ the beam span = 60 in. (Fig. 8). With a 20-in. maximum depth, d = 17.6 in.,

$$k_u = M_u/\phi bd^2 = \frac{1{,}950{,}000}{0.90 \times 60 \times 17.6^2} = 117$$

For negative moment, with a stem width of 14 in. and a depth of 17.6 in.,

$$k_u = M_u/\phi bd^2 = \frac{2{,}835{,}000}{0.90 \times 14 \times 17.6^2} = 727$$

To obtain reinforcement, refer to Table 5. For positive moment, $k_u = 117$, for which ρ is less than the minimum value 200/60,000 = 0.0033. Therefore, $A_s = 0.0033 \times 17.6 \times 60 = 3.52$ sq in.

2 No. 7 truss bars = 1.20 sq in.
2 No. 10 straight bars = 2.54 sq in.
$+A_s$ = 3.74 sq in. (Fig. 9)

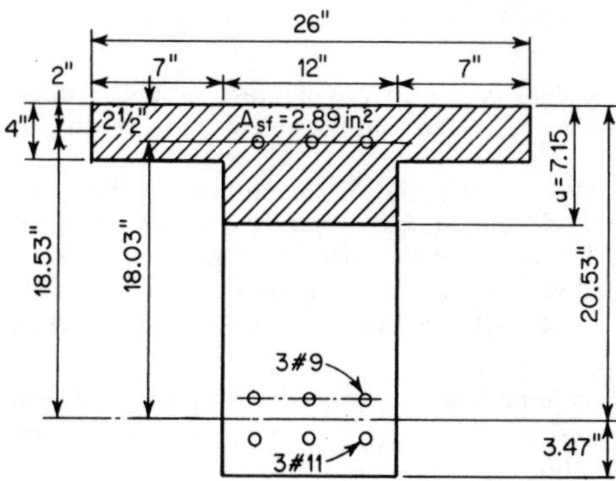

Fig. 7 Example 4.

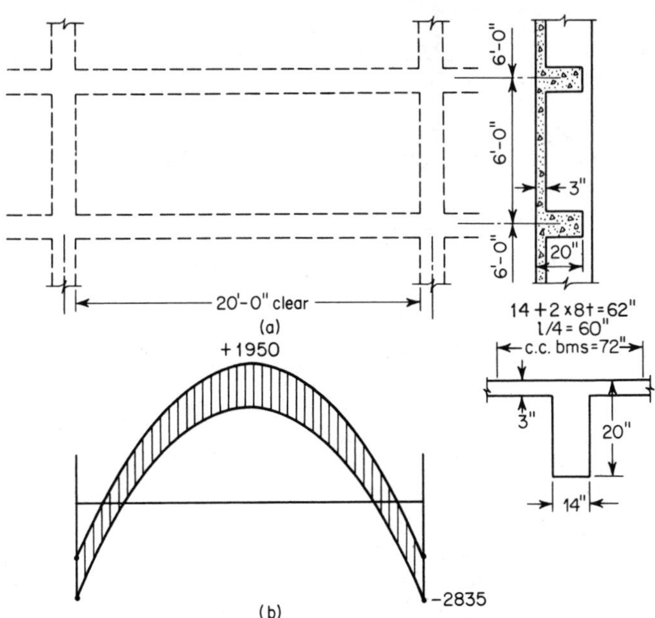

Fig. 8 Example 5.

For negative moment, $k_u = 727$. From Table 5, $\rho = 0.0138$ and $A_s = 0.0138 \times 17.6 \times 14 = 3.40$ sq in.

2 No. 7 this span	=	1.20 sq in.
2 No. 7 continuing span	=	1.20 sq in.
1 No. 10 top	=	1.27 sq in.
	A_s =	3.67 sq in. (Fig. 9)

11. Special Beam Shapes Strength design is adaptable to beams of any shape. In this case, as for rectangular beams, (1) the neutral axis must result in a total tension equal to total compression, (2) the tension steel (at least the bars farthest from the neutral axis) must

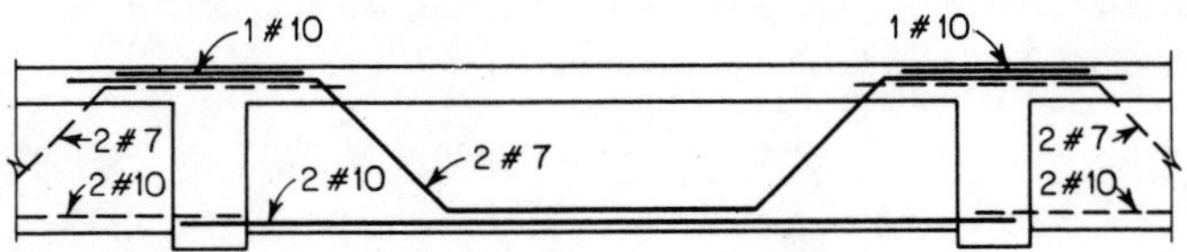

Fig. 9 Reinforcing for beam of Example 5.

be at yield stress (usually above minimum yield strain), (3) the compression strain farthest from the neutral axis must be taken as 0.003, and (4) the compression area can be divided into rectangles or triangles.

12. Shear and Diagonal Tension According to ACI, the mean intensity of shear as a measure of diagonal tension is $v_u = V_u/\phi b_w d$, computed at a distance d from the support, where b_w is the width of the web of a T-beam or rectangular beam. In the absence of torsion, the capacity v_c of an unreinforced web to resist diagonal tension is $2\sqrt{f'_c}$ or, if a more detailed analysis is made, $1.9\sqrt{f'_c} + 2500\,\rho_w Vd/M$ but not to exceed $3.5\sqrt{f'_c}$, where $\rho_w = A_s/b_w d$.

With web reinforcement (and torsion not important) v_u must not exceed $10\sqrt{f'_c}$. Stirrup spacing must not exceed $d/2$ for $v_u \leqq 6\sqrt{f'_c}$ or $d/4$ for $6\sqrt{f'_c} < v_u \leqq 10\sqrt{f'_c}$. Wherever stirrups are required, minimum $A_v = 50 b_w s/f_y$, where A_v is the total area of web reinforcement within the distance s.

The required area A_v of vertical stirrups is

$$A_v = \frac{V_s s}{\phi f_v d} = \frac{v_s b s}{f_y} \tag{8}$$

where $v_s = v_u - v_c$.

For a single bent bar or a single group of parallel bars at the angle α with the longitudinal axis of the member, all bent at the same distance from the support,

$$A_v = \frac{V_s}{\phi f_v \sin \alpha} \tag{9}$$

For a series of parallel bars or groups of bars bent at different distances from the support,

$$A_v = \frac{V_s s}{\phi f_v d\,(\sin \alpha + \cos \alpha)} \tag{10}$$

In these formulas, V_s = shear carried by web reinforcement.

Spacing of stirrups in a triangular shear diagram of base a (Fig. 10) is

$$s = a\,\frac{\sqrt{n} - \sqrt{n - 0.5}}{\sqrt{n}} \qquad a\,\frac{\sqrt{n} - \sqrt{n - 1.5}}{\sqrt{n}} \quad \cdots \quad a\,\frac{\sqrt{n} - \sqrt{0.67}}{\sqrt{n}} \tag{11}$$

where n = number of stirrups in distance a. This locates the stirrups at midlength of the trapezoidal segments of the excess shear prism, except that the last one is at the centroid of the triangular segment. Maximum spacing often controls near the small end of the triangle.

Web reinforcement must be provided to account for three portions of the shear diagram:

1. To continue the size and spacing of such reinforcement as is required at distance d from the face of the support back to the face of the support (the crosshatched area in Fig. 10)
2. To locate a stirrup at or near the centroid of each equal volume into which the excess shear prism (the stippled area in Fig. 10) is divided
3. A length with minimum stirrups beyond area 3 in Fig. 10, within which $v_u \geqq v_c/2$.

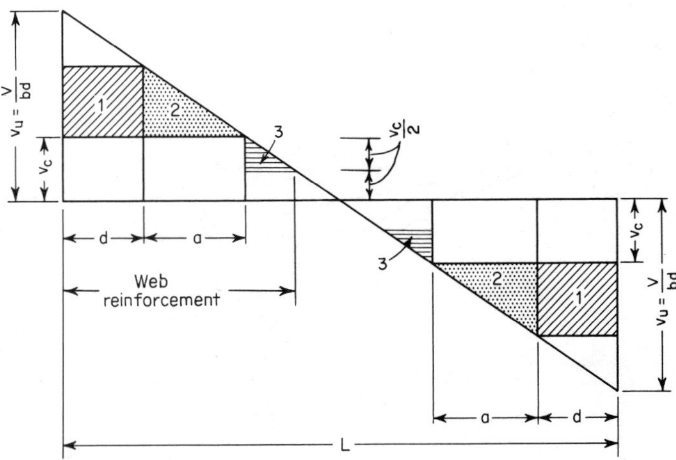

Fig. 10 Location of web reinforcement.

Example 6 For the continuous beam of Example 2, b = 12 in., d = 20.7 in., design interior-span stirrups for f_y = 40,000 psi, neglecting any truss bars.

SOLUTION. Plot the shear envelope curves of Fig. 11*a* from the data of Example 2 and Fig. 4*b*:

CASE I $V_u = wL/2 = (1.4 \times 1 + 1.7 \times 2)10 = 48$ kips

CASE II $V_u = 1 \times 1.4 \times 10 = 14$ kips

CASE III V_{uL} = Case I + $(M_L - M_R)/L = 48 + (-1216 + 2560)/240 = 53.6$ kips

$V_{uR} = 48 - 5.6 = 42.4$ kips

CASE IV V_{uL} and V_{uR} are reversed from Case III.

With a maximum V_L of 53.6 kips, b = 12, and d = 20.7 in.,

$$v_u = V_u/\phi bd = 53{,}600/(0.85 \times 12 \times 20.7) = 254 \text{ psi}$$

At distance d = 20.7 in. = 1.72 ft from the face of the support,

$$v_u = (53.6 - 1.72 \times 4.8)1000/(0.85 \times 12 \times 20.7) = 215 \text{ psi} > 2\sqrt{f_c'} = 126 \text{ psi}$$

Thus web reinforcement is required, and must be extended at least to the point where v_u = 126/2 = 63 psi.

The unit shear diagram, shown in Fig. 11*b*, has a slope of 1.89 psi/in. The shaded areas represent (1) two lengths (AB and BC) where stirrups are required because $v_u > v_c$ and (2) the length CD where minimum stirrups are required because $v_u > v_c/2$.

Assume No. 3 U stirrups (smallest bar size) for length AB and calculate spacing:

$$s = \frac{A_v f_y}{(v_u - v_c)b} = \frac{0.22 \times 40{,}000}{89 \times 12} = 8.23 \text{ in., say 8 in.}$$

as dimensioned in Fig. 11*b*, with the first stirrup at 8/2 = 4 in. from support.*

For length BC, the spacing can be increased as the shear decreases; compute the spacing at the fourth stirrup:

$$v_s = 89 - (28 - 20.7)1.89 = 75 \text{ psi} \qquad s = \frac{0.22 \times 40{,}000}{75 \times 12} = 9.8 \text{ in.}$$

Alternate calculation: s = (89/75)8.23 = 9.8 in.

*See footnote to Example 7, p. 1–19.

Maximum spacing = $d/2$ = 20.7/2 = 10.3 in., say 10 in. The maximum 10-in. spacing governs both here and to within $s/2$ = 5 in. of point D. Next locate D: L_{AD} = 191/1.89 = 101 in.

Use 1 No. 3 U at 4 in., 4 at 8 in., 6 at 10 in., each end.

Example 7 Given the service loads in Fig. 12a, b = 12 in., d = 22.6 in., design stirrups for f'_c = 3750 psi, f_y = 40,000 psi.

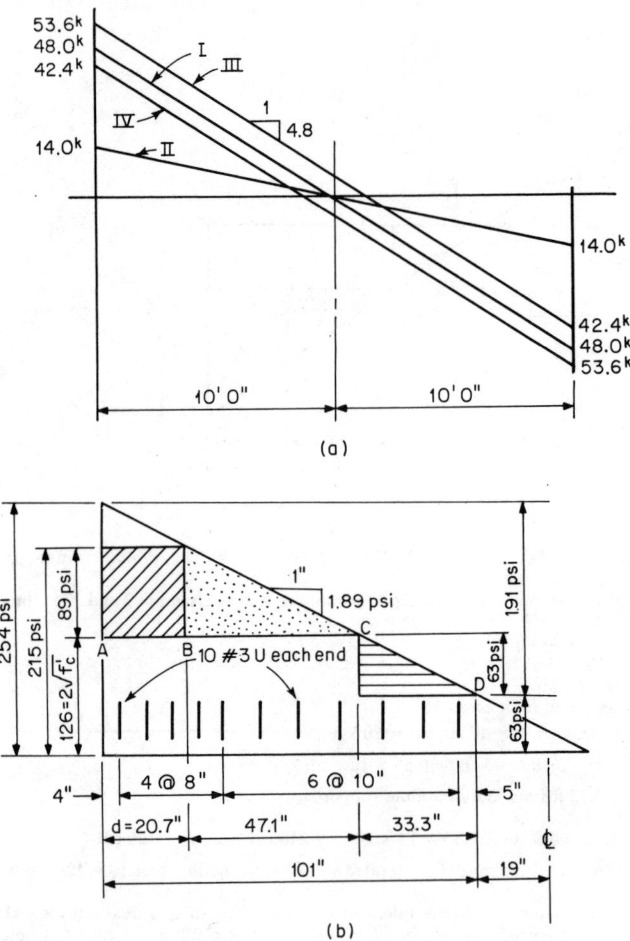

Fig. 11 Example 6.

SOLUTION. The factored-load shear diagram is shown in Fig. 12b. Since the contribution of truss bars is to be disregarded, it will be sufficiently accurate to use the shear diagram of Fig. 12b for either a continuous or simple-span beam. The controlling ultimate unit shears are:

$$v_{\text{end}} = \frac{62{,}000}{0.85 \times 12 \times 22.6} = 270 \text{ psi} \qquad v_c = 2\sqrt{f'_c} = 2\sqrt{3750} = 122 \text{ psi}$$

$$v_d = v_{22.6} = \frac{62{,}000 - 1.88 \times 2800}{0.85 \times 12 \times 22.6} = 246 \text{ psi}$$

$$v_s = v_u - v_c = 246 - 122 = 124 \text{ psi}$$

$$v_c = \frac{34{,}000}{0.85 \times 12 \times 22.6} = 147 \text{ psi}$$

$$v_s = 147 - 122 = 25 \text{ psi}$$

For distance $d = 22.6$ in. from the support (shown crosshatched), try No. 3 U for which $A_v = 0.22$ sq in.

$$s = A_v f_y / v_s b = 0.22 \times 40{,}000/(124 \times 12) = 5.9 \text{ in.}$$

say 5.5 in. for 4 spaces which then reaches to 2.4 in. beyond point B.*
At $2.4 + s =$ say 8.4 in. beyond B, $v_s = 124 - 8.4 \times 1.025 = 115$ psi (where 1.025 is the slope of the shear diagram).

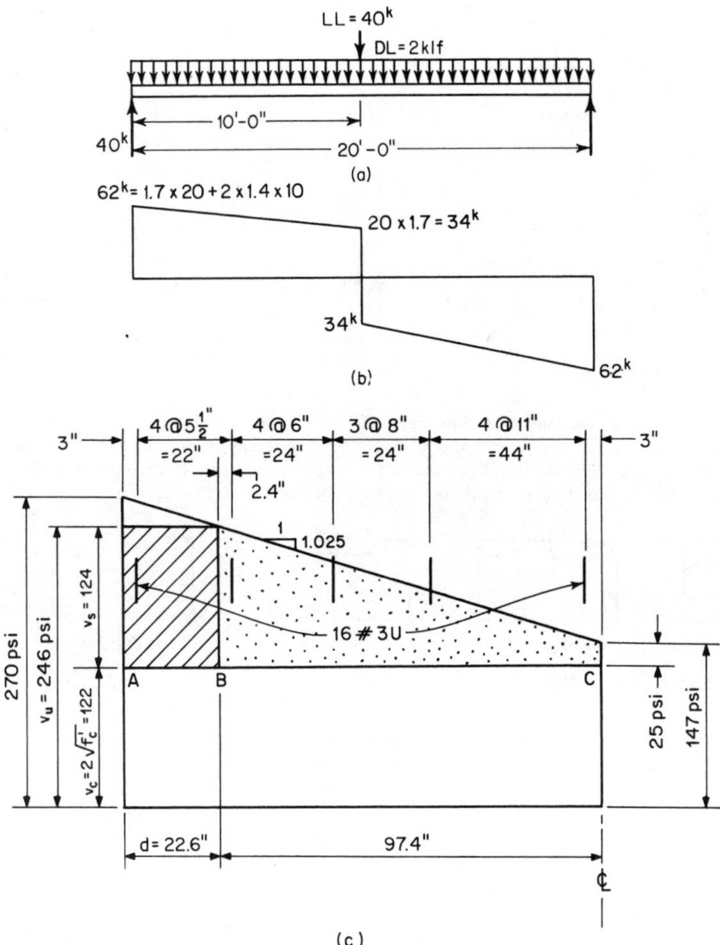

Fig. 12 Example 7.

$$s = 0.22 \times 40{,}000/(115 \times 12) = 6.38 \text{ in.}$$

say 6 in. for 4 spaces to 26.4 in. past B. At 26.4 + 8 (estimated) = 34.4 in. beyond B, $v_s = 124 - 34.4 \times 1.025 = 89$ psi.

Since s varies as $1/v_s$, $s = 6.38 \times 115/89 = 8.2$ in., say 8 in. for 3 spaces, to 26.4 + 24 = 50.4 in. past B. At 50.4 + 11 = 61.4 in. beyond B, $v_s = 124 - 61.4 \times 1.025 = 61$ psi, and $s = 6.38 \times 115/61 = 12.0$ in.,

*Each stirrup theoretically is placed at the centroid of the v_s area it serves. Hence the stirrups just used provide for $s/2$ beyond the last stirrup (to $2.4 + 5.5/2 \approx 5+$ in. beyond B) and the estimated s used next should theoretically be 5.5/2 plus half the next s to be calculated. This is also why on Fig. 12c the first stirrup was placed $s/2 \approx 3$ in. from the support.

which is greater than the maximum $d/2$ = 11.3 in.; say 4 spaces at 11 in. (Note that the 3 in. left to midspan might be made anything from zero to 11/2 in. in similar cases.)

Since stirrups are required over the entire span, requirement No. 3 of Art. 12 does not apply; hence use 16 No. 3 stirrups each end, as shown on Fig. 12*c*.

Example 8 Determine maximum live shears with a movable live load w_L on the beam of Fig. 13, which shows the load position, the corresponding shear, and the fact that live-load shear at midspan is one-quarter the live shear at the end. It also shows that the assumption of a linear variation from end to center of span is on the safe side.

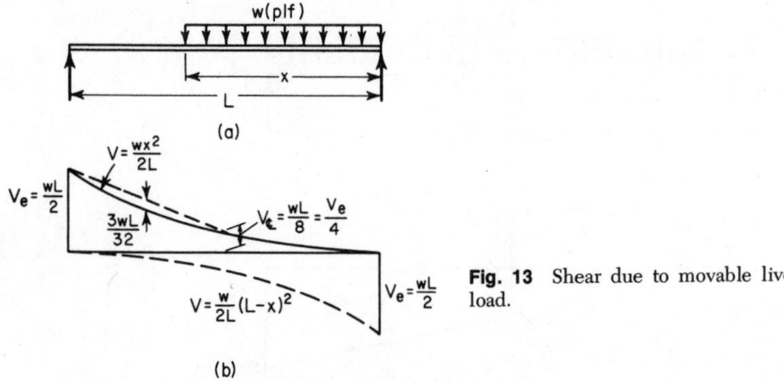

Fig. 13 Shear due to movable live load.

13. Development and Anchorage of Reinforcement The necessary length between the point of maximum stress in a bar and the end of the bar is called its development or anchorage length. An inadequate length can result in a splitting failure of the concrete, the split running along the bar as indicated in the cross sections sketched in Fig. 14. Since

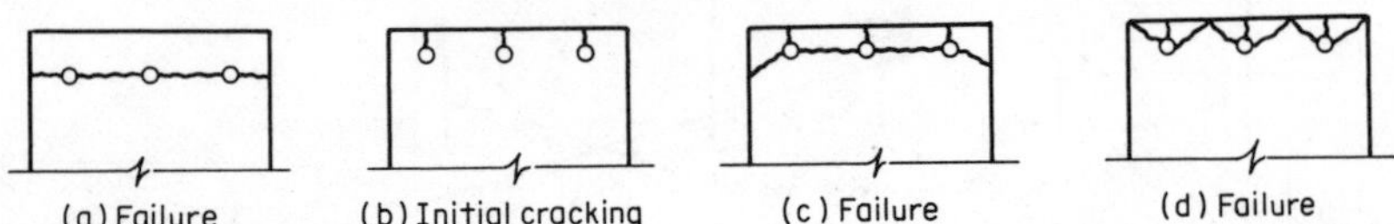

Fig. 14 Development length splitting.

such a failure is normally brittle, conservatism is necessary. The Code development length attempts to assure that such a failure is less probable than some other strength limitation.

Closely spaced bars can split off the cover, as in Fig. 14*a*. If the cover is thin, say less than a half (clear) bar space, the weakness may initially develop as in Fig. 14*b* and bring failure either as shown in Fig. 14*c* for closely spaced bars or in Fig. 14*d* for widely spaced bars.

Neither closely spaced bars (closer than possibly 3 in. clear) nor widely spaced bars in mass concrete are mentioned in the Code; the designer should be aware of the deficiencies of the first and the advantages of the second.

Code development lengths are given in Table 9. These are lengths of straight bars but are also used for trussed bars bent at 45° or flatter. The development length of standard hooks (180, 135, or 90°) is determined by substituting f_h for f_y in Table 9, where $f_h = \xi\sqrt{f'_c}$ and both f_h and $\sqrt{f'_c}$ are in units of psi. Values of ξ are given in Table 10.

The development length may be a combination of the equivalent embedment length of a hook or mechanical anchorage plus additional embedment of the reinforcement.

Research shows that very little tension develops beyond a hook and only the length in front is effective as part of the needed development. The specified extension beyond a hook is primarily to assure that the hook is restrained against straightening under tension.

Hooks in compression bars are ineffective.

TABLE 9 Development Length l_d, in., of Deformed Bars and Deformed Wire*

	Deformed Bars and Deformed Wire in Tension	
(*a*)	l_d = (basic development length) × (modification factor) ≥ 12 in.	
(*b*)	Basic development length†	
	No. 11 bars or smaller	$\geq 0.04 A_b f_y/\sqrt{f'_c}$ and $0.0004\, d_b f_y$
	No. 14 bars	$\geq 0.085\, f_y/\sqrt{f'_c}$
	No. 18 bars	$\geq 0.11\, f_y/\sqrt{f'_c}$
	Deformed wire	$\geq 0.03\, d_b f_y/\sqrt{f'_c}$
(*c*)	Modification factor‡	
	1. Top reinforcement§	1.4
	2. Bars with $f_y > 60{,}000$ psi	$2\text{–}60{,}000/f_y$
	3. Reinforcement located at least 3 in. from side face of member and spaced laterally at least 6 in. on center	0.8
	4. Reinforcement in excess of that required	$A_{s(req)}/A_{s(act)}$
	5. Bars enclosed within a spiral which is not less than ¼ in. in diameter and not more than 4 in. pitch	0.75
	Deformed Bars in Compression	
(*d*)	Development length $l_d = 0.02\, f_y d_b/\sqrt{f'_c} \geq 0.0003\, f_y d_b \geq 8$ in. This length may be modified according to provisions of (*c*) 4 and (*c*) 5	
	Bundled Bars	
(*e*)	l_d = (factor below) × (development length of individual bar)	
	Three-bar bundle	1.2
	Four-bar bundle	1.33

*Adapted from Ref. 6.

†This basic development length must be increased 33 percent for all-lightweight concrete and 18 percent for sand–lightweight concrete. Use linear interpolation for partial sand replacement. Alternatively, the basic development length may be multiplied by $6.7\sqrt{f'_c}/f_{ct} \geq 1$ when f_{ct} is specified and concrete is proportioned in accordance with Code 4.2. The factors in (*c*) also apply.

‡All modification factors can be applicable in a given case, e.g., for top tensile reinforcing in excess of that required, located at least 3 in. from side face and spaced at least 6 in. on center and with $f_y > 60{,}000$ psi, the basic development length from (b) is multiplied by $1.4 \times 0.8 \times (2 - 60{,}000/f_y) A_{s(req)}/A_{s(act)}$.

§Horizontal reinforcement with more than 12 in. of concrete in the member below the bars.

TABLE 10 Values of ξ in $f_h = \xi\sqrt{f'_c}$*†

Bar No.	$f_y = 60$ ksi Top bars	$f_y = 60$ ksi Other bars	$f_y = 40$ ksi All bars
3–5	540	540	360
6	450	540	360
7–9	360	540	360
10	360	480	360
11	360	420	360
14	330	330	330
18	220	220	220

*From Ref. 6.

†Values of ξ may be increased 30 percent where enclosure is provided perpendicular to plane of hook. Enclosure may consist of external concrete or internal closed ties, spirals, or stirrups.

Example 9 A 10 × 18-in. cantilevered beam has 2 No. 7 bars in the top. Determine the embedment required to anchor and fully develop these bars. f'_c = 4000 psi, f_y = 60,000 psi.

SOLUTION. Since these are top bars, the appropriate formula from Table 9 is modified in accordance with (*c*). A_b for the No. 7 bar is 0.6 in.[2] Hence

$$l_d = 1.4\left(\frac{0.04A_bf_y}{\sqrt{f'_c}}\right) = 1.4\frac{0.04 \times 0.6 \times 60{,}000}{\sqrt{4000}} = 31.9 \text{ in.}$$

Part of this embedment will be made up by a hook. This hook for No. 7 to No. 11 bars, inclusive (Table 10) has a value of $360\sqrt{f'_c}$ = 22,800 psi. It needs a lead-in length to care for 60,000 − 22,800 = 37,200 psi, which requires 31.9 × 37,200/60,000 = 19.8 in. This requires an anchorage-member thickness of 19.8 + 3.5 (for hook) + 2 cover = 25.3 in., more than would usually be available unless the beam is one that continues beyond the support. Theoretically, a larger-radius hook should give more resistance, but no Code guidance is available for this. The Code does not forbid a 90° hook plus extra anchorage beyond the bend, but this combination is not recommended, for reasons given in Art. 13. Therefore, the design needs revision.

One solution would be the use of 3 No. 7 bars, equivalent to lowering the bar stress to 40,000 psi. With the same 22,800 psi on the hook the necessary lead-in becomes (17,200/60,000)31.9 = 9.1 in. and the required anchorage-member thickness only 9.1 + 3.5 + 2 = 14.6 in.

Bars could also be made smaller, such as 4 No. 5, furnishing 1.24 in.[2] at a stress of (1.20/1.24)60,000 = 57,100 psi. The hook cares for the same 22,800 psi, and lead-in length for 57,100 − 22,800 = 34,300 psi is $1.4 \times 0.04 \times 0.31 \times 34{,}300/\sqrt{4000}$ = 9.4 in. Total thickness needed would be 9.4 + 2.50 + 2 = 13.9 in.

Example 10 Design the longitudinal reinforcement for a 10 × 18-in. beam carrying a uniform load of 2500 plf live load and 2000 plf dead load on a simple span of 10 ft and determine embedment lengths. f'_c = 4000, f_y = 60,000 psi.

SOLUTION. $d = 18 - 1\frac{1}{2} - \frac{3}{8} - \frac{1}{2} = 15.6$ in. To determine the reinforcing steel:

$$\begin{aligned} DL &= 2000 \times 1.4 = 2800 \text{ plf} \\ LL &= 2500 \times 1.7 = \underline{4250} \text{ plf} \\ &\qquad\qquad\quad 7050 \text{ plf} \end{aligned}$$

$$M = wL^2/8 = 7050 \times (10^2 \times 12)/8 = 1{,}058{,}000 \text{ lb-in.}$$

$$M/\phi bd^2 = \frac{1{,}058{,}000}{0.90 \times 10 \times 15.6 \times 15.6} = 483$$

From Table 5, ρ = 0.0088, A_s = 0.0088 × 10 × 15.6 = 1.37 sq in. Use 1 No. 7 bar and 1 No. 8 bar = 1.39 sq in.

Since the bond resistance is increased by a compressive reaction (Code 12.2.3), the l_d provided must not exceed

$$\begin{aligned} 1.3M_t/V_u + l_a &= 1.3(1.39 \times 60{,}000 \times 15.6)/35{,}250 + l_a \\ &= 48 \text{ in.} + l_a \end{aligned}$$

where l_a is the additional length beyond the theoretical reaction point. For the No. 8 bar (the more critical) from Table 9,

$$l_d = 0.04A_bf_y/\sqrt{f'_c} = 0.04 \times 0.79 \times 60{,}000/\sqrt{4000} = 30 \text{ in.}$$

which is considerably less than the 48-in. limit for M_t/V_u just calculated, even without considering the probable l_a extension of the No. 8 bar past the center of the support.

14. Splices Because reinforcing bars are delivered in manageable lengths, longer runs must be spliced. Splices may be made by lapping (if not larger than No. 11), butt welding, mechanical coupling, or other device. Lapped bars may be in contact or spaced far enough apart to allow more wet concrete between them. The following are Code requirements.

Lap Splices in Compression. The minimum lap length is l_d from (*d*) of Table 9, but not less than $0.0005f_yd_b$ for $f_y \leq 60{,}000$ psi or $(0.0009f_y - 24)d_b$ for $f_y > 60{,}000$ psi, or 12 in., where d_b = diameter of bar. This lap must be increased by one-third for $f'_c < 3000$ psi.

The lap length for spiral compression members, and for tied compression members with ties having an effective area of at least $0.0015hs$ throughout the lap, where h = member thickness and s = tie spacing, may be taken at 0.75 and 0.83, respectively, of the lap length defined above, but not less than 12 in.

Splices in Tension. Lap splices are not permitted in tension tie members; instead, the bars must be fully connected by welding or by mechanical connections, and the splices must be staggered at least $1.7l_d$. Lap splices may be used in other members.

If the area A_s of the steel at the splice is less than twice that required by analysis, lap splices must have a lap of $1.3l_d$ if no more than half the bars are spliced within the lap length and $1.7l_d$ if more than half are spliced, and in either case not less than 12 in.

If the area A_s of the steel at the splice is more than twice that required by analysis, lap splices must have a lap of l_d if no more than three-quarters of the bars are spliced within the lap length and $1.3l_d$ if more than three-quarters are spliced, and in either case not less than 12 in.

The embedment lengths l_d are those for the full f_y (Table 9).

The lap splice of a circumferential tension bar in a round tank need not be considered as being in a tension tie, although it is especially desirable that such splices be well staggered.

Butt-welded or coupled splices must provide positive connections capable of transmitting at least 125 percent of the specified yield strength of the bar. Splices that will be called upon only to transmit compression may have ends of bars sawed square within 1.5° and held in true alignment with a welded sleeve or splice or a mechanical coupler.

Example 11 The tension tie between bottom hinges of a three-hinged arch is 150 ft long and is to consist of deformed bars buried in concrete below the floor, to carry a tension of 220 kips, of which 120 kips is dead load and 100 kips live load. Design the tie, the intermediate splices, and the end anchorages. $f'_c = 4000$ psi, $f_y = 60{,}000$ psi.

SOLUTION.

$$\begin{aligned} \text{DL} &= 120{,}000 \times 1.4 = 168{,}000 \\ \text{LL} &= 100{,}000 \times 1.7 = \underline{170{,}000} \\ & \qquad\qquad\qquad\quad 338{,}000 \end{aligned}$$

This being a tension tie, ϕ equals 0.90 and $\phi f_y = 0.90 \times 60{,}000 = 54{,}000$ psi. Hence the area of bars should be

$$A_s = 338{,}000/54{,}000 = 6.28 \text{ in.}$$
$$\text{Use 4 No. 11 bars} = 6.24 \text{ sq in.}$$

Use mechanical splices, such as Cadweld, or butt-welded splices; lap splices are not allowed by Code. The four bars must be spliced at points staggered $1.7l_d$ along the tie. Since $l_d = 0.04A_b f_y/\sqrt{f'_c} = 0.04 \times 1.56 \times 60{,}000/\sqrt{4000} = 59.2$ in., this requires 101 in.

ANCHORAGE. Length of anchorage is l_d for bottom bars and $1.4l_d$ for top bars, or 59.2 and 83 in., respectively. These lengths give real problems, since little tension can be carried beyond the usual 90° bend. The bars could be extended through the vertical member into an anchorage block, as in Fig. 15a,

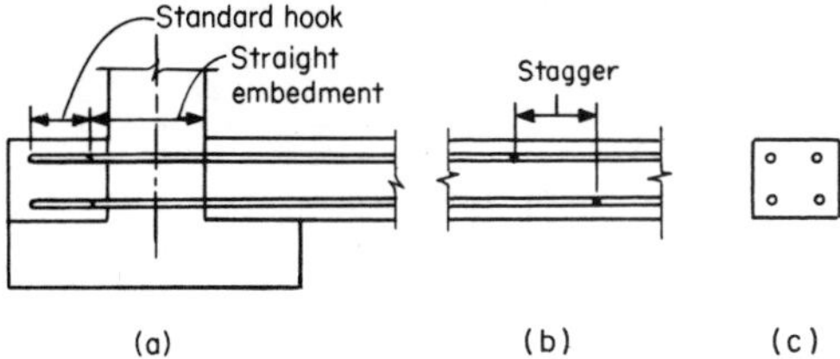

Fig. 15 Example 11.

more bars of smaller diameter could be used to reduce the required l_d, or the bars might be carried through the vertical member and welded to steel anchor plates.

As a further precaution the tie should be buried deep enough or placed on a sufficiently soft layer that loads on the floor above will not damage it from beam action.

Example 12 The wall thickness of the vertical stem of a cantilevered retaining wall varies from 8 to 14 in. in a net height of 12 ft. For $f'_c = 3000$ psi, $f_y = 60{,}000$ psi, design the dowels from the base to the stem and the vertical bars in the stem (Fig. 16). There is a surcharge of 300 psf on the backfill. The service-load moment and shear in the stem are $M = 60(h_x + 3)^3$ and $V = 15(h_x + 3)^2$ per foot length of wall, where h is in feet.

SOLUTION. Computed shears, moments, effective depths, and areas of steel are given in Table 11, using a load factor of 1.7 for horizontal loads. The shears and moments at all levels below 8 ft are given in the schedule. At lesser depths, their values are not critical.

Although the bars shown in the table are all that are required for stress, it would be well to provide at least a mat of No. 4 bars 12 in. c/c horizontal and vertical for temperature reinforcement in the exterior face of the wall.

Development length of No. 5 = $0.04A_bf_y/\sqrt{f'_c}$ = 0.04 × 0.31 × 60,000/$\sqrt{3000}$ = 13.6 in. Therefore, the dowel must extend 14 in. into the footing. However, many designers would prefer to have it terminate in a semicircular hook.

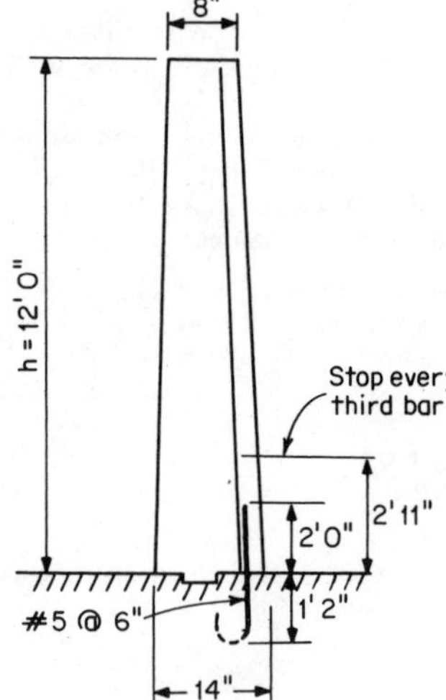

Fig. 16 Example 12.

The projection into the stem acts as a splice, which requires $1.7l_d$ = 1.7 × 13.6 = 23.1 in.

The vertical bars in the wall can theoretically be reduced to No. 5 at 9 in. at a depth of 10 ft from the top, but the Code requires an extension of d = 10.5 in. or $12d_b$ (7.5 in.). Every third bar can then be cut off at 2 ft + 10.5 in. from the top of the base, say 2 ft 11 in. Alternatively, every third bar could be omitted if every third dowel projected 2 ft 11 in. and thus acted as the third bar for the wall.

TABLE 11 Example 12

h_x, ft	V, lb	M, lb-in.	$M/\phi bd^2$	d, in.	ρ Table 5	A_s, in.²	Bars
8	3080	135,800	$\dfrac{135{,}800}{0.90 \times 12 \times 9.5 \times 9.5} = 139$	9.5			
9	3670	176,300	$\dfrac{176{,}300}{0.90 \times 12 \times 10 \times 10} = 163$	10			
10	4310	223,900	$\dfrac{223{,}900}{0.90 \times 12 \times 10.5 \times 10.5} = 188$	10.5	0.0033	0.42	No. 5 at 9
10.5	4650	251,000	$\dfrac{251{,}000}{0.90 \times 12 \times 10.75 \times 10.75} = 201$	10.75	0.0035	0.45	No. 5 at 8
11	5000	280,000	$\dfrac{280{,}000}{0.90 \times 12 \times 11 \times 11} = 214$	11	0.0037	0.49	No. 5 at 7.5
12	5740	344,200	$\dfrac{344{,}200}{0.90 \times 12 \times 11.5 \times 11.5} = 241$	11.5	0.0042	0.58	No. 5 at 6

15. Bar Cutoffs and Bend Points

Example 13 A 10 × 18 in. beam is carrying a factored load of 2550 plf on a simply supported span of 20 ft. Design the tension bars and determine the cutoff for each. f'_c = 4000, f_y = 60,000 psi. (Fig. 17.)

SOLUTION. If the steel is in two layers, d = 18 − 1½ − ⅜ − ½ − ½ = 15.1 in.

$$M_u = wL^2/8 = 2550 \times 20^2 \times 12/8 = 1{,}530{,}000 \text{ in.-lb}$$

$$\frac{M}{\phi b d^2} = \frac{1{,}530{,}000}{0.90 \times 10 \times 15.1 \times 15.1} = 745$$

From Table 5 $\rho = 0.0142$, $A_s = 0.0142 \times 10 \times 15.1 = 2.14$ sq in. Use 5 No. 6 bars = 2.20 sq in.

Code 12.2.1 requires at least one-third of these bars to extend 6 in. into the support, or 2 No. 6 × 21 ft 0 in. Of the remaining bars, one can theoretically be cut off or bent up at $\sqrt{1/5}$ of the distance from midspan to support, i.e., $\sqrt{1/5} \times 10$ ft = 4.47 ft. However, bars should extend the depth $d = 15.1$ in. = 1.26 ft (or 12 bar diameters if greater) past the theoretical cutoff point, which is 4.47 + 1.26 = 5.73 ft from midspan.

The next bottom bar can be spared, cut off or bent up at $\sqrt{2/5}$ of the distance from midspan to support, i.e., 6.32 ft plus the same 1.26 ft for anchorage = 7.58 ft. The third bar can theoretically be spared at $\sqrt{3/5}$ of the distance from midspan, i.e., 7.74 ft + 1.26 ft = 9.00 ft, but might be detailed the same length as the full-length bottom bars.

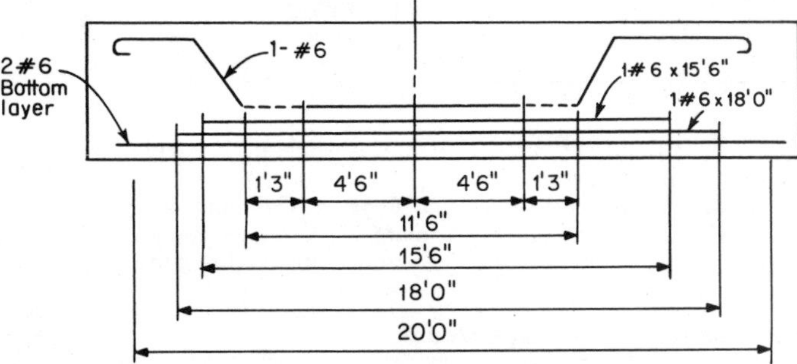

Fig. 17 Example 13.

16. Deflection Safety consists largely in keeping service-load stresses within the maximum allowable or, in strength design, factored loads within stated margins. Serviceability means creating a structure that will not misbehave visibly, i.e., will not settle, crack, warp, or sag to an undesirable degree.

Short-time elastic deflection can be approximated quite adequately by the usual methods. However, due consideration must be given to end restraints (e.g., a simply supported, uniformly loaded beam deflects five times as much as the same beam with ends fully fixed).

ACI limits short-term live-load elastic deflection to $L/180$ for flat roofs not supporting or attached to nonstructural elements likely to be damaged by large deflections, and to $L/360$ for floors under the same restrictions. In roof or floor construction supporting or attached to nonstructural elements not likely to be damaged by large deflections, the deflection which occurs after attachment is limited to $L/240$. This includes long-term deflections due to all sustained load plus the immediate deflection from any additional live load; camber may be deducted. The corresponding limit for members supporting or attached to elements which are likely to be damaged, such as partitions subject to cracking, is $L/480$ except where adequate countermeasures are taken to avoid damage.

Because concrete shrinks as it ages, and creeps or yields plastically under continuous pressure, the compression side of a reinforced-concrete beam shortens with age while the tension side undergoes very little change. Thus, deflections increase with age somewhat in proportion to the amount of compression constantly maintained in the concrete. The amount of such increase depends upon a variety of factors, one of the most important of which is the amount of compressive reinforcement. Table 12 gives fairly workable approximations of the *increase* in the short-term elastic deflection caused by loads that are more or less constantly in place, such as dead load and portions of the live load.

The part of the live load which is likely to be in place most of the time is, to a considerable extent, a question of the designer's judgment. Of the specified live load, only a small part (perhaps a third or so) might be considered permanent for churches or meeting rooms, while practically all would be considered permanent for storage warehouses.

Below certain span-depth ratios, deflections need not be computed. Table 13 lists such ratios, with footnotes indicating modifications required for lightweight concrete or for steel yield strengths other than 60,000 psi.

The ACI procedures are adequate for the control of deflections and may predict actual deflections in monolithic structures reasonably well. However, when tight fits are involved, e.g., glass panels under a spandrel beam, it is prudent to provide clearance or adjustment beyond the estimated deflection.

TABLE 12 Increase in Short-Time Deflection

Duration of load	$A_s' = 0$	$A_s' = A_s/2$	$A_s' = A_s$
1 month..............	0.6	0.4	0.3
6 months..............	1.2	1.0	0.7
1 year.................	1.4	1.1	0.8
5 or more years.........	2.0	1.2	0.8

17. Column Design—Combined Compression and Bending Columns rarely carry axial load alone. Because of loads applied off center, moments introduced at joints, or lateral loading in unbraced frames, column design must be for combined direct load and flexure.

TABLE 13 Minimum Thickness of Beams by Code 9.5*

Member	Min thickness h, in: Simply supported	One end continuous	Both ends continuous	Cantilever
	Members not supporting or attached to partitions or other construction likely to be damaged by large deflections			
Solid one-way slabs†	$l/20$	$l/24$	$l/28$	$l/10$
Beams or ribbed one-way slabs	$l/16$	$l/18.5$	$l/21$	$l/8$

*From Ref. 6.

†The span length l is in inches.

The values given in this table shall be used directly for nonprestressed reinforced-concrete members made with normal-weight concrete (w = 145 pcf) and Grade 60 reinforcement. For other conditions, the values shall be modified as follows:

(*a*) For structural lightweight concrete having unit weights in the range 90 to 120 lb/cu ft, the values in the table shall be multiplied by $1.65 - 0.005w$ but not less than 1.09, where w is the unit weight in lb/cu ft.

(*b*) For nonprestressed reinforcement having yield strengths other than 60,000 psi, the values in the table shall be multiplied by $0.4 + f_y/100{,}000$.

Effective Column Length. Figure 18*a* shows the deflected shape of a column in a braced frame and Fig. 18*b* one in an unbraced frame. When columns are long, the secondary moment $P\Delta$ can be substantial, especially in unbraced frames. Joint rotations are controlled by the relative stiffnesses $\Sigma EI/l_u$ of the columns and $\Sigma EI/l$ of the beams at the joints. The effective length of the column is kl_u, where l_u is the unsupported length of the column.

By Code definition a column in an unbraced frame is short if $kl_u/r < 22$ and in a braced frame if $kl_u/r < 34 - 12M_1/M_2$, where M_1 and M_2 are the column end moments, with $M_1 < M_2$. The ratio M_1/M_2 is positive for members in single curvature and negative for members in double curvature.

Since $k = 1$ and $M_1/M_2 = 1$ are on the safe side for braced frames, $kl_u/r = 34 - 12 \times 1 =$

22, the same as for unbraced frames. Also, since $r = 0.3h$ is a good approximation for rectangular cross sections, $l/h = 0.3 \times 22 = 6.6$ is near the changeover point from short column to long column for braced and unbraced frames.

Design aids. Because of the many variables involved, either interaction diagrams or tables are a practical requirement for the design of compression members. A typical

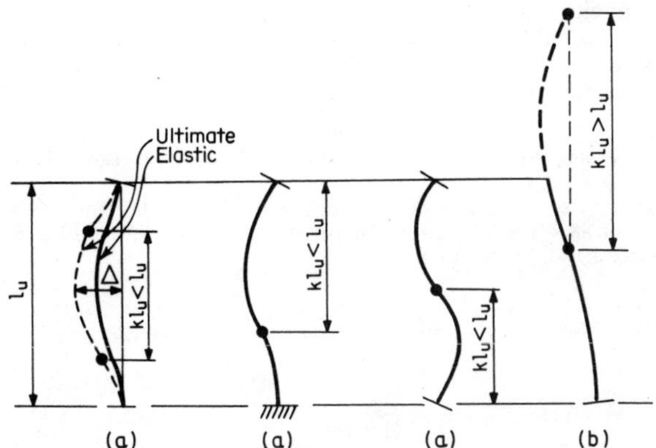

Fig. 18 Effective column length. (*a*) Braced frame; (*b*) unbraced frame.

interaction diagram is shown in Fig. 19. Point P_0 represents concentric compression ($M = 0$) and point M_0 pure bending ($P = 0$). At point a failure is by simultaneous crushing of the concrete and yielding of the steel, which gives a balanced design ($M = M_b$ and $P = P_b$). The segment P_0a represents failure which begins with crushing of the concrete and the portion M_0a failure which begins by yielding of the steel.

The Code specifies $\phi = 0.70$ for tied columns and 0.75 for spiral-reinforced columns.

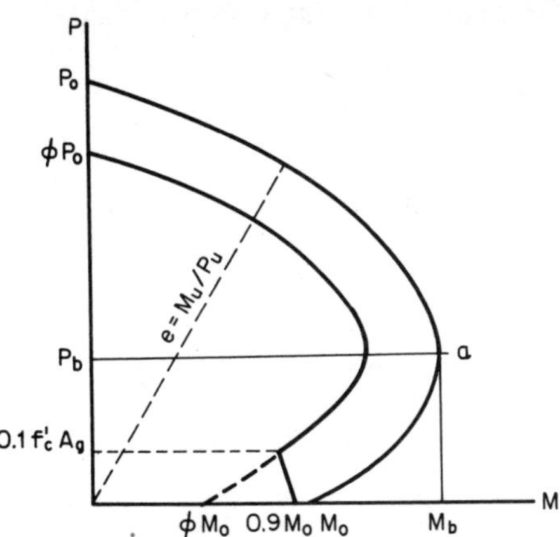

Fig. 19 Column interaction diagram.

Since $\phi = 0.9$ for beams, a linear transition from the value for beams to the value for columns is specified from $P = 0$ to $P = 0.10\,f'_cA_g$ (Fig. 19).

The Code also specifies the following maximum values of the load P_u for concentrically loaded, nonprestressed members:

Members with spiral reinforcement:

$$P_u = 0.85\phi[0.85f'_c(A_g - A_{st}) + f_yA_{st}] \tag{12}$$

Members with tied reinforcement:

$$P_u = 0.80\phi[0.85f'_c(A_g - A_{st}) + f_yA_{st}] \tag{13}$$

A_{st} is the total area of the longitudinal reinforcement. These equations replace the minimum accidental-eccentricity requirements for concentrically loaded members in earlier editions of ACI 318. The reduction in load is about the same as was achieved by the specified minimum eccentricities.

Interaction diagrams for f_y = 60,000 psi and concrete of any strength up to 4000 psi are given in Figs. 20 to 31. They can also be used for steel with other yield strengths without substantial error, at least for preliminary design. In the designation ρq for each curve, $q = f_y/0.85f'_c$.

Example 14 Design a square, tied, short column with bars on two opposite faces for a factored load P_u = 730 kips, M_u = 200 kip-ft, f'_c = 4 ksi, f_y = 60 ksi.

SOLUTION. $e = M_u/P_u = 200 \times 12/730 = 3.29$ in. Assume $\rho = 0.03$, $h = 17$ in., $\gamma = 0.7$. Then $e/h = 3.29/17 = 0.19$. $q = f_y/0.85f'_c = 60/(0.85 \times 4) = 17.6$, $q\rho = 17.6 \times 0.03 = 0.53$. On the chart of Fig. 21 these e/h and $q\rho$ values intersect at $K = 0.84$. Therefore,

$$bh = \frac{P_u}{\phi f'_c K} = \frac{730}{0.7 \times 4 \times 0.84} = 310 \text{ in.}^2$$

Try 18 × 18 in., for which

$$\frac{e}{h} = \frac{3.29}{18} = 0.18 \qquad K = \frac{730}{0.7 \times 4 \times 18^2} = 0.80$$

With these values $q\rho = 0.45$. Therefore, $\rho = 0.45/17.6 = 0.026$, which is close enough to the assumed value 0.03.

Instead of entering Fig. 21 with e/h and $q\rho$, one may assume values of b and h and proceed as follows:

Assume $b = h = 18$ in. and $\gamma = 0.7$.

$$K = \frac{P_u}{\phi f'_c bh} = \frac{730}{0.7 \times 4 \times 17 \times 17} = 0.80$$

$$K\frac{e}{h} = \frac{M_u}{\phi f'_c bh^2} = \frac{200 \times 12}{0.7 \times 4 \times 17 \times 17^2} = 0.15$$

Entering Fig. 21 with these values gives $q\rho = 0.45$, from which $\rho = 0.45/17.6 = 0.026$ as before.

Then $A_s = 0.026 \times 18^2 = 8.42$ in.² Use 4 No. 11 and 2 No. 10, for which $A_s = 8.78$ in.² For simplicity, many designers would use 6 No. 11 = 9.36 in²

$$\gamma = \frac{h - 2d'}{h} = \frac{18 - 2(1.50 + 0.5 + 0.70)}{18} = 0.70, \text{ assumed } 0.70.$$

Table 14 shows that there is no bar-spacing problem.

Use 18 × 18-in. column, 4 No. 11 + 2 No. 10 bars (No. 11 in corners).

Code 7.12.3 requires at least No. 4 ties be used with No. 11 bars, with maximum tie spacing the smallest of $16d_b = 16 \times 1.41 = 22.5$ in., 48 tie diameters = 24 in., or the least dimension of the column = 18 in., the last governing. Also, no bar should be more than 6 in. clear on either side from a braced bar. On a center-to-center basis, these bars are spaced 6.3 in., but not less than 6 in. clear. Crossties are not required, but to be conservative, will be used.

Use No. 4 closed ties at 18 in. around No. 11 corner bars plus No. 4 crossties around No. 10 bars, also at 18-in. spacing (Fig. 32).

Example 15 Design a short round spiral column for a factored load P_u = 1000 kips, M_u = 400 kip-ft, f'_c = 4 ksi, F_y = 60 ksi, reinforcement ratio ρ about 0.04.

SOLUTION. $e = 400 \times 12/1000 = 4.8$. Assume $h = 22$ in. $\quad e/h = 4.8/22 = 0.22$

$q = 60/(0.85 \times 4) = 17.6 \quad \rho q = 0.04 \times 17.6 = 0.70$. Assume $\gamma = 0.8$.

From Fig. 30, $K = 0.60 = \dfrac{1000}{0.75 \times 4h^2} \qquad h = 23.6$ in.

Try $h = 24$, $e/h = 0.20 \qquad K = \dfrac{1000}{0.75 \times 4 \times 24^2} = 0.58$

From Fig. 30, $\rho q = 0.55$. $\rho = 0.55/17.6 = 0.031 \qquad A_s = 0.031 \times 0.785 \times 24^2 = 14.13$ in.²

Use 10 No. 11 = 15.62 in.² or 12 No. 10 = 15.19 in.²

Either can be used, even if the bars are lap-spliced side by side (Table 15).

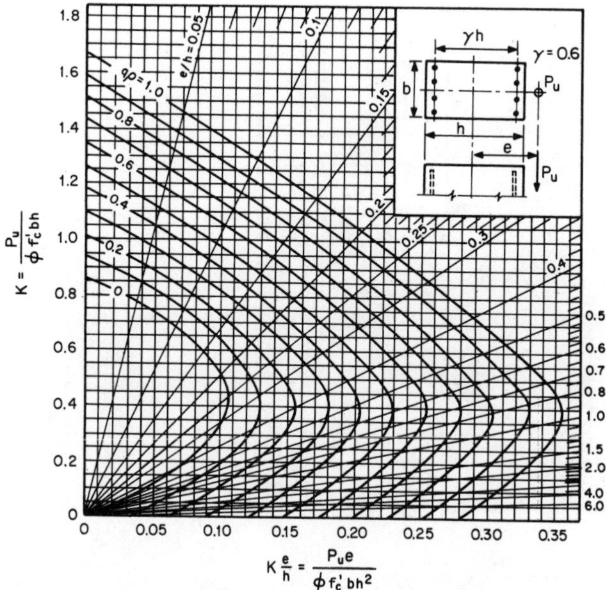

Fig. 20 Column interaction diagram—rectangular section, $\gamma = 0.9$, $f'_c \leq 4$ ksi, $f_y = 60$ ksi. (*Adapted from Ref. 3.*)

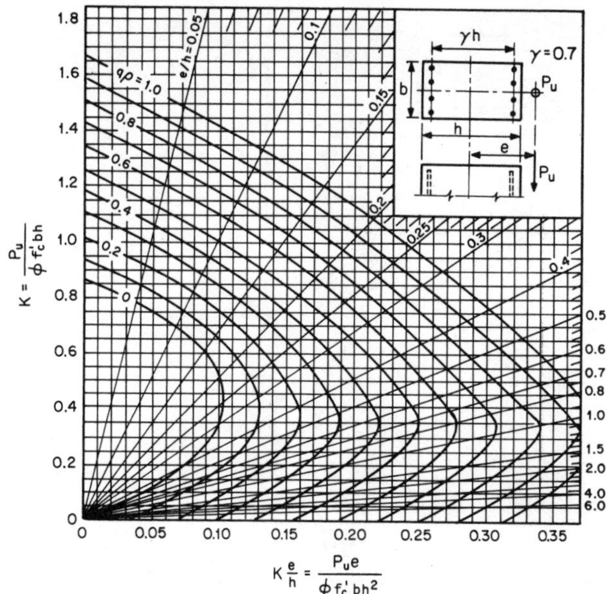

Fig. 21 Column interaction diagram—rectangular section, $\gamma = 0.8$, $f'_c \leq 4$ ksi, $f_y = 60$ ksi. (*Adapted from Ref. 3.*)

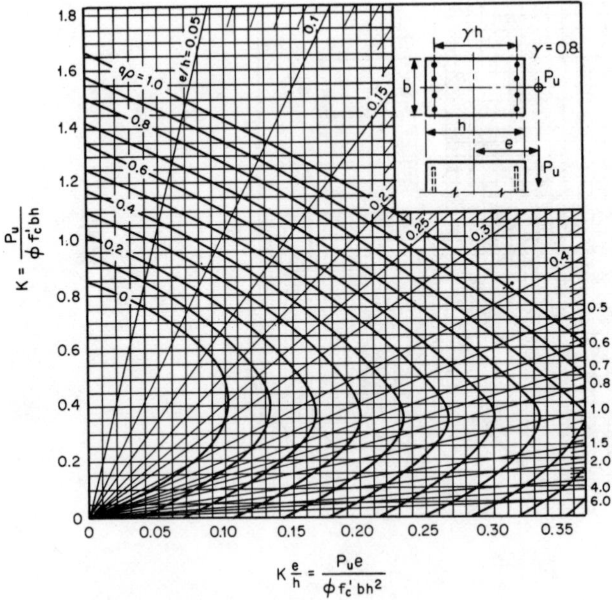

Fig. 22 Column interaction diagram—rectangular section, $\gamma = 0.7$, $f'_c \leqslant 4$ ksi, $f_y = 60$ ksi. (*Adapted from Ref.* 3.)

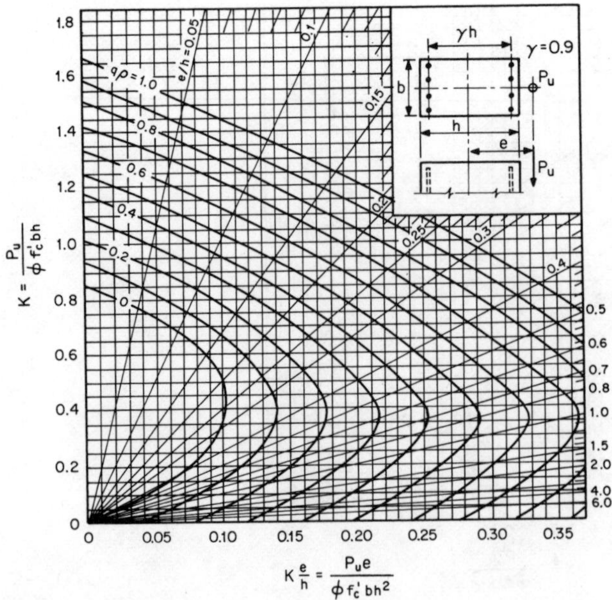

Fig. 23 Column interaction diagram—rectangular section, $\gamma = 0.6$, $f'_c \leqslant 4$ ksi, $f_y = 60$ ksi. (*Adapted from Ref.* 3.)

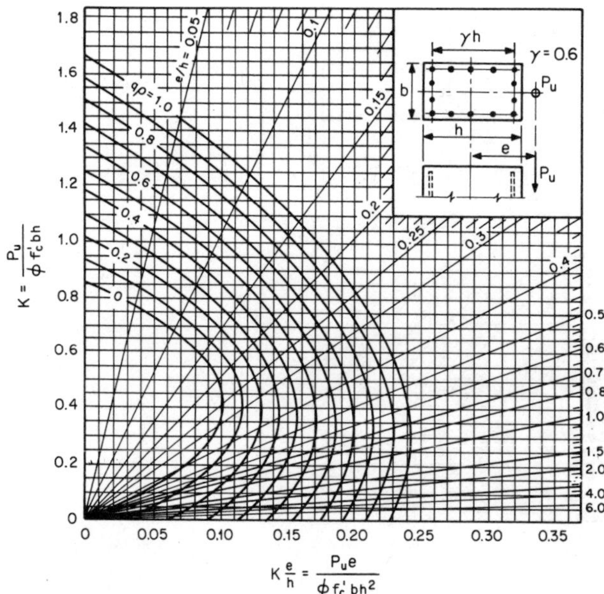

Fig. 24 Column interaction diagram—rectangular section, $\gamma = 0.9$, $f'_c \leqslant 4$ ksi, $f_y = 60$ ksi. (*Adapted from Ref. 3.*)

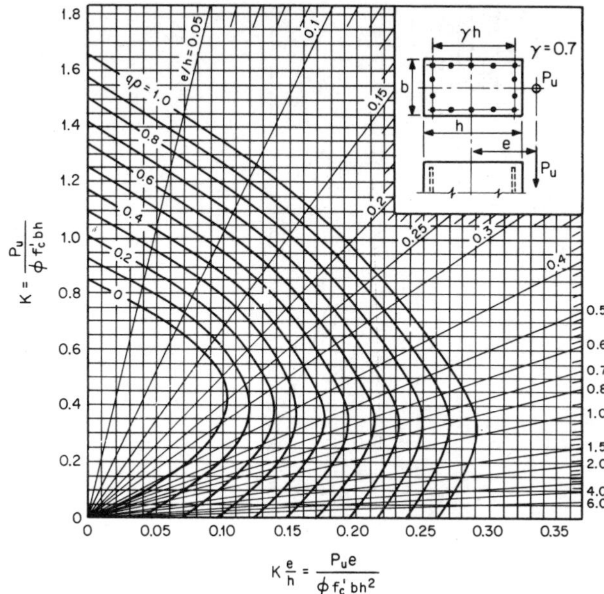

Fig. 25 Column interaction diagram—rectangular section, $\gamma = 0.8$, $f'_c \leqslant 4$ ksi, $f_y = 60$ ksi. (*Adapted from Ref. 3.*)

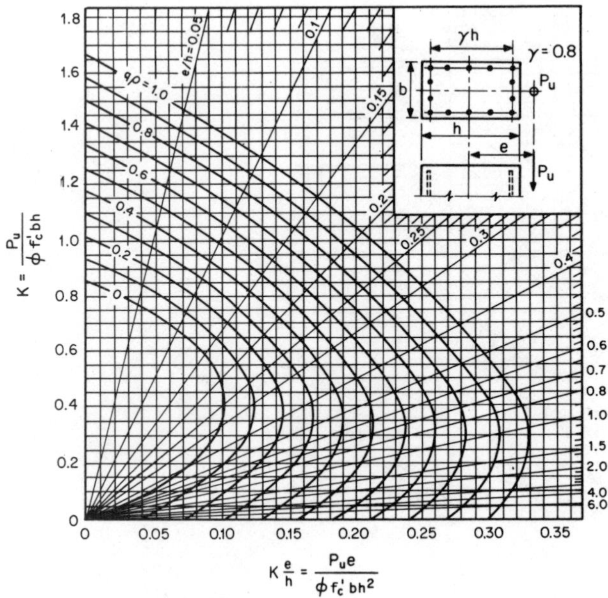

Fig. 26 Column interaction diagram—rectangular section, $\gamma = 0.7$, $f'_c \leqslant 4$ ksi, $f_y = 60$ ksi. (*Adapted from Ref. 3.*)

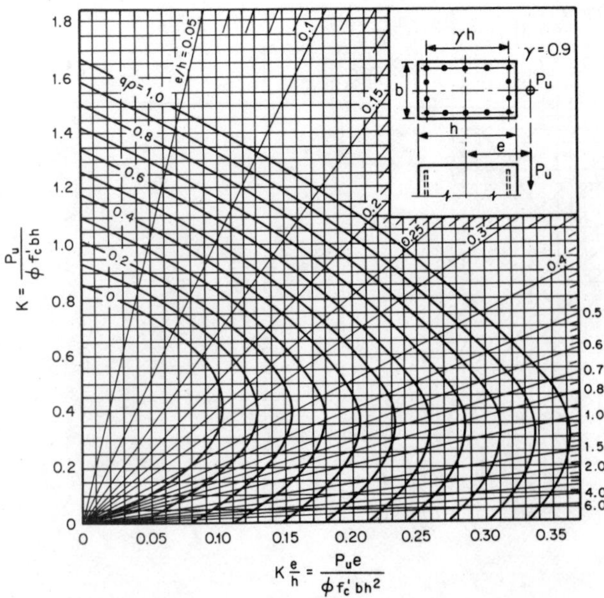

Fig. 27 Column interaction diagram—rectangular section, $\gamma = 0.6$, $f'_c \leqslant 4$ ksi, $f_y = 60$ ksi. (*Adapted from Ref. 3.*)

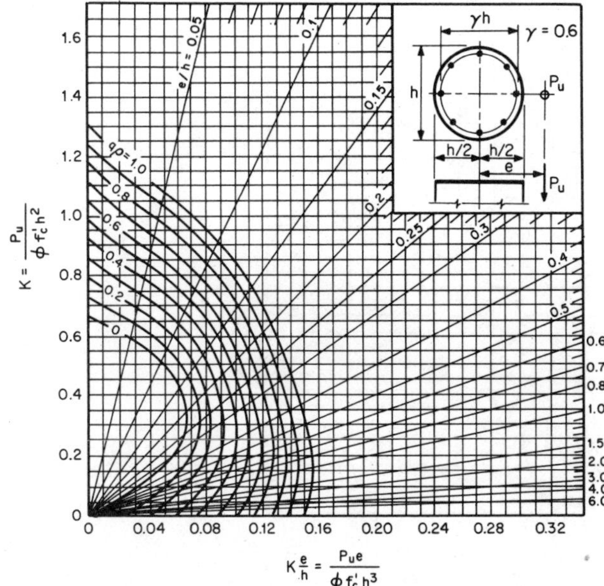

Fig. 28 Column interaction diagram—circular section with spiral, $\gamma = 0.9$, $f'_c \leq 4$ ksi, $f_y = 60$ ksi. (*Adapted from Ref. 3.*)

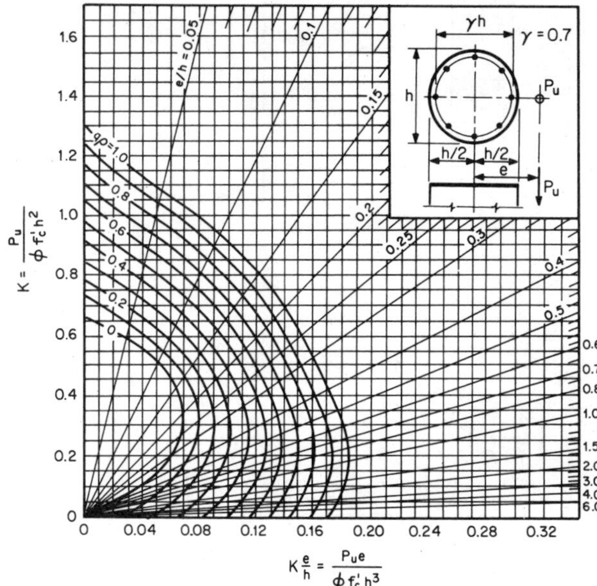

Fig. 29 Column interaction diagram—circular section with spiral, $\gamma = 0.8$, $f'_c \leq 4$ ksi, $f_y = 60$ ksi. (*Adapted from Ref. 3.*)

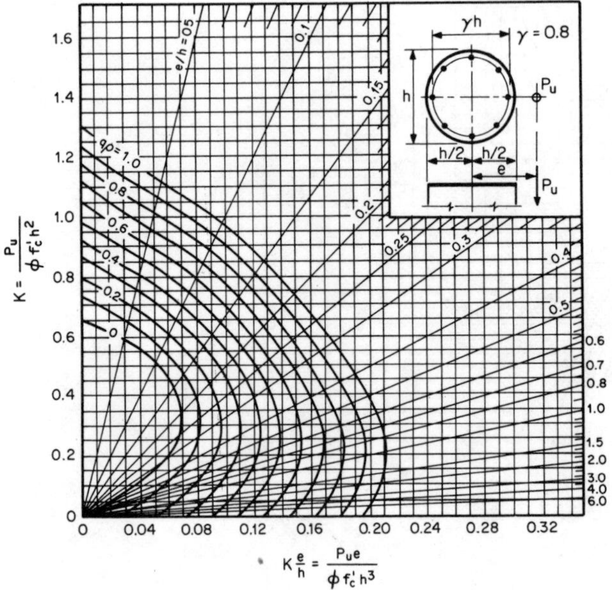

Fig. 30 Column interaction diagram—circular section with spiral, γ = 0.7, $f'_c \leqslant$ 4 ksi, f_y = 60 ksi. (*Adapted from Ref.* 3.)

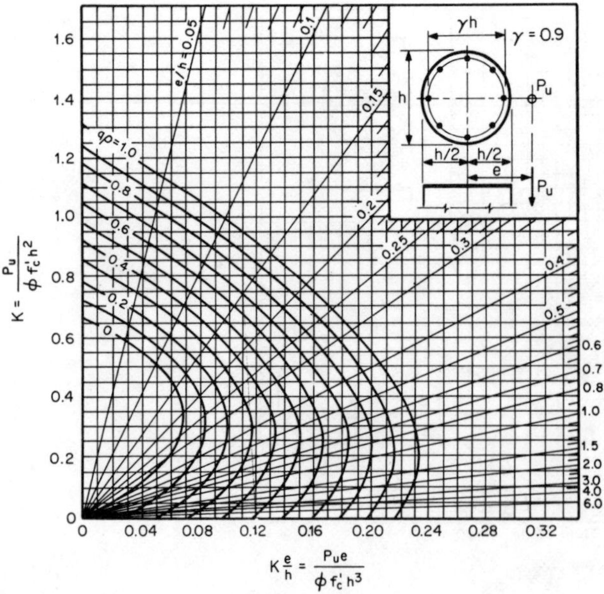

Fig. 31 Column interaction diagram—circular section with spiral, γ = 0.6, $f'_c \leqslant$ 4 ksi, f_y = 60 ksi. (*Adapted from Ref.* 3.)

TABLE 14 Maximum Number of Bars in One Face of Square Tied Columns*

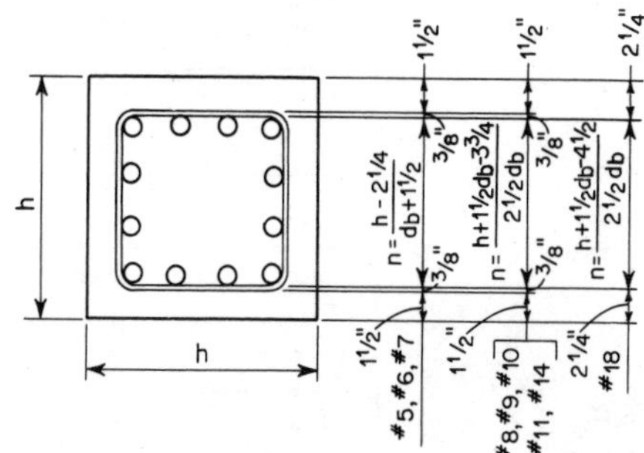

Column size C	No. of bars of a size *n*								
	No. 5	No. 6	No. 7	No. 8	No. 9	No. 10	No. 11	No. 14	No. 18
10	3	3	3	3	2	2	2		
11	4	3	3	3	3	2	2	2	
12	4	4	4	3	3	3	2	2	
13	5	4	4	4	3	3	3	2	
14	5	5	4	4	4	3	3	3†	
15	6	5	5	5	4	4	3	3	2
16	6	6	5	5	4	4	4	3	2
17	6	6	6	5	5	4	4	3	2
18	7	7	6	6	5	5	4	3	2
19	7	7	7	6	6	5	4	4	3†
20	8	7	7	7	6	5	5	4	3
21	8	8	7	7	6	6	5	4	3
22	9	8	8	7	7	6	5	4	3
23	9	9	8	8	7	6	6	5	3
24	10	9	9	8	7	6	6	5	4†
25	10	10	9	9	8	7	6	5	4
26	11	10	10	9	8	7	6	5	4
27	11	11	10	9	8	7	7	6	4
28	12	11	10	10	9	8	7	6	4
29	12	11	11	10	9	8	7	6	4
30	13	12	11	11	9	8	8	6	5
31	13	12	12	11	10	9	8	7	5
32	14	13	12	11	10	9	8	7	5
33	14	13	12	12	10	9	8	7	5
34	14	14	13	12	11	10	9	7	5
35	15	14	13	13	11	10	9	7	6
36	15	15	14	13	12	10	9	8	6
37	16	15	14	13	12	11	9	8	6
38	16	15	15	14	12	11	10	8	6
39	17	16	15	14	13	11	10	8	6
40	17	16	16	15	13	12	10	9	6

*Modified from Ref. 1.

†If this number of bars is used in each of four faces, ρ will exceed 8 percent.

TABLE 15 Maximum Number of Spliced Column Verticals in Various Patterns That Can Be Accommodated in Single Ring within Column Spirals*

Butt-welded

Outside dia. of column
1½"(#8–#14)
2¼"(#18 only)
⅜"φ Spiral all cases
†
1½ bar dia. (#8–#18)
(Bars smaller than #8 seldom welded)

Radially lapped

Outside dia. of column
1½"(#5–#14)
2¼"(#18 only)
⅜"φ Spiral all cases
†
1½"(#5–#8)
1½ bar dia. (#9–#18)

Circumferentially lapped

Outside dia. of column
1½"(#5–#14)
2¼"(#18 only)
⅜"φ Spiral all cases
†
1½"(#5–#8)
1½ bar dia. (#9–#18)

	Butt-welded						Radially lapped									Circumferentially lapped								
Diam of column	No. 8	No. 9	No. 10	No. 11	No. 14	No. 18	No. 5	No. 6	No. 7	No. 8	No. 9	No. 10	No. 11	No.‡ 14	No.‡ 18	No. 5	No. 6	No. 7	No. 8	No. 9	No. 10	No. 11	No.‡ 14	No.‡ 18
10†	6						6									6	6							
11†	7	6					8	7	6							7	7	6						
12†	9	7	6				9	8	7	6						8	7	7	6					
13†	10	9	7	6			10	9	8	7	6					9	8	8	7	6				
14	11	10	8	7			12	11	10	9	8	6				11	9	9	8	7	6			
15	12	11	9	8	6		13	12	11	10	8	7	6			12	11	10	9	8	7	6		
16	14	12	10	9	7		15	13	12	11	9	8	6			13	12	11	10	8	7	6		
17	15	13	11	10	8		16	15	14	12	11	9	7	6		14	13	11	11	9	8	7	6	
18	16	14	12	11	9		18	16	15	14	12	10	8	6		15	14	12	11	10	9	8	6	
19	17	15	13	12	10		19	18	16	15	13	11	9	7		16	15	13	12	11	9	8	7	

20	19	16	14	13	10	6§	21	19	18	16	14	12	10	8		17	16	14	13	12	10	9	7	
21	20	18	15	14	11	7§	22	20	19	17	15	13	11	8		18	17	15	14	12	11	10	8	
22	21	19	16	15	12	7§	23	22	20	18	16	14	11	9		20	18	16	15	13	12	10	8	6
23	22	20	17	15	13	8§	25	23	21	20	17	15	12	10	6	21	19	17	16	14	12	11	9	6
24	24	21	18	16	13	9§	26	24	22	21	18	16	13	11	7	22	20	18	17	15	13	11	9	6
25	25	22	19	17	14	9§	28	26	24	22	19	17	14	11	7	23	21	19	18	16	14	12	10	7
26	26	23	20	18	15	10	29	27	25	23	20	18	15	12	8	24	22	20	19	16	14	13	10	7
27	28	24	21	19	16	11	31	28	26	25	21	19	16	13	8	25	23	21	20	17	15	13	11	8
28	29	25	22	20	16	11	32	30	28	26	22	20	17	13	9	26	24	22	20	18	16	14	11	8
29	30	26	23	21	17	12	33	31	29	27	23	21	17	14	9		25	23	21	19	16	15	12	8
30	31	28	24	22	18	12	35	32	30	28	25	22	18	15	10		26	24	22	20	17	15	12	9
31	33	29	25	23	18	13		34	31	29	26	23	19	16	10		27	25	23	20	18	16	13	9
32	34	30	26	23	19	14			33	31	27	24	20	16	11			26	24	21	19	16	14	10
33	35	31	27	24	20	14			34	32	28	25	21	17	12			27	25	22	19	17	14	10
34	36	32	28	25	21	15				33	29	26	22	18	12				26	23	20	18	15	10
35	38	33	29	26	21	15				34	30	26	23	19	13				27	23	21	18	15	11
36		34	30	27	22	16					31	27	24	19	13					24	22	19	16	11
37		35	31	28	23	16					33	29	25	20	14					25	23	20	16	12
38		37	32	29	24	17					33	29	26	21	14					26	24	21	17	12
39		38	33	30	24	17					35	31	27	22	15					27	24	22	17	12
40			34	31	25	18						31	28	22	15						25	22	18	13
41			35	31	26	19						33	29	23	16						26	23	18	13
42			36	32	27	19						33	30	24	17						26	23	19	14
43			37	33	27	20						35	31	25	17						27	24	19	14
44			38	34	28	20							31	25	18							25	20	14
45				35	29	21							33	26	18							25	20	15
46				36	30	21							33	27	19							26	21	15
47				37	30	22							34	28	19							27	22	16
48				38	31	22							35	28	20							27	22	16

* From Ref. 1.
† ⅜-in. spiral too large to meet all code requirements in 10- to 13-in. columns, but ¼-in. spiral not standard.
‡ Lapped splices not recommended for No. 14 and No. 18 bars.
§ Limited to number of bars that provide a maximum of 8 percent vertical reinforcement.

$$\gamma = \frac{h - 2d'}{h} = \frac{24 - 2 \times 1.5 - 2 \times 0.375 - 1.27}{24} = 0.79, \text{ assumed } 0.8 \qquad \text{O.K.}$$

The spiral must replace the strength of the concrete cover. Code 10.9.2 requires the ratio ρ_s of the volume of spiral steel to the volume of the core out-to-out of spiral to be at least

$$\rho_s = \frac{0.45f_c'(A_g/A_c - 1)}{f_y} = \frac{0.45 \times 4(24^2/21^2 - 1)}{60} = 0.0092$$

Table 16 shows that a 2¼-in. pitch gives $\rho_s = 0.0093$.
Figure 33 shows the column cross section.

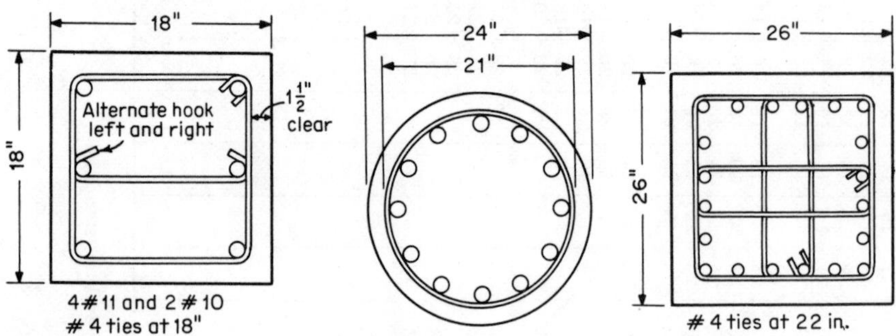

Fig. 32 Example 14. **Fig. 33** Example 15. **Fig. 34** Example 16.

Example 16 Design a square tied column, reinforced on all four faces, for a factored load $P_u = 1600$ kips, $M_u = 800$ kip-ft, $f_c' = 4$ ksi, $f_y = 60$ ksi.

SOLUTION. $e = 6$ in. Assume $h = 28$ in.; then $e/h = 0.21$. Try $\rho = 0.04$ $\qquad q\rho = 17.6 \times 0.04 = 0.70$. Assume $\gamma = 0.8$. From Fig. 26, $K = 0.85$.

$$\frac{P_u}{\phi f_c' bt} = \frac{1600}{0.7 \times 4h^2} = 0.85 \qquad h = 25.9 \text{ in.} \qquad \text{Try 26 in.}$$

$$\frac{e}{h} = \frac{6}{26} = 0.23 \qquad K = \frac{1600}{0.7 \times 4 \times 26^2} = 0.85$$

From Fig. 26, $\rho q = 0.80$ $\qquad \rho = \dfrac{0.80}{17.6} = 0.045.$

$$A_s = 0.045 \times 26^2 = 30.4 \text{ in.}^2 \qquad \text{Use 20 No. 11 or 14 No. 14}$$

Table 15 shows that either six No. 11 or five No. 14 can be placed on one face. Therefore, either arrangement can be used. Use 20 No. 11.

Check γ:

$$d' = 1.5 + 0.5 + 0.71 = 2.71 \text{ in.}$$

$$\gamma = \frac{26 - 2 \times 2.71}{26} = 0.79, \text{ assumed } 0.8 \qquad \text{O.K.}$$

Ties must brace each bar that is as much as 6 in. clear from neighboring braced bars. Distance center to center of corner bars is $26 - 2 \times 1.5 - 1.41 = 20.6$ in. Spacing of six bars is $20.6/5 = 4.12$ in. Therefore, must tie around alternate bars.

Use No. 4 ties at 22 in. ($14d_b = 22.6$ in.) vertically in sets of 3 pieces: one around all bars, one closed around two center bars on each face (Fig. 34).

18. Column Splices If bars must be offset between columns, the slope of the inclined portion with the axis of the column must not exceed 1 in 6, and the portion of the bar above and below the offset must be parallel to the axis of the column. Horizontal support at the offset bends can be provided by the floor construction or or by metal ties or spirals placed not more than 6 in. from the point of bend. The devices must be designed to resist a horizontal thrust equal to one and one-half times the horizontal component of the nominal force in the inclined portion of the bar.

When a column face is offset as much as 3 in., the bars must be spliced with separate

TABLE 16 Spirals as Percentages of Core Volume (Out to Out of Spirals)

Core diam, in.	⅝ in. diam Pitch, in.							½ in. diam Pitch, in.							⅜ in. diam Pitch, in.						
	2	2¼	2½	2¾	3	3¼	3½	2	2¼	2½	2¾	3	3¼	3½	1¾	2	2¼	2½	2¾	3	3¼
11															2.29						
12	5.17							3.33							2.09	1.83					
13	4.77							3.08							1.93	1.69					
14	4.44	3.94						2.86	2.54						1.80	1.57	1.40				
15	4.13	3.68	3.31					2.67	2.37	2.13					1.68	1.47	1.30	1.17			
16	3.88	3.45	3.10					2.50	2.22	2.00					1.57	1.38	1.22	1.10			
17	3.65	3.24	2.92	2.65				2.35	2.09	1.88	1.71				1.48	1.29	1.15	1.03	0.94		
18	3.45	3.06	2.76	2.51	2.30			2.22	1.97	1.78	1.62	1.48			1.40	1.22	1.09	0.98	0.89	0.81	
19	3.26	2.90	2.61	2.37	2.18			2.11	1.87	1.68	1.53	1.40			1.32	1.16	1.03	0.93	0.84	0.77	
20	3.10	2.76	2.48	2.25	2.07	1.91		2.00	1.78	1.60	1.45	1.33	1.23		1.26	1.10	0.98	0.88	0.80	0.73	0.68
21	2.95	2.63	2.36	2.15	1.97	1.82	1.69	1.90	1.69	1.52	1.38	1.27	1.17	1.09	1.20	1.05	0.93	0.84	0.76	0.70	0.64
22	2.82	2.51	2.26	2.05	1.88	1.74	1.61	1.82	1.62	1.45	1.32	1.21	1.12	1.04	1.14	1.00	0.89	0.80	0.73	0.67	0.62
23	2.70	2.40	2.16	1.96	1.80	1.66	1.54	1.74	1.55	1.39	1.26	1.16	1.07	0.99	1.09	0.96	0.85	0.76	0.70	0.64	0.59
24	2.59	2.30	2.07	1.88	1.72	1.59	1.48	1.67	1.48	1.33	1.21	1.11	1.03	0.95	1.05	0.92	0.81	0.73	0.67	0.61	0.56
25	2.48	2.21	1.98	1.80	1.65	1.53	1.42	1.60	1.42	1.28	1.16	1.07	0.98	0.91	1.01	0.88	0.78	0.70	0.64	0.59	0.54
26	2.39	2.12	1.91	1.73	1.59	1.47	1.36	1.54	1.37	1.23	1.12	1.03	0.95	0.88	0.97	0.85	0.75	0.68	0.62	0.56	0.52
27	2.30	2.04	1.84	1.67	1.53	1.41	1.31	1.48	1.32	1.19	1.08	0.99	0.91	0.85	0.93	0.82	0.72	0.65	0.59	0.54	0.50
28	2.21	1.97	1.77	1.61	1.48	1.36	1.27	1.43	1.27	1.14	1.04	0.95	0.88	0.82	0.90	0.79	0.70	0.63	0.57	0.52	0.48
29	2.14	1.90	1.71	1.55	1.43	1.32	1.22	1.38	1.23	1.10	1.00	0.92	0.85	0.79	0.87	0.76	0.67	0.61	0.55	0.51	0.47
30	2.07	1.84	1.65	1.50	1.38	1.27	1.18	1.33	1.19	1.07	0.97	0.89	0.82	0.76	0.84	0.73	0.65	0.58	0.53	0.49	0.45

* Ref. 1.

dowels. Lap splices of No. 14 and No. 18 bars are not acceptable under the Code, in either tension or compression, except for dowels into footings.

For bars carrying compression only, end bearing of bars, held together by an acceptable device, is permitted by Code 7.7.2. However, Code 7.10.5 requires a minimum *tensile* strength at each face of a column equal to 25 percent of the vertical reinforcement area times f_y. Also, Code 7.10.3 requires that bars with stresses ranging between f_y in compression and not more than $f_y/2$ in tension must have splices (or continuing bars) providing a tension resistance of at least twice the computed tension. If the tension calculated exceeds $f_y/2$, lap splices designed for f_y in tension, full-welded splices, or full positive connections are required by Code 7.10.4.

19. Columns with Biaxial Bending For round columns, biaxial bending does not require special design methods because resistance about any axis is the same. One simply designs for $\sqrt{M_x^2 + M_y^2}$, where M_x and M_y are moments about the x and y axes.

For square columns, it is more than safe (somewhat wasteful) to design for the sum of the eccentricities in the two directions, although it is not quite safe to design separately for moment about each axis.

Bresler proposed the following formula for the design of rectangular cross sections under biaxially eccentric load:[7]

$$\frac{1}{P_u} = \frac{1}{P_{x0}} + \frac{1}{P_{y0}} - \frac{1}{P_0} \tag{14}$$

where P_u = ultimate load with eccentricities e_x and e_y
P_{x0} = ultimate load for e_x only
P_{y0} = ultimate load for e_y only
P_0 = ultimate concentric load

Comparisons of the results of this formula with tests, and with ultimate loads for various cross sections computed according to the standard assumptions for ultimate-strength analysis, have shown it to be satisfactory.[8]

Charts and tables to facilitate the design of columns under biaxially eccentric load have been prepared.[2,9]

Figures 20 to 31 can also be used to evaluate Eq. (14).

Example 17 Compute the design load for the column of Fig. 35 for load eccentricities e_x = 12 in. and e_y = 4 in., with f_c' = 4 ksi, f_y = 60 ksi.

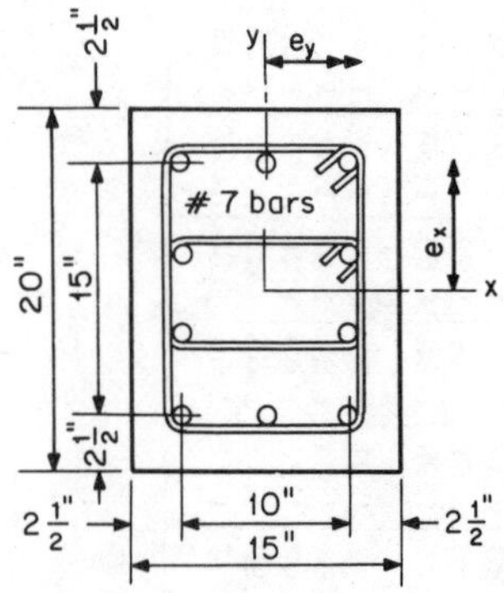

Fig. 35 Example 17.

SOLUTION. $A_s = 10 \times 0.6 = 6$ in.^{2} $\quad \rho = 10/300 = 0.0333$
$q = 60/0.85 \times 4 = 17.6 \quad q\rho = 0.59$

For the x axis:

$$\gamma = 15/20 = 0.75 \qquad e/h = 12/20 = 0.6$$

From Figs. 25 and 26, K = 0.36 and 0.38 for γ = 0.7 and 0.8, respectively. Therefore,

$$P_{x0} = 0.37 \times 4 \times 15 \times 20 = 444 \text{ kips}$$

For the y axis:

$$\gamma = 10/15 = 0.67 \qquad e/b = 4/15 = 0.27$$

From Figs. 24 and 25, $K = 0.62$ and 0.66 for $\gamma = 0.6$ and 0.7, respectively. Therefore,

$$P_{y0} = 0.65 \times 4 \times 15 \times 20 = 780 \text{ kips*}$$

The ultimate concentric load is

$$P_0 = 0.85 \times 4(15 \times 20 - 10) + 60 \times 10 = 1586 \text{ kips}$$

Substituting into Eq. (14) gives

$$\frac{1}{P_u} = \frac{1}{444} + \frac{1}{780} - \frac{1}{1586} = 0.00290 \qquad P_u = 345 \text{ kips}$$
$$P = 0.7 \times 345 = 241 \text{ kips}$$

20. Stairs Concrete stairs are usually constructed after the floors are placed, using dowels to tie them in place. They may be framed in a variety of ways (Fig. 36). They are

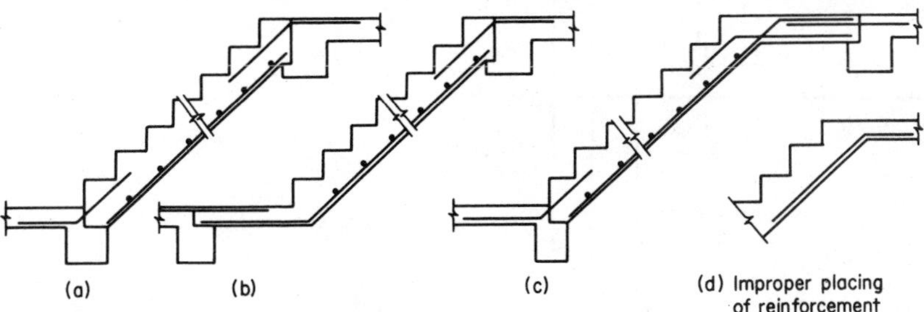

Fig. 36 Stair framing.

usually designed as simple spans, since they are a small part of the structure and the care and expense involved in locating negative-moment bars are not justified. The design is simple. The span is taken as the horizontal distance L between supports and the live load in psf of horizontal projection (Fig. 37). The dead load is calculated as for a slab of length L and thickness z plus one-half the height of riser, plus weight of any finish. Slab dimensions t and d are usually measured at the heel of the step.

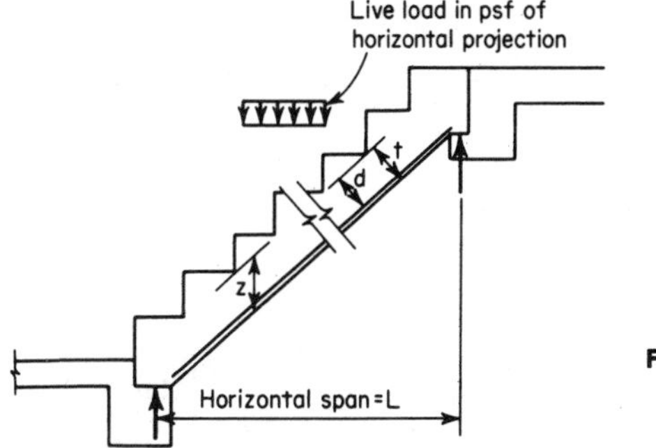

Fig. 37

Typical reinforcement is shown in Fig. 36. Reinforcement at the upper landing should always be spliced and lapped as shown; reinforcing bars in tension should never be bent around a reentrant corner as in *d*.

21. Wall Footings The width of wall footings is determined by the area required to distribute the load. The allowable soil pressure is usually stated for unfactored service

*Comparisons of Eq. (14) with analyses and results of the tests in Ref. 7 were based on P_0 as computed here. According to Eq. (13), P_0 would be 80 percent of this value, which would give a less conservative result ($P = 255$ kips).

loads. Depth is determined by structural requirements for factored loads. Cross reinforcement is required if the flexural stress exceeds the allowable tension for plain concrete. Longitudinal reinforcement is desirable for spacing cross reinforcement.

The bending moment is taken at the face of a concrete wall, column, or pedestal and halfway between the middle and the face for masonry walls.

Example 18 Design a footing for a 12-in. concrete wall carrying a live load of 10,000 plf and a dead load of 5000 plf. Allowable soil pressure = 5000 psf, f'_c = 3000 psi, f_y = 60,000 psi. Neglect weight of footing.

SOLUTION. (Fig. 38). Width of footing = 15,000/5000 = 3 ft. Try a 12-in. depth with d = 8½ in. to allow 3 in. of concrete cover.

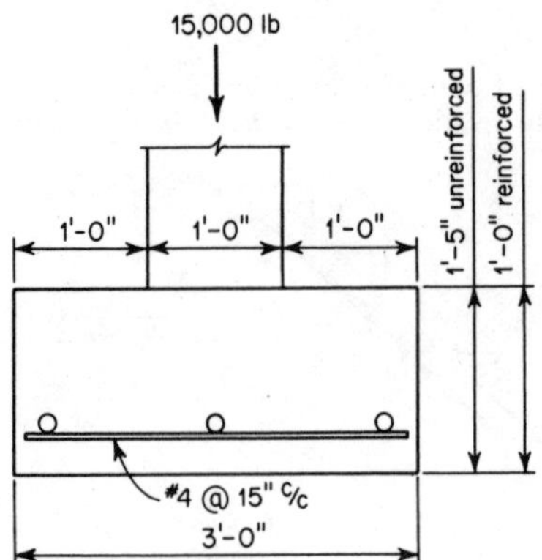

Fig. 38 Example 18.

Factored load w = 1.4 × 5000 + 1.7 × 10,000 = 24,000 plf = 8000 psf.

$$M_u = (8000 \times 1)0.5 = 4000 \text{ lb-ft} = 48{,}000 \text{ lb-in.}$$

Assume internal lever arm = $0.9d$ = 7.65 in.

$$A_s = M_u/(7.65\,\phi f_y) = 48{,}000/(0.9 \times 60{,}000 \times 7.65) = 0.116 \text{ in.}^2/\text{ft}$$

Even though slabs of uniform thickness are excluded in Code 10.5.1, check the minimum reinforcement ratio:

$$\rho_{min} = 200/f_y = 200/60{,}000 = 0.0033 \text{ min} \qquad A_s = \rho bd = 0.0033 \times 12 \times 8.5 = 0.34 \text{ in.}^2$$

This minimum is not required where A_s is over one-third more than required for the given moment. Therefore, use 1.33 × 0.116 = 0.154 in.²/ft.

Use No. 4 at 15 in., A_s = 0.20 × 12/15 = 0.16 in.²/ft. Required development $l_d = 0.04A_b f_y/\sqrt{f'_c}$ = 0.04 × 0.20 × 60,000/$\sqrt{3000}$ = 8.8 in.

From face of wall the bar projects 12 in. minus end cover > 8.8 in., O.K. Shear at d from face of concrete wall (Code 11.2.2) = 8000(12 − 8.5)/12 = 2330 lb. $v = V_u/(\phi bd)$ = 2330/(0.85 × 12 × 8.5) = 27 psi < $2\sqrt{3000}$ = 109 psi. Longitudinal spacer bars may be 2 No. 3 or 2 No. 4. The latter are easier to place and are of the same diameter as moment bars, which may be convenient.

22. Column Footings Individual footings for columns may be square, rectangular, polygonal, round, or irregular to suit the space available (Fig. 39). Square footings are the simplest. Theoretically, footings could taper in depth to a working minimum at the periphery, but as they are frequently constructed by placing a prepared mat of reinforcing bars in a neatly excavated pit and filling to the required level with concrete, such tapering involves more labor than the value of the concrete saved.

Combined footings (Fig. 39*d, e*) may be rectangular or trapezoidal in plan. The trapezoidal footing is used when it is impossible to extend the footing sufficiently beyond the more heavily loaded column. The centroid of the bearing pressure should coincide as nearly as is practicable with that of the loads. In the absence of specific recommendations to the contrary by a soils engineer, the soil pressure may be assumed uniform over the contact surface. Cantilevered footings (Fig. 39g) may be used in place of the combined footing.

The main requirement for footings is that they be safe, simple, and economical. They should be proportioned to produce substantially equal settlements of the supported columns.

Example 19 Design a square footing to support a column load of 200 kips dead + 260 kips live load on 5000-psf soil. $f_c' = 3000$ psi, $f_y = 60{,}000$ psi, column 18 in. square.

SOLUTION. Column load + estimated weight of footing = 460 + 40 = 500 kips. Required size of footing = 500/5 = 100 sq ft = 10 ft square.

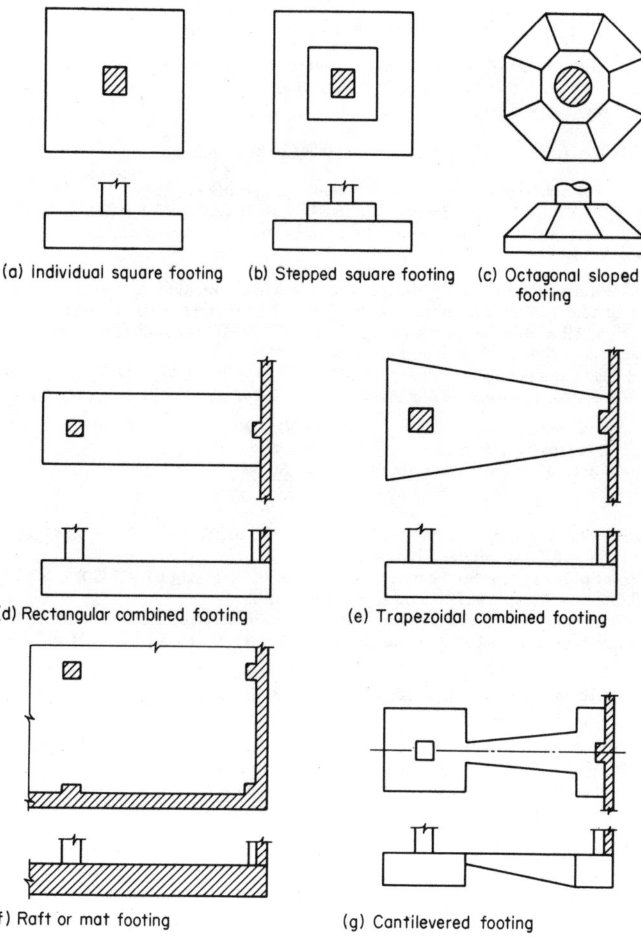

Fig. 39 Typical column footings.

Try a footing 1 ft 10 in. deep with 3 in. clear cover for placement on earth. With bars in two directions, d for the upper layer will be about 17.5 in.

Effective soil pressure $w = (200 \times 1.4 + 260 \times 1.7)/100 = 7.22$ ksf. Except on long narrow footings, punching shear around the column usually determines necessary depth. This is calculated around a perimeter located $d/2$ from the column face (Code 11.10b).

$$\text{Perimeter} = b_0 = 4(18 + 2 \times 17.5/2) = 4 \times 35.5 = 142 \text{ in. or } 2.96 \text{ ft per side}$$

$$V_u = 7.22(10^2 - 2.96^2) = 659 \text{ kips}$$

$$v_u = V_u/(\phi b_0 d) = 659{,}000/(0.85 \times 142 \times 17.5) = 312 \text{ psi}$$

Allowable $v_u = 4\sqrt{3000} = 219$ psi = 0.7 of that needed.

Noting that b_0 also increases as d is increased, try $d = 22.5$ in. = 1.88 ft, thickness = 27 in.

$b_0 = 4(18 + 22.5) = 162$ in. or 3.37 ft per side
$V_u = 7.22(10^2 - 3.37^2) = 640$ kips $\quad v_u = 640{,}000/(0.85 \times 162 \times 22.5) = 207 \text{ psi} < 219$ O.K.

Beam-type shear per foot of width of footing on plane d = 22.5 in. from column:

$$V_u = 7.22 \times 1(5 - 0.75 - 1.87) = 17.2 \text{ kips}$$

$$v_u = 17{,}200/(0.85 \times 12 \times 22.5) = 75 \text{ psi} < 2\sqrt{f_c'} = 109 \text{ psi}$$

Check M_u at column face = $7.22(5 - 0.75)^2/2 = 65.2$ kip ft/ft

$$M_u/(\phi bd^2) = 65{,}200 \times 12/(0.9 \times 12 \times 22.5^2) = 143$$

From Table 5, $\rho = 0.0025$, min $\rho = 200/f_y = 200/60{,}000 = 0.0033 = (4/3)0.0025$.*
$A_s = 0.0033 \times 12 \times 22.5 = 0.90 \text{ in.}^2$
Use 15 No. 7 at 8 in. each way ($A_s = 0.90 \text{ in.}^2$).

$$l_d = 0.04A_bf_y/\sqrt{f'_c} = 0.04 \times 0.60 \times 60{,}000/\sqrt{3000} = 26.3 \text{ in.}$$

Since footing extends over 4 ft from column face, development is no problem.

The assumed weight of 40,000 lb can now be checked as 10 × 10 × 2.25 × 150 = 33,800 lb. If a higher degree of precision seems desirable, a recomputation can be made. No substantial savings over the present design appear likely.

Example 20 Design a combined footing for a 12-in. exterior wall whose outside face is on a property line, 20 ft from the center of a 20 × 20-in. interior column (Fig. 40). The load on the wall is 200 kips and that on the column 325 kips. Allowable soil pressure is 5000 psf, $f_c' = 4000$ psi, and $f_y = 60{,}000$ psi. Assume average load factor is 1.65 on both wall and column.

SOLUTION. At service loads, the centroid of the loads is 0.5 + 325 × 19.5/525 = 12.57 ft from the building line, making the length of a concentric, rectangular footing 2 × 12.57 = 25.14 ft, or, say, 25 ft.

Load on exterior wall	= 200,000
Load on interior column	= 325,000
Footing (at 8 to 10 percent of load)	= 50,000
	575,000 lb

$b = 575{,}000/(5000 \times 25) = 4.6$ ft = 4 ft 8 in. Net soil pressure = $525{,}000/(4.67 \times 25) = 4500$ psf at service loads or $4500 \times 1.65 = 7420$ psf under factored loads.

The distance to the point of zero shear from exterior face is $200{,}000 \times 1.65/(4.67 \times 7420) = 9.52$ ft. Max negative $M_u = -330(9.52 - 0.5) + 34.7 \times 9.52^2/2 = -1410$ kip ft

Clearances or other constraints sometimes demand minimum member size, but economy of the member itself more often goes with ρ smaller than the maximum. Try $\rho = 0.015$, for which $M_u/\phi bd^2 = 781$. Then

$$781 = 1410 \times 1000 \times 12/(0.90 \times 4.67 \times 12d^2)$$
$$d = 20.7 \text{ in}$$

Punching shear seems improbable with this d plus the 20-in. column using up 41 in. of the total 56-in. width. Hence, check flexural shear first. Shear usually controls footing depth, since heavy stirrups in wide members are awkward to place. Flexural shear, by inspection, is maximum near the inside face of the wall or the inside face of the column. Net soil pressure on 56-in. width is 4.67 × 7.42 = 34.7 kips/ft

$$V_{\text{wall}} = 330 - 34.7 \times 1 = 295 \text{ kips}$$

Summing forces from left and including the column,

$$V_{\text{col}} = 34.7(5.0 + 10/12) - 1.65 \times 325 = 202 - 536 = -334 \text{ kips}$$

Critical flexural shear is at d from column or wall.

$$\text{Crit. } V_{\text{col}} = -334 + 34.7 \times 19/12 = -334 + 55 = -279 \text{ kips}$$

$$v_u = V_u/(0.85\,bd) = 279{,}000/(0.85 \times 56 \times 21) = 278 \text{ psi} > 2\sqrt{4000} = 126 \text{ psi}$$

The d of 21 in. could be used with this shear, but the authors prefer not over half the shear to be carried by stirrups in footings, that is, $v_u \eqsim 4\sqrt{f_c} = 252$ psi. Try $d = 24$ in.

$$\text{Crit. } V_{\text{col}} = -334 + (34.7 \times 24/12) = -265 \text{ kips}$$
$$v_u = 265{,}000/(0.85 \times 56 \times 24) = 232 \text{ psi. Stirrups required. (Fig. 40}d\text{.)}$$

Crit. V_{col} on extension beyond column = 34.7(5 − 0.83 − 2) = 75.3 kips < ⅓ × 265 kips (231 psi) on other side of column. No stirrups are needed here.

*Whether footings are excluded (as slabs of uniform thickness) from this minimum ρ might be debated legally, but high shear and low ρ are always a bad combination.

$$\text{Crit. } V_{\text{wall}} = 295 - 34.7 \times 2 = 226 \text{ kips}$$

$$v_u = 226{,}000/(0.85 \times 56 \times 24) = 198 \text{ psi} > 2\sqrt{f_c'}. \text{ Stirrups required.}$$

Punching shear, on the boundary of a square $d/2$ beyond column, i.e., a square of 20 + 24 = 44 in. (3.67 ft) each side gives $b_0 = 4 \times 44 = 176$ in.

$$\text{Crit. } V_u = 325 \times 1.65 - 7.42 \times 3.67^2 = 536 - 100 = 436 \text{ kips}$$

$$v_u = 436{,}000/(0.85 \times 176 \times 24) = 121 \text{ psi} < 4\sqrt{f_c'} = 4\sqrt{4000} = 252 \text{ psi} \qquad \text{O.K.}$$

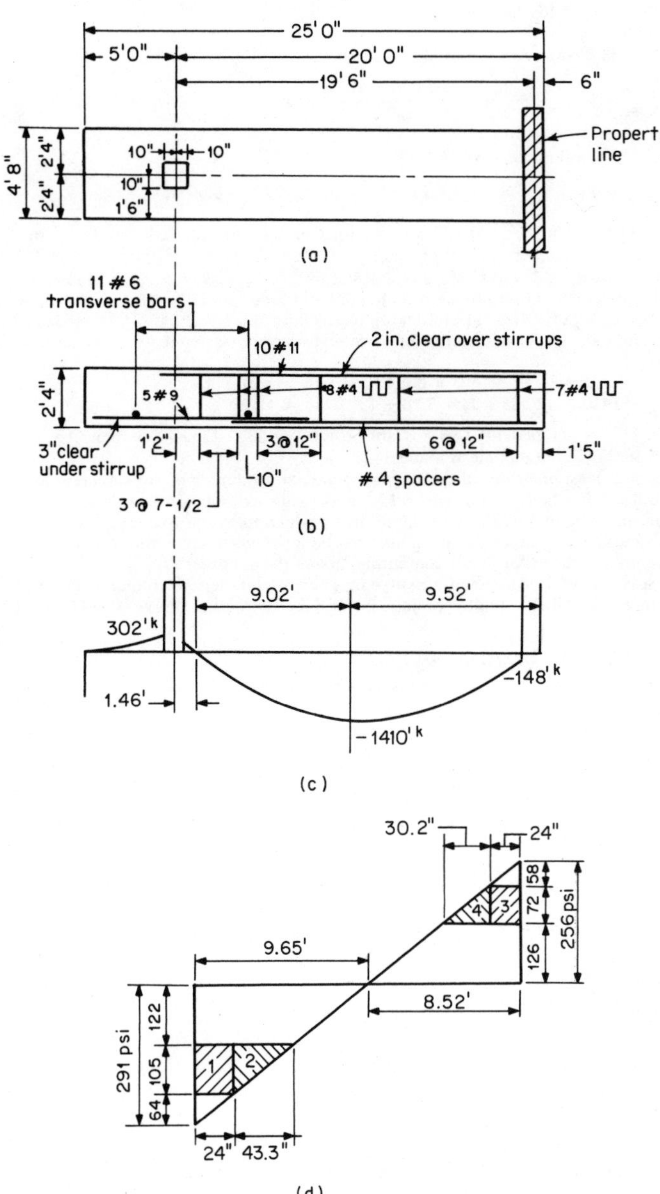

Fig. 40 Example 20.

Stirrups will be designed after flexural reinforcement is established.
Use thickness = 28 in., d = 24 in.
Footing weight = 25 × 4.67 × 2.33 × 150 = 40,800 lb vs. 50,000 lb assumed. O.K.
TOP BARS, LONGITUDINAL.

$$\frac{M_u}{\phi bd^2} = \frac{1{,}410{,}000 \times 12}{0.9 \times 56 \times 24^2} = 583$$

Table 5 gives ρ = 0.0108. A_s = 0.0108 × 56 × 24 = 14.52 in.² Use 10 No. 11 = 15.6 in.²

Required l_d with 1.4 factor for top bars = 1.4 × 0.04 × 156 × $60{,}000/\sqrt{4000}$ = 82.9 in. = 6.9 ft. Continue all bars to end under wall.

BOTTOM BARS, LONGITUDINAL. At exterior face of column, $M_u = 34.7(5 - 0.83)^2/2$ = 302 kip-ft

$$\frac{M_u}{\phi bd^2} = \frac{302{,}000 \times 12}{0.9 \times 56 \times 24^2} = 125$$

Table 5 gives ρ = 0.0027. Minimum ρ = 200/60,000 = 0.0033

$$A_s = 0.0033 \times 56 \times 24 = 4.44 \text{ in.}^2 \qquad \text{Use 5 No. 9} = 5.00 \text{ in.}^2$$

Required $l_d = 0.04 \times 1.00 \times 60{,}000/\sqrt{4000}$= 37.9 in. = 3.1 ft. Continue all No. 9 to left end of footing.

Draw moment diagram to detail bars to right of column (Fig. 40c). At right of column, $M_u = 34.7(5 + 0.83)^2/2 - 1.65 \times 325 \times 0.83 = +145$ kip-ft. M_u at left face of wall = −0.5(330 − 34.7) = 148 kip-ft.

For the top bars, the development requirements at the point of inflection (Fig. 40c) are the same as for positive-moment bars except for the 1.4 factor for top bars. Therefore, $l_a = 1.3M_t/V_u + l_d$, where 1.3 represents the 30 percent increase for reinforcement confined by a compressive reaction (Code 12.2.3).

$M_t \approx M_u$ = 1410 kip-ft
V_u = 34,7 × 9.02 = 313 kips (since V = 0 at point of maximum moment)
Allowable l_d = 1.3 × 1410 × 12/313 + l_a = 70.3 + l_a

Required l_d for No. 11 bars = $1.4(0.04 \times 1.56 \times 60{,}000/\sqrt{4000})$ = 83 in. Development is O.K. if bars are extended l_a = 13 in. beyond the point of inflection.

Code 12.3.3 requires at least one-third of the negative-moment reinforcement to extend at least d beyond the point of inflection, which is 24 in. (since $12d_b$ and 1/16 of clear span are each less). However, the compression from the column modifies the conditions on which this requirement is based, and a designer who understands these various complications can be less conservative, running five bars to the center of the column and the other five to the farther face of the column.

BOTTOM BARS, TRANSVERSE. Use bottom transverse bars centered under the column over a length equal to width of footing = 4.67 ft. The footing projects beyond the face of the columns 0.5(4.67 − 1.67) = 1.50 ft

$$M_u = 4.67 \times 7420 \times 1.50^2/2 = 39{,}000 \text{ lb-ft total}$$

$A_s = M_u/(\phi f_y \times 0.9d)$ = 39,000 × 12/(0.9 × 60,000 × 0.9 × 23) = 0.42 in.²
Min. ρ = 200/60,000 = 0.0033 A_s = 0.0033 × 4.67 × 12 × 23) = 4.29 in.² Use 10 No. 6 = 4.42 in.²

$l_d = 0.04 \times 0.44 \times 60{,}000/\sqrt{4000}$ = 16.7 in. vs. 16 in. available.
Try 11 bars, l_d = (4.29/4.86) × 16.7 = 14.7 in. O.K. Use 11 No. 6 bottom transverse bars.

STIRRUPS. The unit-shear diagram is shown in Fig. 40d. The shaded areas require stirrups. For the d = 24-in. length adjacent to the face of the column,

$$A_v = \frac{105 \times 56 \times 24}{60{,}000} = 2.35 \text{ in.}^2$$

Use 3 No. 4 double-U stirrups at 7.5 in., which gives 4 × 0.2 × 24/7.5 = 2.55 in.
At x = 36 in. from column face,
v_s = 105 −12 × 2.52 = 75 psi
s = 0.8 × 60,000/(75 × 56) = 11.4 in. Use 11 in.
At x = 48 in., v_s = 75 − 11 × 2.52 = 47 psi
Maximum s = d/2 = 12 in. controls

For area 3, with No. 4 double-U stirrups, s = 0.8 × 60,000/(72 × 56) = 11.9 in. Therefore, use maximum allowable (12 in.) for areas 3 and 4.

STIRRUP DEVELOPMENT. Code 12.13.1. A straight leg must have a basic development length l_d above or below middepth of member, with a minimum of 24 stirrup diameters (here 12 in.). Therefore, a hook is required at the top, its 2 in. inside diameter (Code 7.1.3.1) using a total of 1.5 in. plus cover of 2 in. from start of hook, which leaves 24/2 − 2.5 = 8.5 in. for the 0.5 l_d required in addition to the hook.

0.5 $l_d = 0.5(0.04 \times 0.20 \times 60{,}000/\sqrt{4000})$ = 3.79 in. <8.5 in. available

Use No. 4 longitudinal spacers in each corner of stirrups where main longitudinal bars are not available.

Example 21 Design a cantilever footing for a 12-in. exterior wall on the property line, which is 20 ft from the center of a 20-in.-square column (Fig. 41). Wall load = 160 kips, column load = 300 kips, average load factor on each = 1.65, allowable soil pressure = 6000 psf, f'_c = 3750 psi, f_y = 60,000 psi.

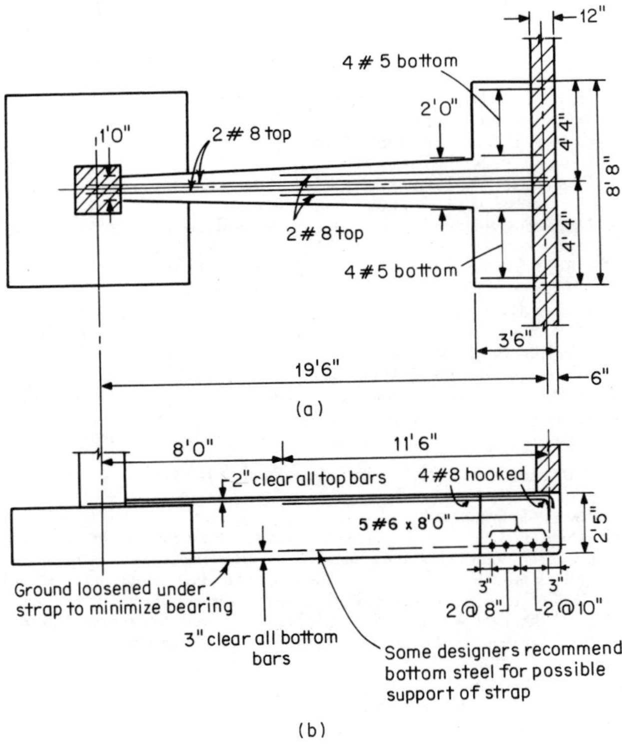

Fig. 41 Example 21.

SOLUTION. Assume the footing under the exterior wall to be 3.5 ft wide. Distance from center of this footing to center of wall is then 1.25 ft and to the center of the interior column is 20 − 1.75 = 18.25 ft. Neglecting weight of the strap beam, the uplift on the interior column (a downward load on the wall footing) is 160,000 × 1.25/18.25 = 10,960 lb.

Wall load	= 160,000 lb
Downward reaction	= 10,960 lb
Footing (assumed)	= 11,000 lb
Total	= 182,000 lb at service loads

Required area of wall footing = 182,000/6000 = 30.3 ft².

Use footing 3 ft 6 in. by 8 ft 8 in. = 30.3 ft²

Net soil pressure = (182,000 − 11,000)/30.3 = 5640 psf at service loads or 5640 × 1.65 = 9310 psf with factored loads.

Assume a strap 2 ft wide where it joins the wall footing, leaving the footing projecting 3.33 ft on each side. If the footing supported a column instead of a wall, it would act as balanced cantilevers beyond each side face of the column and strap beam. The wall, however, delivers its load along the length of the footing but is not interconnected to provide a moment reaction for footing strips perpendicular to it. On the outside strip parallel to the wall it is on the safe side to assume all the net soil pressure carried to the strap beam, creating the worst design shear situation there. This ignores the fact that some reaction is carried directly to the wall because of diagonal strips between wall and strap. It also ignores the torsion developed because footing deflection along the wall is near zero and largest in the corner. If the footing projected much farther from the wall, a more exact analysis of combined shear and torsion would be indicated. Here it is assumed the conservative shear calculation offsets the neglected torsion.

On a 1 ft-strip at the edge* of strap

$$V_u = 9310(4.33 - 1.0) = 31{,}000 \text{ lb}$$

Required $d_v = V_u/\phi v_c b = 31{,}000/(0.85 \times 2\sqrt{3750} \times 12) = 24.8$ in. Use $h = 29$ in., $d = 25.6$ in. The footing may be sloped or stepped down from the strap toward the outer edges; however, a uniform depth will probably be more economical.

$M_u = 9310 \times 3.33^2/2 = 51{,}600$ lb-ft/ft
$A_s = M_u/\phi f_y 0.9d = 51{,}600 \times 12/(0.90 \times 60{,}000 \times 0.9 \times 25.6) = 0.50$ in.²/ft
Minimum $\rho = 200/f_y = 200/60{,}000 = 0.0033$†
Minimum $A_s = 0.0033 \times 25.6 \times 12 = 1.01 \text{ in.}^2 > (4/3)0.50 = 0.67$ in.²/ft
Use 5 No. 6 × 8 ft 0 in. spaced from inner edge as shown on Fig. 41*a*.
$l_d = 0.04 \times 0.44 \times 60{,}000/3750 = 17.2 \text{ in.} <$ available 37 in.

Nominal reinforcing is also desirable perpendicular to wall.

Use on each side of strap beam 4 No. 5 at 10 in., total of 8 bars.
Weight of footing = 3.5 × 8.67 × 2.42 × 150 = 11,000 lb, as assumed.

DESIGN OF STRAP BEAM. Neglecting weight of strap and the limited soil reaction between footings, the point of zero shear is 160,000 × 1.65/(8.67 × 9300) = 3.28 ft from the building line.

$M_u = -(160{,}000 \times 1.65)(3.28 - 0.50) + 8.67 \times 9300 \times 3.28^2/2 = -300{,}200$ lb-ft
$A_s = 300{,}100 \times 12/(0.90 \times 60{,}000 \times 0.9 \times 26.5) = 2.80$ in.², say 4 No. 8 = 3.16 in.²

Development of bars is calculated as explained in Example 20. With 4 No. 8 bars, $M_t = 3.16 \times 60{,}000 \times 0.9 \times 26.5/12 = 377{,}000$ lb-ft. At face of wall $V_u = 160{,}000 \times 1.65 - 8.67 \times 9310 \times 1 = 183{,}000$ lb. The bars extend 4 in. beyond the centerline of the wall; so $l_a = 0.33$ ft. Max usable $l_d = 1.3 \times 377{,}000/183{,}000 + 0.33 = 3.01$ ft = 36 in. Required l_d for No. 8 top bars = $1.4(0.04 \times 0.79 \times 60{,}000/\sqrt{3750}) = 43.3 \text{ in.} > 36$ in. The deficiency in development length available can be made up by adding hooks at the end. An alternate would be to use 5 No. 7 = 3.00 in.² without hooks.

Critical V_u = load transferred from column to wall footing = 10,960 × 1.65 = 18,100 lb.

$$v_u = V_u/\phi bd = 18{,}100/(0.85 \times 24 \times 26.5) = 33.5 \text{ psi} < 2\sqrt{f_c'} = 122 \text{ psi}$$

Strap can be narrowed to 12 in., say at face of column.

Bars may be terminated (Code 12.1.4) when moments permit, provided one of the following conditions is satisfied: (1) shear at the point is not more than half the allowable, (2) extra stirrups are provided, or (3) the continuing bars provide double the area required. The first condition is satisfied over most of the strap beam. Half of the bars can theoretically be cut off where $M_u = 0.5\ M_t$ already calculated = 0.5 × 377,000 = 188,000 lb-ft = $V_u x = 18{,}100x$, where x is the distance from center of column. $x = 10.22$ ft. Code 12.14 also requires the bars to be carried an extra length, the larger of d or $12d_b = 26.5$ in. = 2.20 ft.

Stop 2 No. 8 at 8 ft from center of column.

COLUMN FOOTING. The design of the column footing can be the same as for an isolated footing, usually using the column load reduced by the shear transferred to it by the strap beam. Some modifications could be made in the zone under the beam, but this is usually not done.

If settlement is likely to be such as to compress the soil under the strap, the designer should consider a combined footing.

23. Walls Concrete walls may be plain or reinforced, bearing or nonbearing, cast in forms or in neatly excavated trenches without forms. Design is semiempirical, experience being an even larger factor than tests. The most critical points in a wall are openings or abrupt changes in cross section or outline. Wherever openings occur, there is danger of diagonal tension cracks at reentrant corners (Fig. 42*a*). Reinforcement, equivalent to two or three No. 5 bars in walls of usual thickness, should be supplied as in Fig. 42*b* and *c*.

The capacity of bearing walls is increased considerably by the stiffening effects of intersecting walls, offsets, pilasters, floors, and the like. Unsupported walls, such as the high back wall of a theater stage, require special study. Where loads on reinforced bearing walls fall within the middle third of the thickness, the walls may be proportioned for the allowable compression given by Code 14.2:

$$P_u = 0.55\phi f_c' A_g[(1 - (l_c/40h)^2] \qquad (15)$$

where $\phi = 0.7$ and l_c is the vertical distance between supports and h is the overall thickness, which must be at least 1/25 of the unsupported length or width, whichever is

*Not at d from the face because the lack of an external compressive reaction at the top of the strap makes the shear more critical.

†See footnote on page 1–44.

shorter. Walls in which the load falls outside the middle third must be designed for combined direct compression and flexure.

If the reinforcement is designed, placed, and anchored as for tied columns, the allowable loads may be computed as for tied columns.

The design of walls under tension or tension and flexure is discussed in Sec. 6, Arts. 13 and 14.

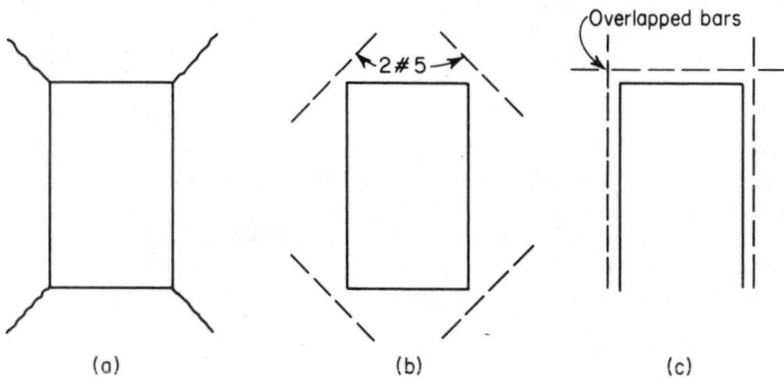

Fig. 42 Reinforcement of wall openings.

24. Slabs One-way solid slabs supported on masonry walls, or on steel or concrete beams, are designed as rectangular beams, isolating a 12-in. width of strip for study.

One-way solid slabs may be required to support concentrated loads rather than uniformly distributed ones. The simplest example is a wheel load. Analysis and tests indicate that if a single wheel is placed on a slab of single span, the width of strip which will support it is about as shown in Fig. 43. The entire slab assumes a saucer shape. Those slab

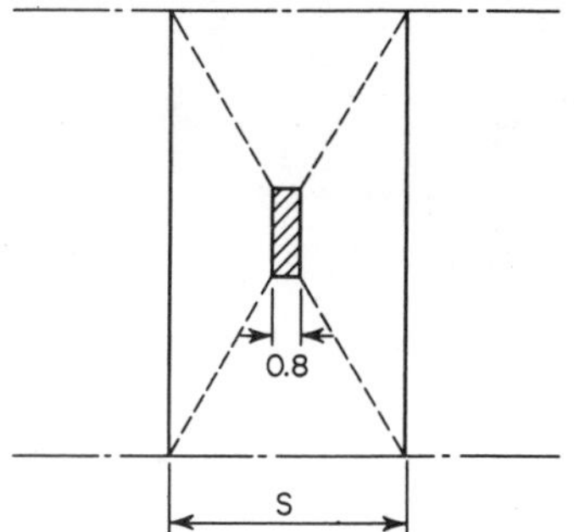

Fig. 43 Concentrated load on slab.

elements directly under the load are most highly stressed, but for a considerable width on either side, elements will participate. A satisfactory scheme is to assume that the radiating lines flare at an included angle of $\tan^{-1} 0.6\pm$. For more precise results, careful analyses are required.

Example 22 A simply supported concrete floor slab spans 10 ft over a factory tunnel. If the slab is designed for a live load of 200 psf, can it carry a truck wheel concentration of 6000 lb?

SOLUTION. The live-load moment on a strip 1 ft wide = $1.7 \times 200 \times 10 \times 11 \times 12/8 = 56{,}100$ lb-in. The corresponding moment of a wheel placed in midspan and assumed distributed over a strip $0.8 + 2 \times 0.6 \times 5 = 6.8$ ft wide is $M = 1.7 \times 6000 \times 11 \times 12/4 \div 6.8 = 49{,}500$ lb-in.; so unless impact and vibration exist, a wheel concentration of $6000 \times 56{,}100/49{,}500 = 6800$ lb would be about equivalent.

Ordinarily, such slabs are cast upon temporary formwork or centering of lumber, plywood, or panels. Sometimes formed metal sheets are used for such supports, the necessary reinforcing bars and concrete placed, and the sheets left permanently in place. In other instances, the formed sheets are bonded to the concrete with welded cross wires or other devices so as to replace the bars and become the slab reinforcement.

Another variation is to key or anchor the supporting beams, whether of steel or precast concrete, to the slab with shear developers (composite construction). Precast-concrete slabs or planks, with or without voids, are available.

Holes in floors are common because of stairs, elevators, ducts, pipes, and the like. Reentrant corners are points of weakness and require supplementary reinforcement in the way of corner bars or overlapped side and end bars. One rough rule is to make the tensile resistance of the added bars at least equal to that of the concrete removed by the opening. Obviously, the stress bars must not be cut off, or the capacity is reduced; rather the full amount of required steel must be grouped, some at either side of the opening. In addition, bars at right angles are needed to divert the stresses around the openings.

Two-Way Solid Slabs. As distinguished from two-way flat slabs or flat plates, two-way solid slabs span in two directions (usually at right angles to each other) onto supporting beams or girders, possibly of steel or precast concrete, but frequently of concrete cast monolithically with the slab. The nearer the panels are to being square, the more effective they become. The moment and shear in Appendix A of the 1963 ACI Code are recommended as giving adequate information for the design of two-way solid slabs. The comments in the previous paragraphs regarding holes and openings apply equally well here.

Two-Way Flat Slabs. The two-way flat slab (Fig. 44) is supported on columns, usually

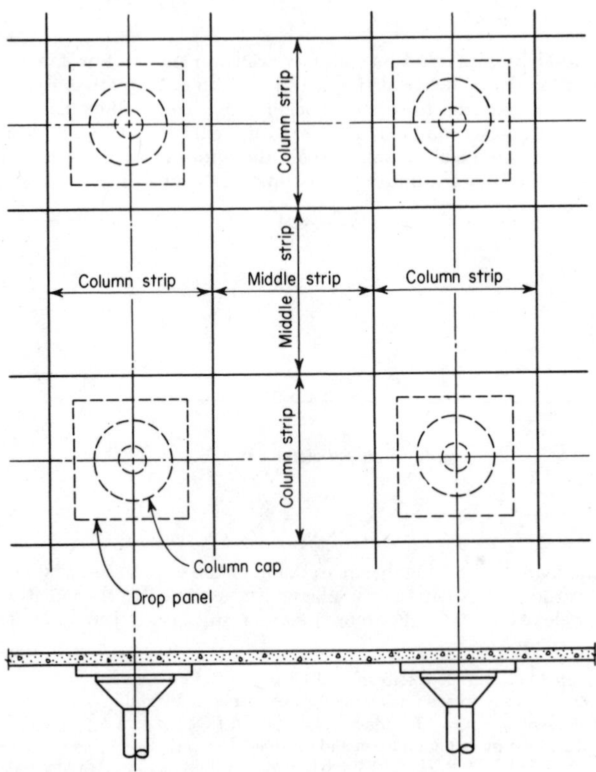

Fig. 44 Two-way flat slab.

arranged in square or rectangular patterns. Flaring heads on the columns and plinthlike dropped panels are an integral part of the system. Methods for design are set out in detail in ACI 318. Numerous variations, such as columns in triangular pattern, four-way reinforcement of approximately square panels, and structural steel or heavy bar shear heads in lieu of column capitals, have been successfully built.

Two-Way Flat Plates. Two-way flat plates are quite similar to flat slabs except that the drop panels and column capitals are omitted. Columns may be increased in size to keep the shear within allowable limits. Flat plates are much used in high-rise apartment or hotel construction, especially when the underside of the plate can also serve as the ceiling of the room below. Columns may be irregularly spaced to suit the location of corners of closets or bathrooms. Columns are usually uniform in size from bottom to top, with the amount of reinforcement varied from maximum to minimum. The design is an adaptation of ACI 318.

One-Way Concrete Ribbed Slabs. Since the concrete in a solid slab below and near the neutral axis is greatly in excess of what is needed to resist shear, much of it can be saved, and dead weight eliminated, by forming voids with permanent steel forms or removable forms of metal, wood, plastic, or other material (Fig. 45). The joists are

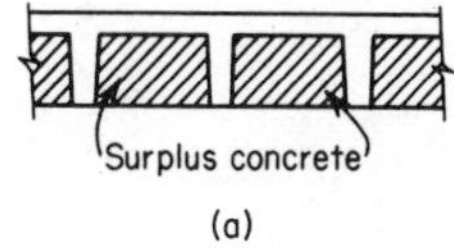

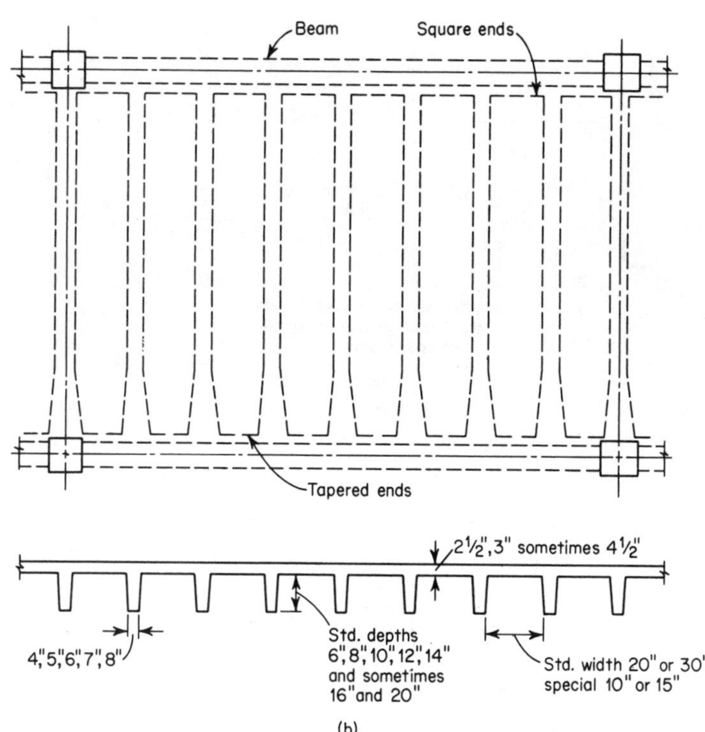

Fig. 45 One-way ribbed slab.

designed as regular tee beams. It is important that form depths, joist widths, tapered ends, etc., all conform to the standard forms that it is proposed to use (see U.S. Department of Commerce Simplified Practice Recommendation R87). Maintaining the same sizes throughout a project makes for maximum economy because of the many reuses of a minimum amount of form material.

Two-Way Joist Construction. Much of the unnecessary concrete in flat-slab construction can be eliminated by using waffle slabs or two-way joist construction (Fig. 46). ACI 318 recommends designing for the shears and moments used in flat-slab design.

25. Structural Framing Systems One of the first decisions to be made by the designer is the selection of a suitable framing system for any structure. The determining factors are many. Compliance with all laws, codes, and ordinances (or a special permit to depart for

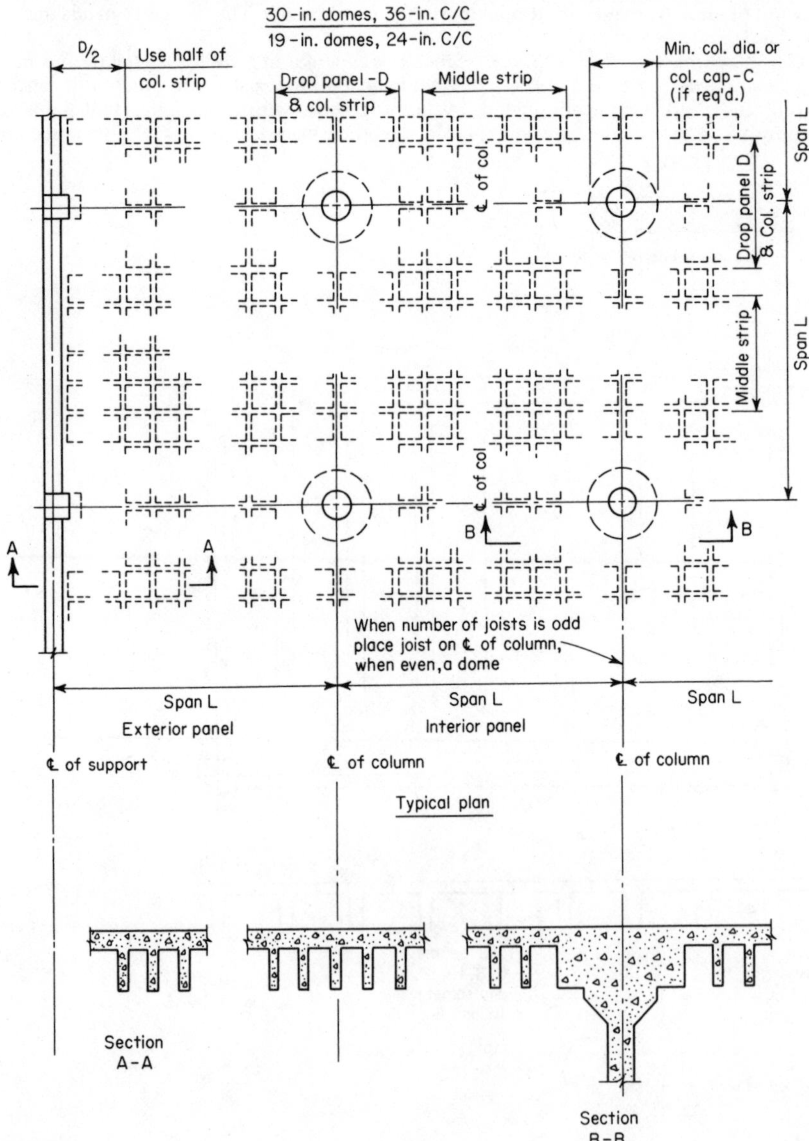

Fig. 46 Two-way joist system.

good reason) is the first essential. Economy, not merely in the structural frame itself but, more importantly, overall economy in the finished structure, including mechanical equipment, is the second most important factor. Following well-understood local practices and customs results in better bids. After these factors are duly considered, studies are usually made of various possibilities, but the following observations are pertinent. Two-way flat slabs are appropriate for heavy loads on moderate spans. Loads as low as 100 or 150 psf can be accommodated, but as the loads rise to 300 or 400 psf or higher, the more suitable the flat slab. While column spacings can be as low as 15 or 18 ft, those of 20, 25, and even 30 ft are common and economical. Waffle slabs are appropriate in lieu of flat slabs when the loads are modest, say 100 to 200, or even 300 psf, or the spans considerable, say 25 to 40 or more feet, because dead weight becomes an increasingly important consideration. Both flat slabs and waffle slabs are particularly suitable for industrial plants, garages, and shopping centers.

Flat plates are most natural for relatively light loads, say 20 or 30 up to 80 or 100 psf on spans of 15 or 16 up to 20 or 25 ft. By elimination of drops and caps, they provide much more acceptable framing in apartments and hotels.

Solid slabs have worked particularly well for relatively low, several-storied dormitories spanning 15 to 18 ft from exterior walls to corridor bearing wall and across the corridor. Required wall thicknesses have limited such structures to five or six stories. The underside of the slab can be painted to become the ceiling of the room below, but all ducts, pipes, etc., must be run in vertical shafts or in suspended ceilings over the corridors.

Whatever system is chosen, rough framing sketches with approximate sizes should be checked repeatedly with all interested parties, for clearances, story heights, interferences, economy, simplicity, and likelihood of obtaining favorable bids before a final choice is made, because resulting computational work makes later changes costly.

REFERENCES

1. "Design Handbook," vol. II, Concrete Reinforcing Steel Institute, Chicago, 1965.
2. "CRSI Handbook," Concrete Reinforcing Steel Institute, Chicago, 1975.
3. Everard, Noel J., and Edwin Cohen: "Ultimate Strength Design of Reinforced Concrete Columns," American Concrete Institute Publication SP-7, Detroit, 1964.
4. "Manual of Standard Practice for Detailing Reinforced Concrete Structures," ACI 315-74, American Concrete Institute, Detroit, 1974.
5. Ferguson, Phil M.: "Reinforced Concrete Fundamentals," 3d ed., John Wiley & Sons, Inc., New York, 1973.
6. "Building Code Requirements for Reinforced Concrete," ACI 318-77, American Concrete Institute, Detroit, 1975.
7. Bresler, B.: Design Criteria for Reinforced Concrete Columns under Axial Load and Biaxial Bending, *J. ACI,* November 1960.
8. Ramamurthy, L. N.: "Investigation of Ultimate Strength of Square and Rectangular Columns under Biaxial Eccentric Loads, Symposium on Reinforced Concrete Columns," ACI Publication SP-13, 1966.
9. Parme, A. L., J. M. Nieves, and A. Gouwens: Capacity of Rectangular Columns Subject to Biaxial Bending, *J. ACI,* September 1966
10. "Design Handbook," ACI Publication SP-17, 1973.

Section 2

Design of Prestressed-Concrete Structural Members*

T. Y. LIN

Professor Emeritus of Civil Engineering, University of California, Berkeley

PAUL ZIA

Professor of Civil Engineering, North Carolina State University

NOTATION

a = lever arm between centers of C and T
A = sectional area
A_c = net cross-sectional area of concrete
A_{ps} = area of steel
A_t = transformed area
C = total compressive force in concrete
C' = ultimate compressive force in concrete
c_b = distance from c.g.c. to bottom fiber
C_c = coefficient of creep = δ_t/δ_i
c.g.c. = centroid of concrete
c.g.s. = centroid of steel
c_t = distance from c.g.c. to top fiber
e = eccentricity of c.g.s.
E_c = modulus of elasticity of concrete
E_s = modulus of elasticity of steel
F = effective prestress (after losses)
F_1, F_2 = total prestress at points 1 and 2, respectively
F_i = total initial prestress before transfer
F_0 = total prestress just after transfer
f = unit stress in general
f_r = modulus of rupture of concrete
f_c = unit stress in concrete
f'_c = compressive strength of concrete, 28 days

*Much of the material for this section was taken from T. Y. Lin, "Design of Prestressed Concrete," 2d ed., John Wiley & Sons, Inc., New York, 1963.

f'_{ci} = compressive strength of concrete at transfer
$f'_{c\infty}$ = strength of concrete at time infinity
f_{se} = effective prestress after deducting all losses
f_i = initial prestress before transfer
f_0 = prestress just after transfer
f_s = steel stress in general
f_{pu} = ultimate unit stress in steel
f_{ps} = stress in steel at ultimate load on section
f_{py} = yield strength of prestressing steel
f'_t, f'_b = tensile stress at top (bottom) fiber
f_y = yield strength of nonprestressed reinforcement
h = sag of cable, overall depth
I_t = moment of inertia of transformed section
K = wobble coefficient, per ft
k = coefficient for depth of compression
k' = k at ultimate load
k_b, k_t = kern distances from c.g.c. to bottom (top)
M_u = ultimate resisting moment
m = load factor or factor of safety
n = E_s/E_c
ρ_p = A_{ps}/bd
r = radius of gyration of cross section
S_t = principal tensile stress
T = total tension in prestressing steel
T' = ultimate tension in prestressing steel
V_c = total shear carried by concrete
V_s = total shear carried by steel
δ = unit elastic shortening due to transfer of prestress
δ_i = initial unit strain in concrete due to elastic shortening
δ_s = unit strain in steel due to shrinkage of concrete
δ_t = final unit strain in concrete including effect of creep but not of shrinkage
Δ_a = deformation of anchorage
Δf_s = loss of prestress
μ = coefficient of friction between tendon and surrounding material

MATERIALS

1. Concrete Higher-strength concrete is usually required for prestressed than for reinforced work. In the United States, 28-day cylinder strength of 4000 to 5000 psi is generally specified and is often the most economical mix for prestressed concrete, while a strength of 6000 psi is commonly used in the plant. A low-slump 5000-psi concrete can be obtained with a water-cement ratio of about 0.45 by weight and a cement content under 7 sacks per cu yd of concrete, provided good internal or external vibration is used.

High early strength is desirable for fast turnover in a pretensioning plant, or fast removal of formwork at the jobsite. With steam or hot-air curing, a transfer strength of 3500 psi is often attained in less than 24 hr. A transfer strength of over 3000 psi can be attained in 3 or 4 days by using high-early-strength cement without special curing.

Shrinkage and creep characteristics affect the behavior and efficiency of prestressed concrete. The total amount of creep strain at the end of 20 years ranges from one to five times the instantaneous elastic deformation under load (averaging about three times), the low values occurring for moist storage and for limestone aggregates. Of the total amount of creep strain, about one-fourth takes place within the first 2 weeks after application of prestress, another one-fourth within 2 to 3 months, another one-fourth within a year, and the last one-fourth in the course of many years. Upon removal of the sustained stress, roughly 80 to 90 percent of the creep will be recovered during the same length of time that it has developed.

Shrinkage is primarily dependent on time and on moisture conditions, but not on stresses. The magnitude of shrinkage strain varies with many factors; it may range from 0.0000 to 0.0010 and beyond. At least a portion of the shrinkage strain resulting from drying is recoverable upon the restoration of lost water. The amount of shrinkage is somewhat proportional to the amount of water employed in the mix. Larger size of aggregates needing a smaller amount of cement paste and harder and denser aggregates produce smaller shrinkage. The chemical composition of the cement also affects the amount of shrinkage. For the purpose of design, shrinkage strain is assumed to be about 0.0002 to 0.0004 for the usual concrete mixtures employed in prestressed construction.

The rate of shrinkage depends chiefly on weather conditions. If the concrete is left dry, most of the shrinkage will take place during the first 2 or 3 months. When stored in air at 50 percent humidity and 70°F, the rate of shrinkage is comparable with that of creep.

Lightweight concrete has been successfully used in prestressed-concrete construction. It has a lower modulus of rupture than normal-weight concrete and slightly less favorable shrinkage and creep characteristics. However, with the better aggregates these properties are comparable with those of normal-weight concrete.

2. Steel High-tensile steel for prestressing usually takes one of three forms: wires, strands, or bars. Wires for prestressing generally conform to ASTM Specification A421 for Uncoated Stress-Relieved Wire for Prestressed Concrete. They are made from rods produced by the open-hearth or electric-furnace process. After being cold-drawn to size, wires are mechanically straightened and stress-relieved by a continuous low-temperature (about 700°F) heat treatment to produce the prescribed mechanical properties.

The tensile strength and the minimum yield strength (measured by the 1.0 percent total-elongation method) are prescribed in Table 1 for the various sizes of wires. Currently, the ¼-in. wire is the most commonly used.

TABLE 1 Tensile and Yield Strength for Prestressing Wires*

Nominal diam, in.	Remarks	Area, sq in.	Min tensile strength, psi		Min yield strength, psi	
			Type WA	Type BA	Type WA	Type BA
0.192	Gage 6	0.02895	250,000		200,000	
0.196	5 mm	0.03017	250,000	240,000	200,000	192,000
0.250	¼ in.	0.04909	240,000	240,000	192,000	192,000
0.276	7 mm	0.05983	235,000		188,000	

*Type WA for wedge-type anchorage; type BA for button-type anchorage.

A typical stress-strain curve for a stress-relieved ¼-in. wire conforming to ASTM A421 is shown in Fig. 1, with a typical modulus of elasticity between 28,000,000 and 30,000,000 psi. The specified minimum elongation in 10 in. is 4 percent. Typical elongation at rupture is likely to be from 5 to 6 percent.

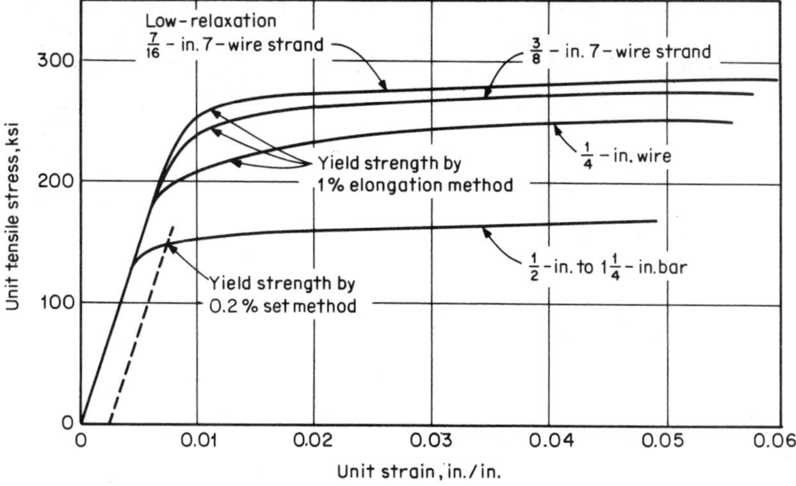

Fig. 1 Typical stress-strain curve for prestressing steels.

Strands for prestressing generally conform to ASTM Specification A416 for Uncoated Seven-Wire Stress-Relieved Strand for Prestressed Concrete. While these specifications were intended for pretensioned, bonded construction, they are applicable to posttensioned construction, whether of the bonded or the unbonded type. These strands have a guaranteed minimum ultimate strength of 250,000 psi, with the properties listed in Table 2. A higher-strength steel known as 270K grade, which is more commonly used, has a guaranteed minimum ultimate strength of 270,000 psi (Table 2).

A typical stress-strain curve for a stress-relieved ⅜-in. seven-wire strand (ASTM A416 grade) is shown in Fig. 1, which is also typical for strands of all sizes. For approximate calculations, a modulus of elasticity of 27,000,000 psi is often used for ASTM A416 grade and 28,000,000 psi for 270K grade. The specified minimum elongation of the strand is 3.5 percent in a gage length of 24 in. at initial rupture, although typical values are usually in the range of 6 percent. When these strands are galvanized, they are about 15 percent weaker.

TABLE 2 Seven-Wire Uncoated Stress-Relieved Strands

Nominal diam, in.	Weight per 1000 ft, lb	Approx area, in.^{2}	Ultimate strength, lb	Yield strength, lb
		250K grade		
¼	122	0.036	9,000	7,650
5/16	197	0.058	14,500	12,300
⅜	272	0.080	20,000	17,000
7/16	367	0.108	27,000	23,000
½	490	0.144	36,000	30,600
0.6	737	0.216	54,000	45,900
		270K grade		
⅜	290	0.085	23,000	19,550
7/16	390	0.115	31,000	26,350
½	520	0.153	41,300	35,100
0.6	740	0.217	58,600	49,800
		270K grade (low relaxation)		
⅜	292	0.085	23,000	20,700
7/16	400	0.115	31,000	27,900
½	532	0.153	41,300	37,170
0.6	737	0.215	54,000	48,600
		Dyform strand		
5/16	230	0.069	20,000	17,000
⅜	330	0.099	28,000	23,800
7/16	450	0.134	38,000	32,300
½	600	0.174	47,000	39,950
0.6	860	0.253	65,000	55,250

Another type of seven-wire strand of 270K grade is the "stabilized" strand, which is produced by a combined process of low-temperature heat treatment and high tension. Because of this special process, the yield strength of the strand is raised and its relaxation is substantially reduced. This type is also called low-relaxation strand.

A "Dyform" seven-wire stress-relieved strand, originated in Great Britain, differs from the regular seven-wire strand in that the outer wires of the Dyform strand are deformed, being run through a die after the stranding operation. It has the advantage of having a greater steel area than that of a regular strand with the same nominal diameter, thus resulting in a larger ultimate tensile strength.

Three-wire stress-relieved strands up to ⅜ in. in diameter and four-wire stress-relieved strand of 7/16 in. in diameter are also available. These strands have the same diameter, steel area, and ultimate-strength requirements as the corresponding seven-wire strands.

High-strength alloy bars for prestressing generally conform to ASTM Specification A722. These bars are usually proof-stressed (cold-stretched) to at least 80 percent of the guaranteed ultimate strength, which has a minimum value of 150,000 psi. A typical stress-strain curve for these bars is shown in Fig. 1, which shows that a constant modulus of elasticity exists only for a limited range (up to about 80,000 psi) with a value between 25,000,000 and 28,000,000 psi.

The yield strength of high-tensile bars is often defined by the 0.7 percent extension method or the 0.2 percent offset method, as indicated in Fig. 1. Most specifications call for a minimum yield strength of 85 percent of the guaranteed ultimate strength, though actual values are often higher. Minimum elongation at rupture in 20 diameters length is specified at 4 percent, with minimum reduction of area at 20 percent. Common sizes and properties of high-tensile bars for prestressing are listed in Table 6.

3. Grouting For bonding the tendons to the concrete after tensioning (in the case of posttensioning), cement grout is injected, which also serves to protect the steel against corrosion. Entry for the grout into the cableway is provided by means of holes in the anchorage heads and cones, or pipes buried in the concrete members. The grout can be injected at one end of the member until it is forced out of the other end. For longer members, it can be applied at both ends until forced out of a center vent. Either ordinary portland cement or high-early-strength cement may be used for the grout. Coarse sand is preferred for bond and strength, but sufficient fineness is necessary considering the limited space through which the grout has to pass. To ensure good bond for small conduits, grouting under pressure is desirable; however, care should be taken to ensure that the pressure on the walls of the cable enclosure can be safely resisted. Machines for mixing and injecting the grouts are commercially available.

Where larger space between the wires is obtained, such as in a Magnel cable, a 1:1 cement-sand mix is often used with a water-cement ratio of about 0.5 by volume, and a pressure of a few psi may be sufficient for short cables. Where the space is limited, as in a Freyssinet or Prescon cable, neat cement paste with about the same water-cement ratio should be employed. Admixtures are generally used to increase workability, reduce bleeding and shrinkage, or provide expansion. When it is desired to save cement, fine sand of 1/64-in. grain size can be added. The water:cement:sand proportion should be about 1.0:1.3:0.7 by volume with water-cement ratio of no more than 0.45 by weight. Grouting pressure generally ranges from 80 to 100 psi. After the grout has discharged from the far end, that end is plugged and the pressure is again applied at the injecting end to compact the grout. It is also good practice to wash the cables with water before grouting is started, the excess water being removed with compressed air. When tendons are unbonded, they must be properly greased and wrapped to prevent corrosion.

A minimum grout temperature of 60°F is generally recommended. The temperature of members at time of grouting should be maintained above 35°F until job-cured 2-in. cubes of grout reach a minimum compressive strength of 800 psi. Test cubes should be cured under temperature and moisture conditions as close as possible to those of the grout in the member. During mixing and pumping, the grout temperature should not exceed 90°F. Otherwise, difficulties may be encountered in pumping.

When tendons are unbonded, they must be properly greased and wrapped to prevent corrosion. Plastic shielded wires and strands up to 0.6 in. in diameter, prepacked with corrosion inhibitor, are commercially available for posttensioned work. Such tendons are claimed to have a very low coefficient of friction during tensioning.

METHODS AND SYSTEMS OF PRESTRESSING

4. Tensioning Methods Methods of tensioning tendons can be classified into four groups: (1) mechanical prestressing by means of jacks, (2) electrical prestressing by application of heat, (3) chemical prestressing by means of expansive cements, (4) others.

Mechanical stressing of the tendons is by far the commonest method for both posttensioning and pretensioning. In posttensioning, hydraulic jacks are used to pull the steel against the hardened concrete; in pretensioning, to pull it against bulkheads or molds. The capacity of these jacks varies from about 3 tons up to 200 tons or more. The Clifford-Gilbert system in England employs a small screwjack weighing about 20 lb pulling one wire at a time. The B.B.R.V. and Prescon systems employ jacks of various capacities to fit cables of different sizes. The Leonhardt system in Germany employs reinforced-concrete jacks tensioning hundreds of wires at one time. In all cases, both the jack gage pressure and the tendon elongation are measured to determine the amount of prestress.

Tendons can be lengthened by heating with electricity. Originated in the United States for the posttensioning process, it has not proved to be commercially applicable. A combination of electrical and mechanical stressing, known as the electrothermal method, has been developed in the U.S.S.R. for pretensioning.

Chemical prestressing utilizes expansive cements that expand chemically after setting and during hardening. When these cements are used to make concrete with embedded steel, the steel is elongated by the expansion of the concrete. Thus, the steel is prestressed in tension, which in turn produces compressive prestress in the concrete, resulting in chemically prestressed or self-stressed concrete. Modern development of expansive cement started in France in 1940 (Lossier cement). Its use for self-stressing has been investigated intensively in the U.S.S.R. since 1953. At the University of California, Berkeley, the use of calcium sulfoaluminate admixtures for expansive cements was developed by A. Klien in 1956. Since 1963, a number of structures have been built using his shrinkage-compensating cement, which has an expansion of about 0.05 to 0.10 percent, intended to compensate the expected amount of shrinkage strain.

The Preflex method in Belgium consists of prebending a high-tensile steel beam and encasing its tensile flange in concrete: releasing the bending places the concrete under compression, thus enabling it to take tension.

5. Pretensioning Pretensioning in the United States is usually accomplished in the plant by the long-line process. Tendons are stretched between two bulkheads held against the ends of a stressing bed several hundred feet long. Concrete is then placed along the bed between steel, timber, or concrete forms. When the concrete has set sufficiently to carry the prestress, the tendons are freed from the bulkheads and the prestress is transferred to the members, generally through bond between steel and concrete. Since strands anchor themselves much better than wires, they are widely used.

Devices for gripping the tendons to the bulkheads are usually made on the wedge and friction principle. Quick-release grips for holding strands are employed.

In order to improve the behavior of prestressed beams, their tendons are often bent to given profiles. In the long-line process, this is achieved by deflecting the tendons up and down along the length of the bed, known as harping or draping. When individual molds are used for pretensioning, complicated patterns of tendon arrangement can be accomplished, such as are carried out in the U.S.S.R. by their continuous prestressing process whereby the tendons are mechanically fed under a controlled tension force and woven around pegs fixed to the mold.

6. Posttensioning Systems There are hundreds of patents and systems for posttensioning. A partial list is given in Table 3. Other systems are described in publications of the Post-tensioning Institute, Chicago. Patent royalties are indirectly included in the bid price for supplying tendons and anchorages. The bid sometimes includes the furnishing of equipment and technical supervision of jacking operations. Tables 4 through 7 give data for a few prestressing systems commonly used in the United States. Typical end anchorages are shown in Fig. 2.

Loss of Prestress

Loss of prestress results from the following:

1. Immediate elastic shortening of concrete under compression
2. Creep of concrete under sustained compression
3. Shrinkage of concrete due to drying
4. Relaxation in steel under tension
5. Slippage and slackening of tendons during anchoring
6. Frictional force between tendon and concrete during tensioning

7. Elastic Shortening of Concrete As prestress is transferred to the concrete, shortening of the concrete results in loss of prestress. Considering the axial shortening of concrete produced by pretensioning, we have

$$\text{Unit elastic shortening } \delta = \frac{f_c}{E_c} = \frac{F_0}{A_c E_c}$$

where F_0 is the total prestress just after transfer, that is, after the shortening has taken place. Loss of prestress in steel is

TABLE 3 Linear Prestressing Systems

Type	Classification	Description		Name of system	Country of origin
Pretensioning	Methods of stressing	Against buttresses or stressing beds		Hoyer	Germany
		Against central steel tube		Shorer, Chalos	U.S., France
		Continuous stressing against molds		Continuous wire winding	U.S.S.R.
		Electric current to heat steel		Electrothermal	U.S.S.R.
	Methods of anchoring	During prestressing	Wires	Various wedges	
			Strands	Strandvise, Supreme	U.S.
		For transfer of prestress	Bond, for strands and small wires		Europe, U.S.
			Corrugated clips, for big wires	Dorland	U.S.
Posttensioning	Methods of stressing	Steel against concrete		Most systems	
		Concrete against concrete		Leonhardt	Germany
				Billner	U.S.
		Expanding cement		Lossier	France
		Electrical prestressing		Billner	U.S.
		Bending steel beams		Preflex	Belgium
	Methods of anchoring	Wires, by frictional grips		Freyssinet	France
				Magnel	Belgium
				Morandi	Italy
				Holzmann	Germany
				Preload	U.S.
				Kelly	U.S.
				WCS	U.S.
		Wires, by bearing		B.B.R.V.	Switzerland
				INRYCO	U.S.
				WCS	U.S.
				Prescon	U.S.
				Texas P.I.	U.S.
		Wires, by loops and combination of methods		Billner	U.S.
				Monierbau	Germany
				Huttenwerk Rheinhausen	Germany
				Leoba	Germany
				Leonhardt	Germany
		Bars, by bearing and by grips		Lee-McCall	England
				Stressteel	U.S.
				Stress rods	U.S.
				Finsterwalder	Germany
				Dywidag	Germany
				Karig	Germany
				Polensky and Zollner	Germany
				Wets	Belgium
				Bakker	Holland
		Strands, by bearing		Roebling	U.S.
				Wayss and Freytag	Germany
		Strands, by friction grips		CCL	England
				Freyssinet	U. S., France
				Anderson	U.S.
				Atlas	U.S.
				VSL	Switzerland, U.S.
				Prescon	U.S.
				CCS	U.S.
				CONESCO	U.S.
				Continental Structures	U.S.
				CONA	Switzerland, U.S.
				PTS/Howlett	U.S.
				PTI	U.S.
				Kelly	U.S.
				WCS	U.S.

$$\Delta f_s = E_s\delta = \frac{E_sF_0}{A_cE_c} = \frac{nF_0}{A_c} \tag{1}$$

where $n = E_s/E_c$.

The value of F_0 may not be known exactly. However, the value of the initial prestress F_i before transfer is usually known; hence another solution can be obtained. Using the transformed-section method, with $A_t = A_c + nA_{ps}$,

$$\delta_i = \frac{F_i}{A_cE_c + A_{ps}E_s}$$

$$\Delta f_s = E_s\delta_i = \frac{E_sF_i}{A_cE_c + A_{ps}E_s} = \frac{nF_i}{A_c + nA_{ps}} \tag{2}$$

$$\Delta f_s = \frac{nF_i}{A_t}$$

TABLE 4 Typical Tendons for B.B.R.V. System*

No. of ¼-in. wires	1	14	28	40
Section area of wires, in.^{2}	0.04909	0.687	1.3744	1.963
Max force after anchoring (70% of ultimate), lb	8250	115,500	231,000	330,000
Max jacking force (80% of ultimate), lb	9420	131,880	263,760	376,800
Ultimate strength, lb	11,780	164,920	329,840	471,200
Baseplate, B.B.R.V., in.		6¾ × 6¾	9¼ × 9¼	11 × 11
Baseplate, Prescon, in.		6 × 8½	7 × 12	

*Almost any number of ¼-in. wires up to about 192 for either system.

TABLE 5 Freyssinet System. Cable Characteristics

Wires (0.196 and 0.276 in. diameters)

Cable size	12/0.196	18/0.196	12/0.276
Nominal steel area, in.^{2}	0.362	0.543	0.718
Ultimate strength, lb	90,000	135,000	168,500
Max jacking force, lb (80% ultimate)	72,000	108,000	135,000
Max force after anchoring, lb (70% ultimate)	63,000	94,500	117,950
Cable weight—sheath not included, lb/ft	1.23	1.85	2.45
Recommended hole diam, in.	1⅛	1½	1½
Anchorage diam, in.	3⅞	4⅞	4⅞
Anchorage length, in.	4	4⅞	4⅞

12-strand (½-in. 7-wire Strands)

Nominal steel area, in.^{2}	1.73
Ultimate cable strength (1.73 × 250,000 psi), lb	432,000
Max jacking force (80% of ultimate), lb	345,600
Max force after anchoring (70% of ultimate), lb	302,400
Cable weight (sheath not included) lb/ft	5.93
Recommended hole diam (ID), in.	2⅝
Anchorage diam, in.	8¼

TABLE 6 Prestressing Bars

			Ultimate strength guaranteed min		Initial tensioning load, $0.7f_{pu}$*		
			Regular†	Special‡	Regular	Special	Anchorage plate, in.
			All values in units of 1000 lb				
			Stressteel bars (smooth)				
¾	1.50	0.442	66	71	46	50	4 × 4
⅞	2.04	0.601	90	96	63	67	4½ × 5
1	2.67	0.785	118	126	82	88	5 × 5½
1⅛	3.38	0.994	149	159	104	111	6 × 6
1¼	4.17	1.227	184	196	129	137	6 × 7
1⅜	5.05	1.485	223	238	156	166	7 × 7½
			Dywidag bars (threaded)				
⅝	0.98	0.28	43.5		30.5		3 × 3
1	3.01	0.85	127.8	136.3	89.5	95.4	5 × 5½
1¼	4.39	1.25	187.5	200.0	131.0	140.0	6 × 7
1⅜	5.56	1.58	234.0	163.8	249.6	174.7	7 × 7½

*Losses due to creep, shrinkage of concrete, and steel relaxation should be deducted from this value. Overtension to $0.8f_{pu}$ is permitted to account for friction loss and/or wedge seating loss.

†Regular is for minimum ultimate strength of 150,000 psi.

‡Special is for minimum ultimate strength of 160,000 psi.

TABLE 7 VSL Posttensioning System Using ½-in. 7-Wire Strands of 270K Grade

Unit	No. of strands	Steel area, in.^{2}	Weight, lb/ft	Max temp. force, kips	Initial force, kips	Sheath diam, in. Flexible tubing	Sheath diam, in. Rigid tubing	Bearing plate, in.
E5-3	2	0.31	1.05	66	58	1¼	1½	5¼ × 5¼
	3	0.46	1.58	99	87	1½	1¾	
E5-4	4	0.61	2.10	132	116	1⅝	1¾	6⅛ × 6⅛
E5-7	5	0.77	2.63	165	145	1¾	2 1/16	8 × 8
	6	0.92	3.15	198	174	1⅞	2 1/16	
	7	1.07	3.68	231	202	2	2¼	
E5-12	8	1.22	4.20	264	231	2	2¼	10½ × 10½
	9	1.38	4.73	297	260	2⅛	2 7/16	
	10	1.53	5.25	330	289	2¼	2 7/16	
	11	1.68	5.78	363	318	2⅜	2 7/16	
	12	1.84	6.30	397	347	2½	2 13/16	
E5-19	13	1.99	6.83	430	376	2⅝	3	13¼ × 13¼
	14	2.14	7.35	463	405	2⅝	3	
	15	2.30	7.88	496	434	2¾	3 3/16	
	16	2.45	8.40	529	463	2⅞	3 3/16	
	17	2.60	8.93	562	492	3	3 9/16	
	18	2.75	9.45	595	520	3	3 9/16	
E5-22	19	2.91	9.98	628	549	3⅛	3 9/16	14⅜ × 14⅜
	20	3.06	10.50	661	578	3¼	3¾	
	21	3.21	11.03	694	607	3¼	3¾	
	22	3.37	11.55	727	636	3⅜	3¾	
E5-31	23	3.52	12.08	760	665	3½	3 15/16	17 × 17
	24	3.67	12.60	793	694	3½	3 15/16	
	25	3.83	13.13	826	723	3⅝	3 15/16	
	26	3.98	13.65	859	752	3⅝	3 15/16	
	27	4.13	14.18	892	781	3¾	4 5/16	
	28	4.28	14.70	925	810	3⅞	4 5/16	
	29	4.44	15.23	958	838	3⅞	4 5/16	
	30	4.59	15.75	991	867	4	4½	
	31	4.74	16.28	1024	896	4	4½	
E5-55	55	8.42	28.88	1818	1590	5½	6	23 × 23

For posttensioning, the problem is different. If we have only a single tendon in the member, the concrete shortens as that tendon is jacked against the concrete. Since the force in the cable is measured after the elastic shortening of the concrete has taken place, no loss due to that shortening need be accounted for.

If we have more than one tendon and the tendons are stressed in succession, then the prestress is gradually applied to the concrete, the shortening of concrete increases as each

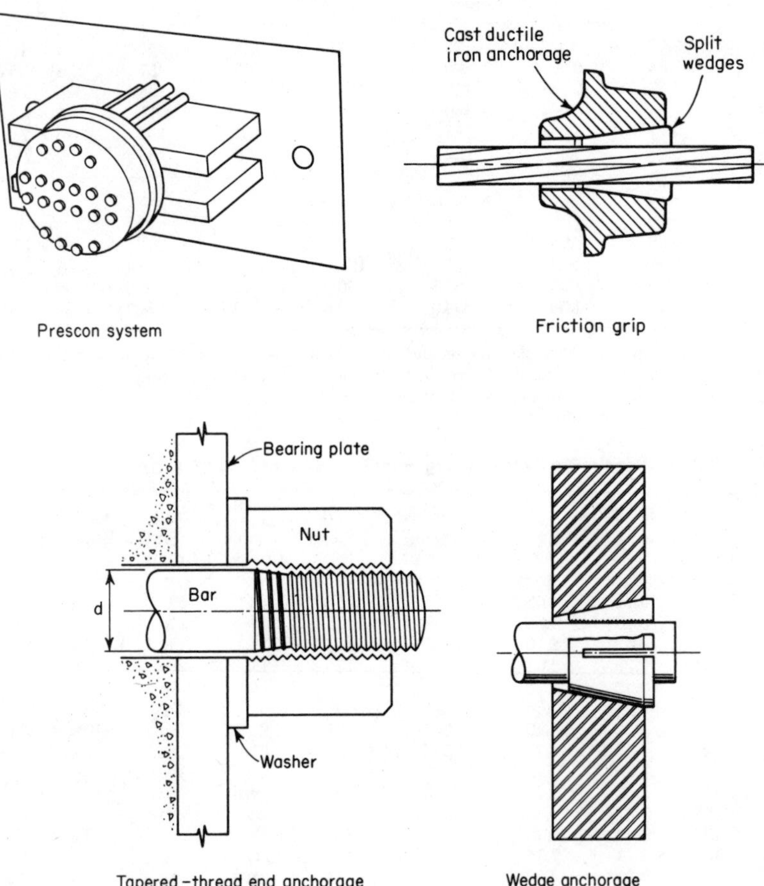

Fig. 2 End anchorages.

cable is tightened against it, and the loss of prestress due to elastic shortening differs in the tendons. The tendon that is first tensioned would suffer the maximum amount of loss due to the shortening of concrete by the subsequent application of prestress from all the other tendons. The tendon that is tensioned last will not suffer any loss due to elastic concrete shortening. For practical purposes, it is accurate enough to determine the loss for the first cable and use half that value for the average loss of all the cables. If each tendon is tensioned to a value above the specified initial prestress by the magnitude of the expected loss, no loss from elastic shortening need be considered.

8. Creep The shortening due to creep ranges from one to five times the instantaneous elastic shortening, averaging about three times for normal-weight concrete. A simple method to estimate creep is therefore to use the elastic shortening and multiply it by a suitable creep coefficient C_c, realizing that this coefficient depends on many factors. Thus, the loss of prestress Δf_s due to creep is given by

$$\Delta f_s = C_c n f_c \tag{3a}$$

where f_c = compressive stress in the concrete.

To estimate the creep coefficient the approach of the European Concrete Committee may be used. For the usual range of mixes at stress-strength ratios not exceeding 0.35,

$$C_c = \eta_b \eta_c \eta_d \eta_e \eta_t \tag{3b}$$

Values of η are given in Table 8.

TABLE 8a Values of η_b in Eqs. (3b) and (4b)

W/C ratio	Cement content in mix, lb/yd³ 340	505	675	840
0.3				0.65
0.4		0.6	0.8	1.0
0.5	0.6	0.88	1.2	1.5
0.6	0.75	1.13	1.6	
0.7	1.0	1.5		

TABLE 8b Values of η_c* in Eq. (3b)

Relative humidity %	40	50	60	70	80	90	100
η_c	3.1	2.85	2.6	2.3	1.9	1.5	1.0

TABLE 8c Values of η_d* in Eq. (3b)

Concrete age at loading, days	1	3	7	28	90	360
Normal cement	1.8	1.6	1.4	1.0	0.7	0.5
High-early-strength cement	1.7	1.4	1.1	0.7	0.5	0.3

*For concrete hardening at a constant temperature of 68°F. If the concrete hardens at a temperature other than 68°F, the age at loading should be computed as $D = \Sigma[\Delta t(T - 14)/54]$ in which Δt is the number of days during which hardening has taken place at T°F.

TABLE 8d Values of η_e in Eq. (3b)

Volume/surface ratio, in.	1	2	4	6	9	10
η_e	1.2	1.0	0.85	0.75	0.72	0.7

9. Shrinkage Loss of prestress resulting from shrinkage of the concrete is given by

$$\Delta f_s = \delta_s E_s \tag{4a}$$

The shrinkage strain δ_s is commonly taken to be 0.0003. If a closer estimate is required, one may use the approach of the European Concrete Committee,

$$\delta_s = \eta_b \eta_t \eta_e' \eta_p \epsilon_h \tag{4b}$$

In the above expression, $\eta_p = 1/(1 + n\rho_p)$, in which ρ_p is the percentage of longitudinal bonded reinforcement A_{ps} with respect to the cross-sectional area A_c of member. The values of η_b and η_t are the same as for predicting creep, and the values of η'_e and ϵ_h are given in Table 9.

10. Relaxation in Steel Stress relaxation in steel is the loss of stress when it is maintained at a constant strain for a period of time. Relaxation varies with steels of different compositions and treatments, but its approximate characteristics are known for most of the prestressing steels. In general, the percentage increases with increasing stress, and when a steel is under low stress, relaxation is negligible. Typical curves giving the relation between relaxation and initial stress level in three types of steel wires are shown in Fig. 3.

TABLE 8e Values of η_t in Eqs. (3b) and (4b)

Loading duration	Volume/surface ratio, in. 1	2	4	8	16
3 days	0.2	0.1			
7	0.3	0.18	0.1		
14	0.4	0.25	0.15		
28	0.5	0.36	0.2	0.1	
90	0.7	0.6	0.4	0.2	0.1
180	0.85	0.75	0.55	0.33	0.15
1 year	0.95	0.85	0.7	0.5	0.25
2	0.98	0.92	0.85	0.68	0.35
5	0.99	0.95	0.95	0.85	0.65
10	1.0	0.99	0.99	0.95	0.85
20		1.0	1.0	0.99	0.95

Relaxation in stress-relieved seven-wire strands has characteristics similar to those of stress-relieved wires, and can be expressed as

$$\Delta f_s = f_{si}\left[\frac{\log_{10} t}{10}\left(\frac{f_{si}}{f_{py}} - 0.55\right)\right]$$

where f_{si} is the initial stress in the prestressing strand, f_{py} is the yield strength of the strand at 1 percent elongation, and t is the time in hours. An expression for stabilized strands is

$$\Delta f_s = f_{si}\left[\frac{\log_{10} t}{45}\left(\frac{f_{si}}{f_{py}} - 0.55\right)\right]$$

Both formulas are applicable only for $f_{si} \geqslant 0.60 f_{py}$.

For high-tensile bars stressed to about $0.60 f_{pu}$, relaxation is about 3 percent, which is also the average loss assumed for relaxation of most tendons stressed to the usual allowable values.

TABLE 9a Values of η'_e in Eq. (4b)

Volume/surface ratio, in.	1	2	4	6	8	10
η'_e	1.20	1.00	0.80	0.65	0.55	0.50

TABLE 9b Values of ϵ_h in Eq. (4b)

Relative humidity %	40	50	60	70	80	90	100
ϵ_h, 10^{-6} in./in.	420	380	330	275	210	115	0

11. Slippage of Tendons during Anchoring For most systems of posttensioning, when the jack is released and the prestress transferred to the anchorage, the tendon tends to slip slightly. The amount of slippage depends on the type of wedge and the stress in the wires, an average value being around 0.1 in. For direct bearing anchorages, the heads and nuts are subject to a slight deformation at the release of the jack, an average value for such deformation being about 0.03 in. If long shims are required to hold elongated wires in place, there will be a deformation in the shims. A shim 1 ft long may deform 0.01 in.

A general formula for computing the loss of prestress due to deformation Δ_a at anchoring is

$$\Delta f_s = \frac{\Delta_a E_s}{L} \tag{5}$$

where L is the length of tendon in inches. Since this loss of prestress is caused by a fixed total amount of shortening, the percentage of loss is higher for short wires than for long

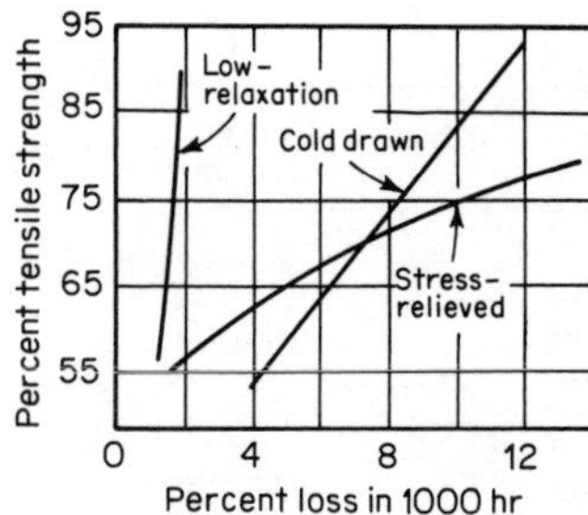

Fig. 3 Relaxation of prestressing wires.

ones. Hence it is quite difficult to tension short wires accurately, especially for systems of prestressing whose anchorage losses are relatively large. On the other hand, in the long-line process of pretensioning, this type of loss is insignificant and is not taken into consideration in design.

12. Friction The stress in a tendon tends to decrease with distance from the tensioning end, because there is friction between the tendon and its surrounding concrete or sheathing. This frictional loss can be conveniently considered in two parts: the length effect and the curvature effect. The length effect is the amount of friction that would be encountered if the tendon were straight. Since in practice the duct for a straight tendon will not be perfectly straight, some friction will exist between the tendon and its surrounding material. This is sometimes described as the wobbling effect of the duct and is dependent on the length of and stress in the tendon, the coefficient of friction, and the workmanship and method used in aligning and forming the duct.

The loss of prestress due to curvature effect results from the intended curvature of the tendons. This loss is also dependent on the coefficient of friction and the pressure exerted by the tendon on the concrete. The coefficient of friction depends on the nature of the surfaces in contact, the amount and nature of lubricants, and sometimes the length of contact. The pressure between the tendon and concrete depends on the stress in the tendon and the change in angle.

The force F_2 at any point on the tendon is given by

$$F_2 = F_1 e^{-\mu\theta - Kx} \tag{6}$$

where F_1 = force at the jacking end
θ = angle between F_1 and F_2, radians
x = length between points 1 and 2, ft
μ = coefficient of friction between tendon and surrounding material
K = wobble coefficient, per ft

When $\mu\theta + Kx$ is less than 0.3, the following approximate formula can be used:

$$F_2 = F_1(1 - \mu\theta - Kx) \tag{7}$$

Table 10 may be used to estimate values of μ and K. Actual values may differ greatly from those given and can only be obtained from experience. For example, values depend a great deal on the care exercised in construction. Tendons well greased and carefully wrapped in plastic tubes will offer little friction, but if mortar leaks through openings in the tube, the tendons may become tightly stuck.

There are several methods for overcoming the frictional loss in tendons. One method is to overtension them. Jacking from both ends is another. Lubricants can be used to advantage for unbonded tendons. For bonded tendons, water-soluble oils can be used to reduce friction while tensioning; the lubricant is flushed off with water afterward.

13. Effective Prestress The effective prestress is obtained by deducting the losses from the initial prestress. The loss of prestress varies with many factors. For a close estimate it is necessary to consider the amount of various losses in successive time intervals such as before transfer of prestress, during transfer of prestress, first year after transfer of prestress, and from first year to the end of service life of the structure. However, for the average case, the values in Table 11 are representative.

TABLE 10 Coefficients for Frictional Loss

	Wobble coefficient, K	Curvature coefficient, μ
Grouted tendons in metal sheathing:		
Wire tendons	0.0010–0.0015	0.15–0.25
High-strength bars	0.0001–0.0006	0.08–0.30
7-wire strand	0.0005–0.002	0.15–0.25
Unbonded tendons:		
Mastic-coated		
Wire tendons	0.001–0.002	0.05–0.15
7-wire strand	0.001–0.002	0.05–0.15
Pregreased		
Wire tendons	0.0003–0.002	0.05–0.15
7-wire strand	0.0003–0.002	0.05–0.15

TABLE 11 Prestress Losses

	Pretensioning, %	Posttensioning, %
Elastic shortening of concrete	4	2
Creep of concrete	6	5
Shrinkage of concrete	7	6
Relaxation in steel	6	6
Total loss not including frictional loss	23	19

14. Elongation of Tendons If a tendon has uniform stress F along its entire length L, the total elongation is given by

$$\Delta_s = \frac{FL}{E_s A_{ps}} \tag{8}$$

For a tendon with uniform curvature, considering frictional loss throughout its length L, the total elongation is given by

$$\Delta_s = \frac{F_2 L}{E_s A_{ps}} \frac{e^{\mu\theta + KL} - 1}{\mu\theta + KL} \tag{9}$$

For an approximate solution,

$$\Delta_s = \frac{F_1 + F_2}{2} \frac{L}{E_s A_{ps}} \tag{10}$$

ANALYSIS FOR FLEXURE

15. Basic Concepts Three different concepts may be used to explain and analyze the behavior of prestressed concrete.

1. Consider a simple rectangular beam, eccentrically prestressed by a tendon (Fig. 4). The prestress F produces a stress f at any cross section:

$$f = \frac{F}{A} \pm \frac{Fey}{I} \tag{11}$$

where A = area of concrete section
e = eccentricity of tendon
y = distance from centroidal axis (c.g.c.)
I = moment of inertia of the section

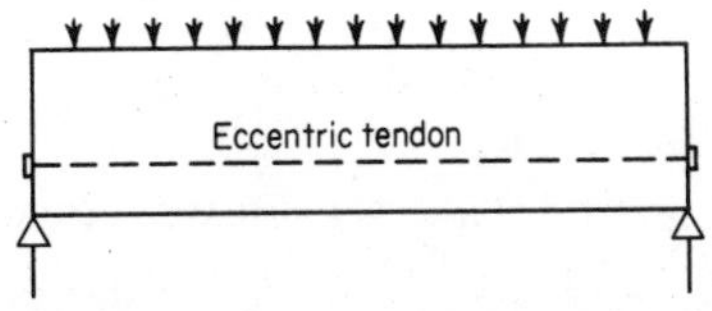

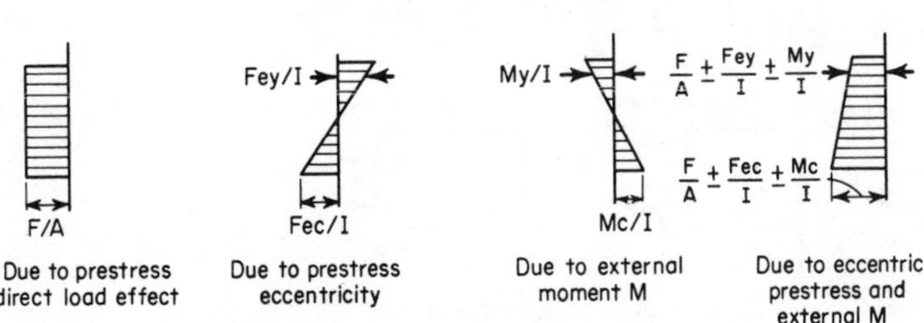

Fig. 4 Stress distribution across an eccentrically prestressed concrete section.

If M is the moment at a section due to external load, the stress at any point on that section is

$$f = \frac{F}{A} \pm \frac{Fey}{I} \pm \frac{My}{I} \tag{12}$$

Fig. 5 Example 1.

If prestress eccentricities and external moments exist on both principal axes

$$f = \frac{F}{A} \pm \frac{Fe_y y}{I_x} \pm \frac{Fe_x x}{I_y} \pm \frac{M_x y}{I_x} \pm \frac{M_y x}{I_y} \tag{13}$$

Example 1 A prestressed-concrete rectangular beam 20 × 30 in. has a simple span of 24 ft and is loaded by a uniform load of 3 kips/ft which includes its own weight (Fig. 5). The prestressing tendon is located as shown and produces an effective prestress of 360 kips. Compute fiber stresses in the concrete at the midspan section.

SOLUTION. $F = 360$ kips, $A = 20 \times 30 = 600$ in.^{2} (neglecting any hole due to the tendon), $e = 6$ in., $I = bd^3/12 = 20 \times 30^3/12 = 45{,}000$ in.^{4}; $y = 15$ in. for extreme fibers.

$$M = 3 \times \frac{24^2}{8} = 216 \text{ kip-ft}$$

$$\begin{aligned} f &= \frac{-360{,}000}{600} \pm \frac{360{,}000 \times 6 \times 15}{45{,}000} \pm \frac{216 \times 12{,}000 \times 15}{45{,}000} \\ &= -600 + 720 - 864 = -744 \text{ psi for top fiber} \\ &= -600 - 720 + 864 = -456 \text{ psi for bottom fiber} \end{aligned}$$

The resulting stress distribution is shown in Fig. 5.

2. The internal resisting couple C-T is shown in Fig. 6. When the prestress F and the

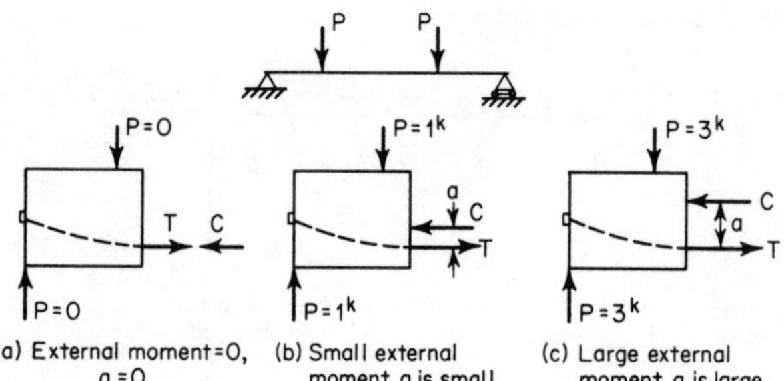

Fig. 6 Variation of a.

external moment M are known and since $C = T = F$, the lever arm a, which locates the center of the compressive force C, is given by

$$a = \frac{M}{F} \tag{14}$$

Once C is located, the stress at any point on the section is given by

$$f = \frac{C}{A} \pm \frac{Ce'y}{I} \tag{15}$$

where e' = eccentricity of C with respect to c.g.c.

Various stress distributions in the concrete can be obtained, depending upon the location of C (Fig. 7). Thus, if C is at the bottom kern point, the triangular distribution b results. Similarly, if C is at the top kern point, distribution e is obtained. The kern distances k_b and k_t are given by

$$k_b = \frac{r^2}{c_t} \qquad k_t = \frac{r^2}{c_b}$$

where r = radius of gyration of the cross section
c_t = distance from c.g.c. to top fiber
c_b = distance from c.g.c. to bottom fiber

Example 2 Same data as Example 1.

SOLUTION. $M = 3 \times 24^2/8 = 216$ kip-ft. The internal couple furnished by the forces $C = T = 360$ kips (Fig. 8) has the lever arm

$$a = \frac{216}{360} \times 12 = 7.2 \text{ in.}$$

Since T acts 9 in. from the bottom, C lies 16.2 in. from the bottom, and $e' = 16.2 - 15 = 1.2$ in. Then

$$f = \frac{-360{,}000}{600} \pm \frac{360{,}000 \times 1.2 \times 15}{45{,}000}$$
$$= -600 \pm 144$$
$$= -744 \text{ psi for top fiber}$$
$$= -456 \text{ psi for bottom fiber}$$

3. By this concept, one visualizes prestressing as an attempt to balance a portion of the external loads on a beam. In its simplest form, one assumes a parabolic tendon in a simple beam prestressed so as to exert a uniform upward force on the beam. If this beam supports an external downward load of equal intensity, then the net transverse load is zero, so that there is a uniform compressive stress $f = F/A$ at any cross section. If the external load is not balanced by the upward force, the moment M of the *unbalanced* load produces an additional stress $f = My/I$.

The prestress F required to balance a uniform load w lb/ft is given by

$$F = \frac{wL^2}{8h}$$

where L = span of beam, ft
h = sag of cable, ft

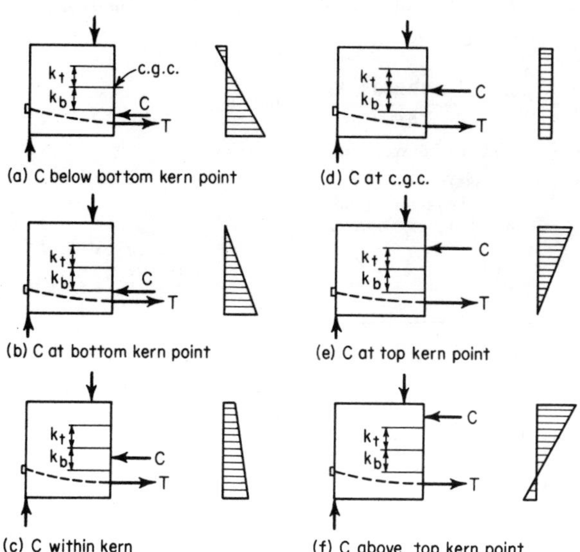

Fig. 7 Stress distribution (elastic theory).

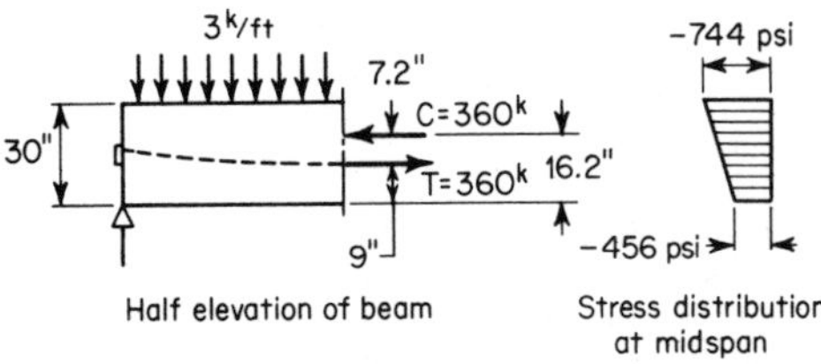

Fig. 8 Example 2.

Example 3 A 20 × 30-in. concrete beam is prestressed with a parabolic cable located as shown in Fig. 9. Compute the prestress required to balance an external load of 2.5 kips/ft.
SOLUTION. From Fig. 9, $h = 6$ in.

$$F = \frac{wL^2}{8h} = \frac{2.5 \times 24^2}{8 \times 0.5} = 360 \text{ kips}$$

Under this prestress, the beam will be uniformly stressed to 360,000/600 = 600 psi. (The horizontal component of the prestress force should be used if greater accuracy is desired.)

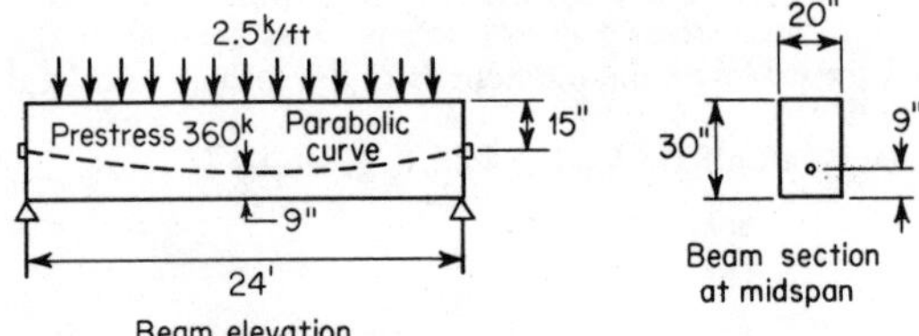

Fig. 9 Example 3.

The beam will be under no bending when subjected to an external load of 2.5 kips/ft. If the external load is 3 kips/ft (Examples 1 and 2), the unbalanced load is 0.5 kip/ft, and the midspan moment

$$M = \frac{wL^2}{8} = \frac{0.5 \times 24^2}{8} = 36 \text{ kip-ft}$$

$$f = \pm\frac{My}{I} = \frac{36 \times 12{,}000 \times 15}{45{,}000} = \pm 144 \text{ psi}$$

The stresses resulting from prestress and the external load of 3 kips/ft are

$$\begin{aligned} f &= -600 \pm 144 \\ &= -744 \text{ psi for top fiber} \\ &= -456 \text{ psi for bottom fiber} \end{aligned}$$

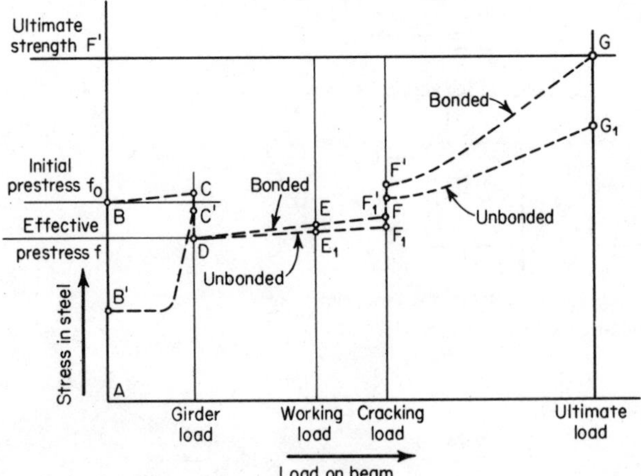

Fig. 10 Variation of steel stress with load.

16. Stress in Steel Variation in stress in the steel of a prestressed-concrete beam is shown in Fig. 10. For a posttensioned beam, as the tendons are tensioned, the steel stress increases from A to B. Simultaneously, the prestress is transferred to the beam. If the beam is heavy, its full dead load will not come into play until after its falsework is removed, causing a slight increase in steel stress from B to C. If the beam is relatively light, it usually begins to camber before the steel stress reaches B, and its dead load comes into play immediately. Thus the steel stress may vary from an intermediate point, say B' to C'. Because of the camber the tendons shorten slightly so that their stress is slightly lower, as represented by C'. In a pretensioned beam, the steel stress generally varies from B to C' upon transfer of prestress.

Assuming the losses of prestress take place before the application of superimposed dead and live loads, the steel stress is reduced from C or C' to D. Only minor changes in the range DE are induced by superimposed dead and live loads.

At cracking, stress in the steel at the crack jumps from F to F', after which it continues to increase until the ultimate load G is reached. Unbonded tendons would be forced to slip except for frictional resistance. This slippage would allow any strain in the unbonded tendon to distribute throughout its entire length. Consequently, as load increases, the tendon stress will increase more slowly than that in a bonded tendon. The line $DE_1F_1F_1'G_1$ shows the stress variation in unbonded tendons, assuming the same effective prestress before addition of external load.

17. Cracking Moment The external moment producing first hair cracks in a prestressed-concrete beam is known as the cracking moment. It is a measure of the serviceability of the beam. It occurs when the tensile stress in the extreme fiber of concrete reaches its modulus of rupture f_r (Fig. 11).

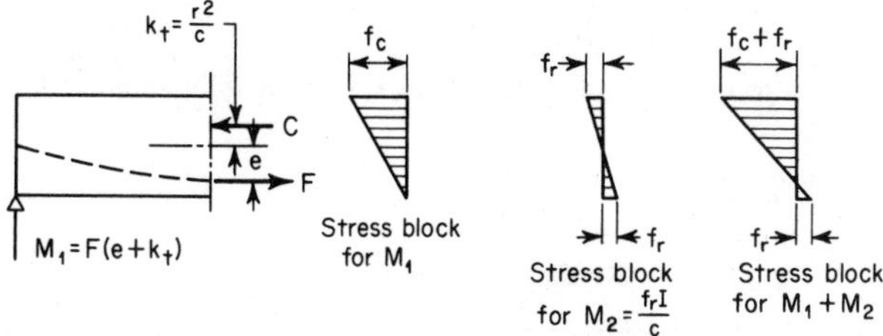

Fig. 11 Cracking moment.

$$-\frac{F}{A} - \frac{Fec}{I} + \frac{Mc}{I} = f_r$$

Transposing, we have the value of cracking moment,

$$M = Fe + \frac{FI}{Ac} + \frac{f_r I}{c} \tag{16a}$$

This can also be written

$$M = F\left(e + \frac{r^2}{c}\right) + \frac{f_r I}{c}$$

$$= F(e + k_t) + \frac{f_r I}{c} \tag{16b}$$

which is shown as $M = M_1 + M_2$ in Fig. 11. Unless the value of f_r is known from tests, it is commonly specified as $7.5\sqrt{f_c'}$ psi.

18. Ultimate Moment *Underreinforced Beams.* The steel in an underreinforced bonded beam is usually stressed to its ultimate strength f_{pu} under the action of the ultimate moment. Thus, the ultimate tension force is $T' = A_{ps}f_{pu}$. The compression force $C' = T'$ can be located approximately using a rectangular stress block, with an average stress of $0.85f_c'$ and a depth of $k'd$ (Fig. 12) such that the area of concrete A_c' within the depth $k'd$ satisfies the relation $A_c' = C'/0.85f_c'$. Equating T' and C' results in the equation

$$k' = \frac{A_{ps}f_{pu}}{0.85f_c'db} \tag{17}$$

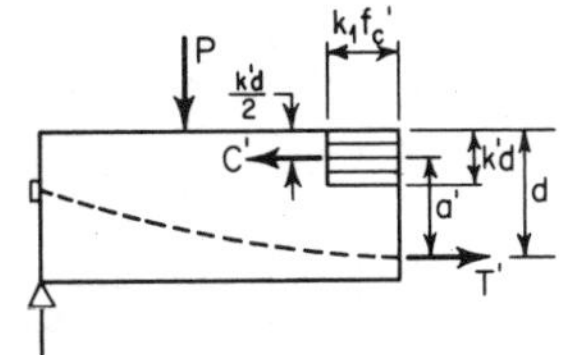

Fig. 12 Ultimate moment.

These formulas apply even though the cross section is not rectangular, provided that the compression flange has a uniform width b for the depth $k'd$.

The ultimate resisting moment $T'a'$ (Fig. 12) is

$$M_u = A_{ps} f_{pu} d \left(1 - \frac{k'}{2}\right) \tag{18}$$

Most building codes and bridge specifications give a slightly different and more conservative formula for rectangular sections

$$M_u = A_{ps} f_{ps} d \left(1 - 0.6 \frac{\rho_p f_{ps}}{f'_c}\right) \tag{19}$$

where f_{ps} = stress in steel at failure and $\rho_p = A_{ps}/bd$.

For flanged sections in which the neutral axis falls outside the flange (usually where the flange thickness t is less than $1.4d\rho_p f_{ps}/f'_c$),

$$M_u = A_{sr} f_{ps} d \left(1 - \frac{k'}{2}\right) + 0.85 f'_c (b - b')\, t \left(d - \frac{t}{2}\right) \tag{20}$$

where b is the effective width of the flange, b' is the width of the web, and

$$A_{sr} = A_{ps} - A_{sf}$$
$$A_{sf} = 0.85 f'_c (b - b') t / f_{ps}$$

Unless the value of f_{su} is determined from detailed analysis, the following is often specified:

Bonded members: $$f_{ps} = f_{pu} \left(1 - 0.5 \frac{\rho_p f_{pu}}{f'_c}\right) \tag{21a}$$

Unbonded members: $$f_{ps} = f_{se} + 10{,}000 + \frac{f'_c}{100\rho_p} \leq f_{se} + 60{,}000 \leq f_{py} \tag{21b}$$

where f_{se} is the effective stress in the prestressing steel after losses, in psi, and f_{py} is the specified yield strength of the prestressing steel.

In unbonded members, it is desirable to use a moderate amount of bonded nonprestressed reinforcement which will help distribute cracks and improve the postcracking stiffness and the ultimate strength of the beam. The minimum amount of such reinforcement required by the ACI Code is $A_s = 0.004A$, where A is the area of that part of the cross section between the flexural tension face and the center of gravity of the cross section.

Overreinforced Beams. It is safe to assume that a rectangular section is underreinforced if $\rho_p f_{ps}/f'_c < 0.30$. If $\rho_p f_{ps}/f'_c > 0.30$, the beam is usually considered to be overreinforced, and it is often specified that the ultimate flexural strength be taken not greater than $M_u = 0.25 f'_c bd^2$. More accurate predictions can be made with the following procedure, considering both equilibrium and strain compatibility.

The maximum compressive strain in the concrete at failure is taken to be 0.003. Assuming plane sections remain plane and using a trial value of $k'd$, the strain in the steel at rupture of the beam is given by $e_s = e_{s1} + e_{s2}$, where e_{s1} is the strain at the time when the concrete compressive strain on the top fiber is zero, and $e_{s2} = 0.003\,(1 - k')/k'$ (Fig. 13). The value of f_{ps} can then be obtained from the stress-strain diagram. If f_{ps} so determined is near the ultimate value f_{pu}, the section is not overreinforced. However, if f_{ps} is appreciably lower than f_{pu}, the section is overreinforced.

Knowing the value of f_{ps}, the equilibrium condition $C' = T'$ should be checked. If the condition is not satisfied, a new trial value of $k'd$ should be assumed and the procedure repeated until $C' = T'$. The moment capacity is then computed by Eq. (19) or Eq. (20).

Example 4 RECTANGULAR SECTION. Given a beam of rectangular cross section 12 in. wide by 24 in. deep. The c.g.s. of the prestressing wires is 4 in. above the bottom of the beam. Area of wires is 1.5 in.^{2}, f_{pu} = 240,000 psi, f_0 = 150,000 psi, f'_c = 5000 psi. Compute the ultimate resisting moment.

SOLUTION. Assuming that the beam is underreinforced, the stress in the steel at the ultimate moment is $f_{ps} = f_{pu}$ = 240,000 psi. The depth $d = 24 - 4 = 20$ in., and from Eqs. (17) and (18).

$$k' = \frac{1.5 \times 240{,}000}{0.85 \times 5000 \times 20 \times 12} = 0.353$$

$$M_u = 1.5 \times 240{,}000 \times 20\left(1 - \frac{0.353}{2}\right) = 5{,}930{,}000 \text{ in.-lb}$$

Using the more conservative formula, Eq. (19), the reinforcement ratio $\rho_p = 1.5/(12 \times 20) = 0.00625$, and from Eq. (21*a*)

$$f_{ps} = f_{pu}\left(1 - 0.5\frac{\rho_p f_{pu}}{f'_c}\right) = 240{,}000\left(1 - 0.5\frac{0.00625 \times 240}{5}\right)$$
$$= 204{,}000 \text{ psi}$$

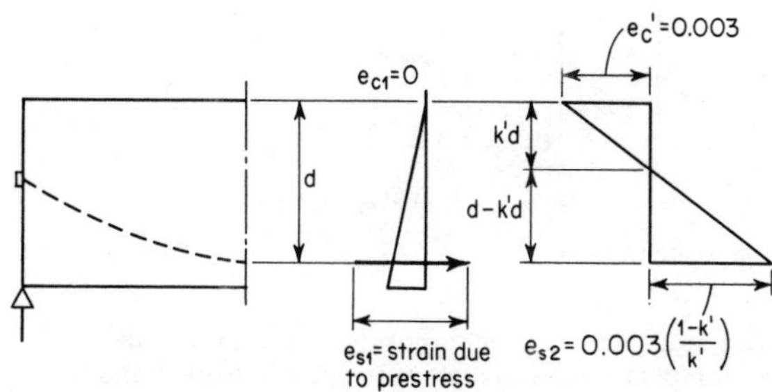

Fig. 13 Strains at rupture.

From Eq. (19),

$$M_u = 1.5 \times 204{,}000 \times 20\left(1 - 0.6 \times \frac{0.00625 \times 204}{5}\right) = 5{,}184{,}000 \text{ in.-lb}$$

This result is about 13 percent smaller than the previous value. Now

$$\rho_p \frac{f_{ps}}{f'_c} = 0.00625 \times \frac{204{,}000}{5000} = 0.255$$

which, since it is less than 0.3, suggests that the beam is underreinforced as assumed. If a more accurate value of M_u is desired, assume $k' = 0.31$ and the maximum strain in the concrete to be 0.003 (Fig. 13). Then

$$e_{s2} = 0.003 \times \frac{1 - 0.31}{0.31} = 0.0067$$

With $f_0 = 150{,}000$ psi, assume the effective prestress $f_e = 125{,}000$ psi. The corresponding strain $f_e/E_s =$ 0.0042 and the total strain at failure is 0.0067 + 0.0042 = 0.0109. From Fig. 1, this corresponds to a stress of 210,000 psi. Therefore,

$$k' = \frac{1.5 \times 210{,}000}{0.85 \times 5000 \times 20 \times 12} = 0.309$$

which is very close to the originally assumed value of 0.31. Therefore, with $f_{ps} = 210{,}000$ psi,

$$M_u = 1.5 \times 210{,}000 \times 20\left(1 - 0.6 \times \frac{0.00625 \times 210}{5}\right)$$
$$= 5{,}308{,}000 \text{ in.-lb}$$

which is about 10 percent less than the first computed value.

T-SECTION. Assume a T-section 24 in. deep whose flange is 20 × 3 in. and web is 5 in. thick. The c.g.s. of the prestressing wires is 4 in. above the bottom of the beam. Area of wires is 1.5 in.², $f_{pu} =$ 240,000 psi, $f_0 = 150{,}000$ psi, $f'_c = 5000$ psi. Compute the ultimate resisting moment.

SOLUTION. The required area in compression is (assuming $f_{ps} = f_{pu}$)

$$\frac{1.5 \times 240{,}000}{0.85 \times 5000} = 85 \text{ in.}^2$$

The flange furnishes 60 in.², leaving 25 in.² to be supplied by the web. Therefore, the neutral axis is 25/5 = 5 in. below the bottom of the flange.

For an approximate solution compression in the web can be neglected and the center of compression assumed at middepth of the flange. The lever arm is 24 − 4 − 1.5 = 18.5 in. and M_u = 360,000 × 18.5 = 6,660,000 in.-lb.

For a more exact solution, the centroid of the compressive area, including the web, can be determined. Thus, the distance from middepth of flange to the centroid is 4 × 25/(25 + 60) = 1.2 in., and the lever arm 18.5 − 1.2 = 17.3 in. The resulting moment is M_u = 360,000 × 17.3 = 6,220,000 in.-lb., which is about 6 percent less than the approximate value.

19. Composite Sections Figure 14 shows a composite section at the midspan of a simply

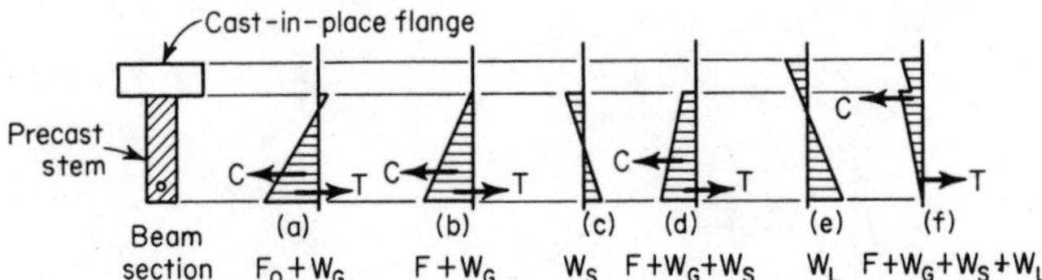

Fig. 14 Stress distribution for a composite section.

supported beam, whose stem is precast and lifted into position with the top slab cast in place directly on the stem. If no temporary intermediate support is furnished, the weight of both the slab and the stem will be carried by the stem acting alone. After the slab concrete has hardened, the composite section will carry live or dead load that may be added to it. The following stress distributions are shown for various stages of working-load conditions.

a. Owing to the initial prestress F_0 and the weight of the stem W_G there will be heavy compression in the lower fibers and possibly some small tension in the top fibers. The tensile force T in the steel and the compressive force C in the concrete form a resisting couple with a small lever arm between them.

b. After losses have taken place, the effective prestress F together with the weight of the stem will result in a slightly lower compression in the bottom fibers and some small tension or compression in the top fibers. The C-T couple will act with a slightly greater lever arm.

c. Addition of the slab of weight W_s produces additional moment and stresses as shown. Stresses resulting from differential creep and shrinkage between the slab and the stem are neglected.

d. Adding b to c, smaller compression is found to exist at the bottom fibers and some compression at the top fibers. The lever arm for the C-T couple increases further.

e. Stresses resulting from live load W_L are shown, the moment being resisted by the composite section.

f. Adding d to e, we have the stress block f. The C-T couple now acts with an appreciable lever arm.

The cracking moment and ultimate moment can be determined using methods similar to those previously described for noncomposite sections. An illustration of composite design is given in Example 12.

DESIGN FOR FLEXURE

20. Preliminary Design Preliminary design for flexure cn be based on the C-T couple. Under working load, the lever arm a varies between $0.30h$ and $0.85h$, where h = total depth of section, and averages about $0.60h$. Hence, the required effective prestress F can be estimated from the equation

$$F = T = \frac{M_T}{0.60h} \tag{22}$$

where M_T = total external moment produced by the working load.

The depth h for a prestressed section varies between 50 and 80 percent of that of an equivalent reinforced-concrete section, and may be taken at 70 percent for a first trial. Having estimated the force F, the area of steel is computed by

$$A_{ps} = \frac{F}{f_s} \tag{23}$$

where f_s depends on the steel but usually equals about 150 ksi.

The area of concrete required is estimated by

$$A_c = \frac{F}{f_{av}} \tag{24}$$

where f_{av}, the average precompression in the concrete, varies from 700 to 1300 psi for I- and T-beams and from 250 to 500 psi for solid slabs.

The load-balancing method, Concept 3 of Art. 15, can also be used for preliminary design (Art. 23).

21. Elastic Design Concept 1 of Art. 15 can be used for design by the elastic theory. However, it is often more convenient to use Concept 2. Thus:

Case 1. Girder moment $M_G = 0$. Allowing no tension in the concrete, either at transfer or at maximum load, the permissible moment is (Fig. 15*a*)

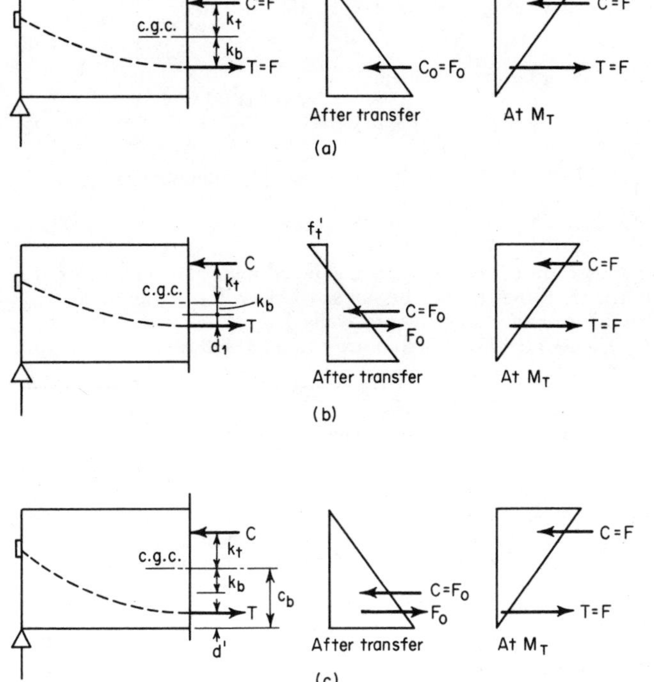

Fig. 15 Elastic design for different ratios of M_G/M_T: (*a*) M_G = O; (*b*) small M_G/M_T; (*c*) large M_G/M_T.

$$M_T = F(k_t + k_b) \tag{25}$$

where k_t and k_b are the kern distances defined in Art. 15. If tensile stress f_b' is allowed for the bottom fibers, an additional moment $f_b'I/c_b$ can be carried:

$$M_T = F(k_t + k_b) + \frac{f_b'I}{c_b} \tag{26}$$

If tensile stress f_t' is allowed for the top fibers at transfer, another additional moment $f_t'I/c_t$ can be carried:

$$M_T = F(k_t + k_b) + \frac{f_b' I}{c_b} + \frac{f_t' I}{c_t} \tag{27}$$

Case 2. When M_G is small, so that the c.g.s. cannot be located at its lowest possible position, as determined by the required concrete protection d', the distance d_1 from the bottom kern point to the c.g.s. is given by

$$d_1 = \frac{M_G + f_t' I / c_t}{F_0} \tag{28}$$

where f_t' is the allowable tension in the top fiber at transfer. The permissible moment is (Fig. 15*b*)

$$M_T = F(k_t + k_b + d_1) \tag{29}$$

If tensile stress f_b' is allowed for the bottom fibers

$$M_T = F(k_t + k_b + d_1) + \frac{f_b' I}{c_b} \tag{30}$$

Case 3. When M_G is large so that c.g.s. is located at the lowest possible position as determined by the required concrete protection d' (Fig. 15*c*),

$$M_T = F(k_t + c_b - d') \tag{31}$$

If tensile stress f_b' is allowed for the bottom fibers

$$M_T = F(k_t + c_b - d') + \frac{f_b' I}{c_b} \tag{32}$$

After the prestress F has been determmed the area of steel is computed by

$$A_s = \frac{F}{f_s}$$

and the extreme fiber stresses in the concrete are computed under M_G and under M_T, using the formulas in Art. 15. If the stresses are not satisfactory, the section is revised. Direct design formulas and computer programs are available.

Allowable Stresses. The ACI318-77 allowable stresses are as follows:

18.4. Permissible stresses in concrete—flexural members

18.4.1. Flexural stresses immediately after transfer, before losses, shall not exceed the following:

(a) Compression $0.60f_{ci}'$

(b) Tension except as permitted in (c) $3\sqrt{f_c'}$

(c) Tension at ends of simply supported members $6\sqrt{f_c'}$

Where the calculated tension stress exceeds this value, reinforcement shall be provided to resist the total tension force in the concrete computed on the assumption of an uncracked section.

18.4.2. Stresses at service loads, after allowance for all prestress losses, shall not exceed the following:

(a) Compression $0.45f_c'$

(b) Tension in precompressed tensile zone $6\sqrt{f_c'}$

(c) Tension in precompressed tensile zone in members, other than in two-way slab systems, where computations based on the transformed cracked section and on bilinear moment-deflection relationships show that immediate and long-term deflections comply with requirements of ACI Section 9.5 $12\sqrt{f_c'}$

18.4.3. The permissible stresses in Sections 18.4.1 and 18.4.2 may be exceeded when it is shown experimentally or analytically that performance will not be impaired.

18.5. Permissible stresses in steel

18.5.1. Due to jacking force $0.80f_{pu}$ or $0.94f_{py}$ whichever is smaller, but not greater than the maximum value recommended by the manufacturer of the steel or of the anchorages

18.5.2. Pretensioning tendons immediately after transfer, or posttensioning tendons immediately after anchoring $0.70f_{pu}$

---

22. Ultimate Design The ultimate moment capacity of the section must be not less than the working load moment multipled by a load factor m, usually 1.8 for buildings and 2.0 for bridges. For underreinforced sections, the ultimate lever arm will be around $0.9d$, where d = effective depth. The area of steel required is

$$A_{ps} = \frac{mM_T}{0.9df_{pu}} \tag{33}$$

Assuming that the concrete on the compressive side is stressed to $0.85f'_c$, the required area under compression is

$$A_c = \frac{mM_T}{0.9d \times 0.85f'_c} \tag{34}$$

When designed by the above formulas, the section should be checked by the equations in Art. 18. In addition, compressive stresses at transfer must be investigated for the tension flange, usually by the elastic theory, and checks for excessive camber, deflection, and cracking may be required.

23. Balanced-Load Design Concept 3 of Art. 15 can be conveniently used for design. Balancing of a uniformly distributed load by a parabolic cable was described in Art. 15. Figure 16 illustrates the balancing of a concentrated load by bending the tendon at

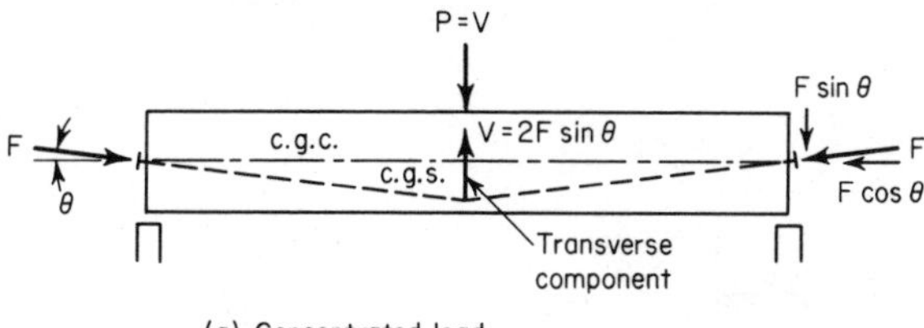

(a) Concentrated load

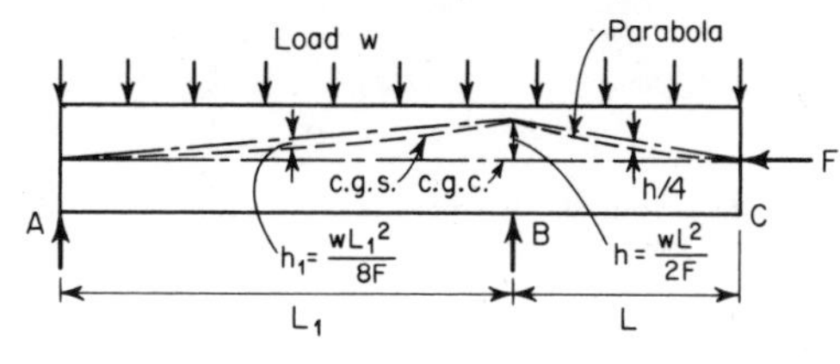

(b) Cantilever beam

Fig. 16 Load balancing for beams.

midspan, creating an upward component $V = 2F \sin \theta$. If this V exactly balances a concentrated load P, the fiber stress in the beam at any section (except for local stress concentrations) is given by $f = (F \cos \theta)/A_c = F/A_c$ for small values of θ. Any loading in addition to P will cause bending in an elastic homogeneous beam (up to point of cracking), and the additional stresses can be computed by $f = Mc/I$, where M is the moment produced by the additional load.

Now consider a cantilever beam (Fig. 16*b*). The conditions for load balancing become slightly more complicated, because any vertical component at the cantilever end C will upset the balance unless there is an externally applied load at that end. To balance a uniformly distributed load w, the tangent to the c.g.s. at C must be horizontal. The parabola for the cantilever portion is located by computing $h = wL^2/2F$, and the parabola for the anchor arm by $h_1 = wL_1^2/8F$.

Example 5 A double cantilever beam is to be designed so that its prestress will exactly balance a total uniform load of 1.6 kips/ft on the beam (Fig. 17*a*). Design the beam using the least amount of prestress, assuming that the c.g.s. must have a concrete protection of at least 3 in. If a concentrated load $P = 14$ kips is added at midspan, compute the maximum fiber stresses.

SOLUTION. In order to balance the load in the cantilever, the c.g.s. at the tip must be located at the c.g.c. with a horizontal tangent. To use the least amount of prestress, the eccentricity over the support should be a maximum, that is, $h = 12$ in. or 1 ft. The prestress required is

$$F = \frac{wL^2}{2h} = \frac{1.6 \times 20^2}{2 \times 1} = 320 \text{ kips}$$

In order to balance the load on the center span, using the same prestress, the sag for the parabola must be

$$h_1 = \frac{wL_1^2}{8F} = \frac{1.6 \times 48^2}{8 \times 320} = 1.44 \text{ ft} = 17.3 \text{ in.}$$

Hence the c.g.s. is located as shown in Fig. 17*b*.

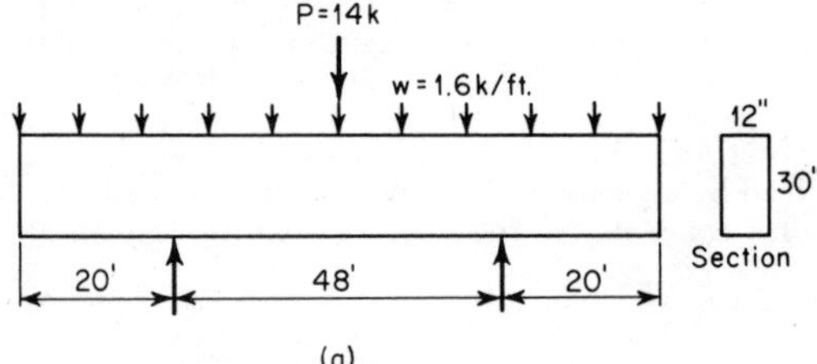

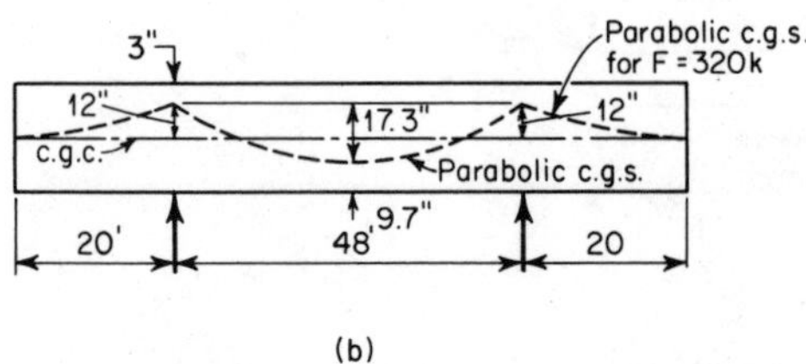

Fig. 17 Example 5.

Under the combined action of the uniform load and the prestress, the beam has no deflection anywhere and is under uniform compressive stress of

$$f = \frac{F}{A_c} = \frac{320{,}000}{360} = -889 \text{ psi}$$

Owing to $P = 14$ kips, the moment M at midspan is

$$M = \frac{PL}{4} = \frac{14 \times 48}{4} = 168 \text{ kip-ft}$$

and the extreme fiber stresses are

$$f = \frac{Mc}{I} = \frac{6M}{bd^2} = \frac{6 \times 168 \times 12{,}000}{12 \times 30^2} = \pm 1120 \text{ psi}$$

The resulting stresses at midspan are

$$f_{top} = -889 - 1120 = -2009 \text{ psi compression}$$
$$f_{bot} = -889 + 1120 = +231 \text{ psi tension}$$

Note that the actual cable placement may not possess the sharp bend shown over the supports, and the effect of any deviation from the theoretical position must be investigated accordingly. Also note that $F = 320$ kips is the effective prestress, so that under the initial prestress there will be a slight camber at midspan and either a camber or a deflection at the tips which can be computed.

For better stress conditions under the load P, it would be desirable to relocate the c.g.s. so that it would have more sag at midspan. Then a balanced condition would not exist under the uniform load w.

24. Deflections While controlled deflections resulting from prestress can be advantageously used to produce desired cambers and to offset load deflection, excessive camber can cause serious trouble. Deflection owing to prestress can be computed as in Example 6.

Example 6 A concrete beam of 32-ft simple span (Fig. 18) is posttensioned with 1.2 sq in. of high-tensile steel to an initial prestress of 140 ksi immediately after prestressing. Compute the initial deflection at midspan due to prestress and the beam's own weight, assuming E_c = 4,000,000 psi.

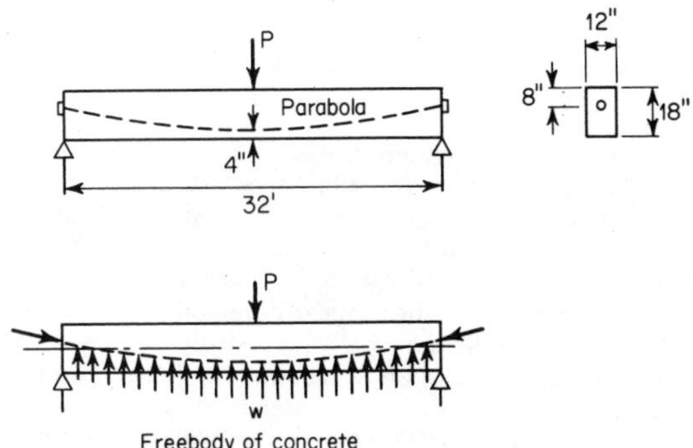

Fig. 18 Example 6.

Estimate the deflection after 3 months, assuming a creep coefficient C_c = 0.8 and an effective prestress of 120 ksi at that time.

SOLUTION. The parabolic tendon with 6-in. midordinate is replaced by a uniform load acting along the beam with intensity

$$w = \frac{8Fh}{L^2} = \frac{8 \times 140{,}000 \times 1.2 \times 6}{32^2 \times 12} = 655 \text{ plf}$$

The prestress force is eccentric 1 in. at each end of the beam, producing a moment of 140,000 × 1.2 × 1/12 = 14,000 ft-lb.

Since the weight of the beam is 225 plf, the net uniform load on the concrete is 655 − 225 = 430 plf, which produces an upward deflection at midspan given by

$$\Delta = \frac{5wL^4}{384EI} = \frac{5 \times 430 \times 32^4 \times 12^3}{384 \times 4{,}000{,}000 \times (12 \times 18^3)/12} = 0.434 \text{ in.}$$

The end moments produce a downward deflection given by

$$\Delta = \frac{ML^2}{8EI} = \frac{140{,}000 \times 1.2 \times 1 \times 32^2 \times 12^2}{8 \times 4{,}000{,}000 \times (12 \times 18^3)/12} = 0.133 \text{ in.}$$

Thus the net deflection due to prestress and beam weight is

$$0.434 - 0.133 = 0.301 \text{ in. upward}$$

To calculate the estimated deflection at 3 months, the prestress deflection is reduced proportionately for loss of prestress, and the resulting net deflection is multiplied by 1.8 to allow for creep. The camber due to initial prestress is 0.434 × 655/430 − 0.133 = 0.528 in. The initial deflection due to beam weight is 0.434 × 225/430 = 0.227 in. Thus the required deflection is

$$\Delta = 1.8\left(0.528 \times \frac{120}{140} - 0.227\right) = 0.407 \text{ in. upward}$$

Deflections resulting from external load are calculated in the usual manner for homogeneous beams, provided the concrete is not cracked. If the beam is bonded, the moment of inertia should be computed on the basis of the transformed section, but it can be approximated by using the gross concrete section. If the beam is unbonded, it is close enough for practical purposes to use the moment of inertia of the gross section of concrete. For computing instantaneous deflection due to live load, the following approximate values of E_c may be used:

Age of concrete	E_c, ksi	
	Hard rock	Lightweight
1 day	4,000	2,500
7 days	4,500	3,000
30 days	5,000	3,400
1 year	5,500	3,800

If live load is of long duration, creep must be considered. Also, if the load produces cracking, the elastic theory can be used only as an approximation. Accurate data concerning deflection after cracking are not available. However, the ultimate deflection can be computed accurately if moment-curvature relationships are known.

It is difficult to predict camber, because it varies not only with E_c and creep of concrete, but also with age, support conditions, temperature and shrinkage differential between top and bottom fibers, and variations in properties of the concrete. It is usually necessary to have experience with the product of a particular plant before accurate prediction can be made. Lacking such experience, camber computations for 1-day strength of 4000 psi may be based on E_c = 4,000,000 psi for hard rock and 2,500,000 psi for lightweight concrete. These values may then be modified for loss of prestress and creep, which can be approximated roughly by the following table:

Age of concrete	Ratio of effective to initial steel stress, %	Coefficient of flexural creep	
		Hard rock	Lightweight
1 day	94	1.0	1.0
7 days	89	1.6	1.3
30 days	86	2.0	1.5
1 year	83	2.5	1.8

It is good practice to balance the dead-load deflection by camber whenever possible. If this is done, flexural creep and variation in E_c will have little effect on camber or deflection.

Formulas for the calculation of prestress camber are given in Fig. 19. In these formulas, moments M_1 and M_2 are determined by multiplying the prestress F (more accurately, its horizontal component) by the corresponding ordinate y.

SHEAR, BOND, AND BEARING

25. Principal Tension Under service-load conditions a prestressed-concrete beam is generally uncracked and behaves elastically. The principal tensile stress in the web is computed as follows:

1. From the total external shear V at the section, deduct the shear V_s carried by the tendon to obtain the shear V_c carried by the concrete

$$V_c = V - V_s \tag{35}$$

Occasionally, though rarely, $V_c = V + V_s$; this happens when the cable inclination is such that it adds to the shear on the concrete.

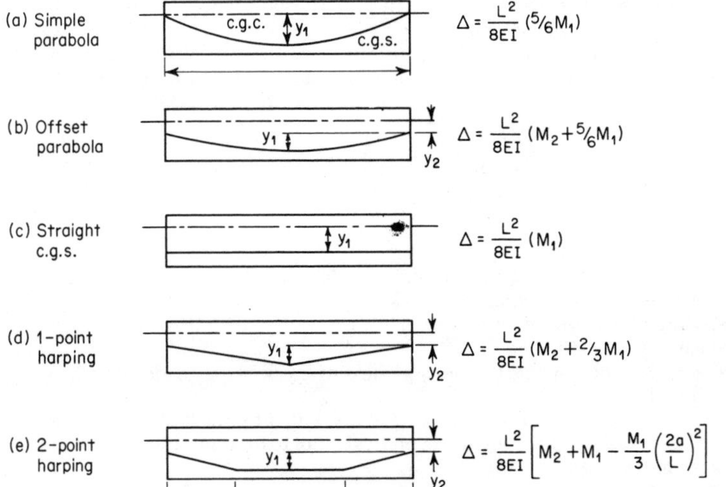

Fig. 19 Formulas for computing midspan camber due to prestress (simple beams).

2. Compute the distribution of V_c by

$$v = \frac{V_c Q}{Ib} \tag{36}$$

where v = shearing unit stress at any given level
Q = statical moment of the cross-sectional area above (or below) that level about the centroidal axis
b = width of section at that level

3. Compute the fiber stress distribution for that section due to external moment M and the prestress F

$$f_c = \frac{F}{A} \pm \frac{Fec}{I} \pm \frac{Mc}{I} \tag{37}$$

4. The maximum principal tensile stress S_t is given by

$$S_t = \sqrt{v^2 + \left(\frac{f_c}{2}\right)^2} - \frac{f_c}{2} \tag{38}$$

The greatest principal tensile stress does not necessarily occur at the centroidal axis, where the maximum vertical shearing stress exists. At some point, where f_c is diminished, a higher principal tension may exist even though v is not a maximum.

For sections without web reinforcement, S_t is often limited to $2\sqrt{f'_c}$ or less.

In composite construction the shearing stress v between precast and in-place portions is computed from Eq. (36) with V = the total shear applied after the in-place portion has been cast. The allowable value of v varies from about 40 psi for a smooth surface without ties to 160 psi for a roughened surface with adequate ties. Ties are not usually required for composite slabs or panels with large contact areas. For beams with a narrow top flange composite with in-place slabs, ties are almost always needed.

Push-off tests at the Portland Cement Association indicate an ultimate stress for composite action of about 500 psi for a rough bonded surface and 300 psi for a smooth bonded surface. About 175 psi may be added for each 1 percent stirrup reinforcement crossing the joint.

26. Web Reinforcement Tension cracks which may develop under combined flexure and shear reduce moment capacity unless web reinforcement is provided. ACI318-77 gives an empirical method, based on ultimate-strength test results at the University of Illinois, for design of web reinforcement. The yield strength of the stirrups extending over the length d is assumed to be effective in transmitting the shear (Fig. 20). Thus

$$\frac{A_v f_y d}{s} = V_u - V_{ci} \tag{39}$$

$$\frac{A_v f_y d}{s} = V_u - V_{cw} \tag{40}$$

where A_v = area of one stirrup
f_y = yield stress of stirrup, not more than 60,000 psi
d = effective depth of section, in.
s = spacing of stirrups, in.
V_u = shear at ultimate load
V_{ci} = shear at section when the vertical flexure crack starts to develop into an inclined one
V_{cw} = shear at section when web cracking starts without prior flexural cracking

The critical section is taken at $h/2$ from the theoretical point of maximum shear, where h is the overall depth of the beam.

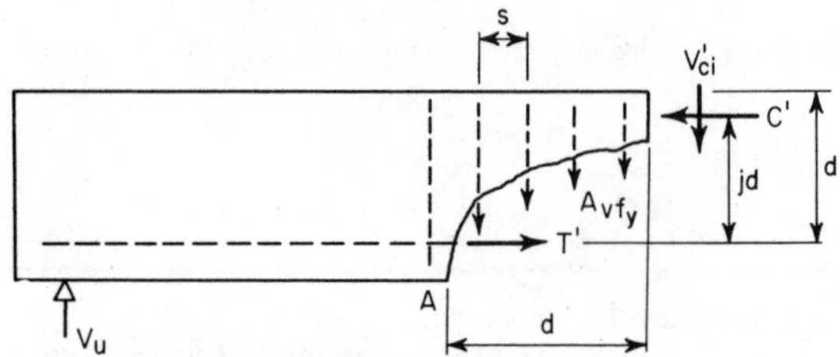

Fig. 20 Ultimate design for combined moment and shear.

V_{ci} in Eq. (39) is given by

$$V_{ci} = \frac{M_{cr}}{M/V} + V_d + 0.6 b_w d \sqrt{f'_c} \tag{41}$$

where M_{cr} = cracking moment [Eq. (43)]
M = maximum moment at section due to superimposed loads
V = shear coincident with M
V_d = dead-load shear at section
b_w = width of web
d = distance from extreme fiber in compression to centroid of prestressing tendons or $0.8h$, whichever is larger

This equation is identical in form with ACI318-77, Eq. (11-11). The last term accounts for the additional shear required to cause the vertical flexural crack to develop into an inclined one. Test results show that V_{ci} need not be taken less than $1.7 b_w d \sqrt{f'_c}$.

V_{cw} in Eq. (40) is given by

$$V_{cw} = (3.5\sqrt{f'_c} + 0.3 f_{pc}) b_w d + V_p \tag{42}$$

where f_{pc} = compressive stress in concrete at centroid of section (or at junction of web and flange if centroid is in flange) due to effective prestress
V_p = vertical component of effective prestress at section

This is Eq. (11-13) of ACI318-77. Instead of the value given by this equation, V_{cw} may be taken as the shear at the section for the multiple of dead load plus live load that produces a principal tensile stress of $4\sqrt{f'_c}$ at the centroid, or at the junction of the web and flange if the centroid is in the flange.

The cracking moment is given by

$$M_{cr} = \frac{I}{c_t}(6\sqrt{f_c'} + f_{pe} - f_d) \tag{43}$$

where I = moment of inertia of beam section
c_t = distance from c.g.c. to extreme fiber in tension
f_{pe} = effective prestress at extreme fiber which is in tension due to superimposed load
f_d = dead-load stress at extreme fiber which is in tension due to superimposed load

Procedures for design of web reinforcement are the same as for nonprestressed members (Sec. 1, Art. 12) except that, if the effective prestress force is equal to at least 40 percent of the tensile strength of the flexural reinforcement, the minimum area of web reinforcement may be determined by

$$A_v = \frac{A_{ps}}{80}\frac{f_{pu}}{f_y}\frac{s}{d}\sqrt{\frac{d}{b_w}} \tag{44}$$

instead of the formula in Sec. 1.

Example 7 Assume for the cantilever beam shown in Fig. 21: f_c' = 5000 psi, f_y = 40,000 psi (stirrups), effective prestress = 390 kips. Design the web reinforcement by ACI318-77.

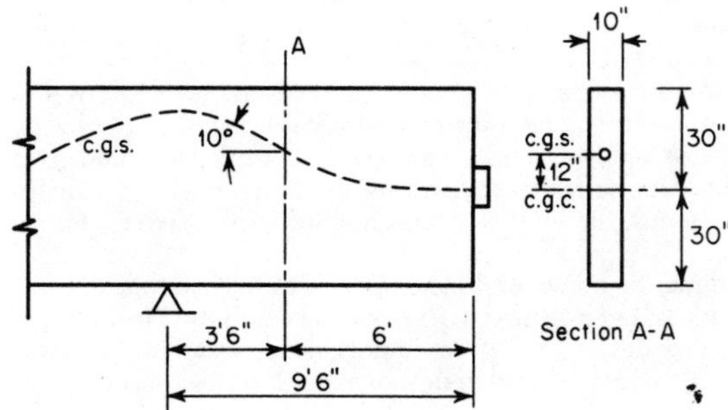

Fig. 21 Example 7.

SOLUTION. The weight of the beam is 10 × 60 × 150/144 = 625 lb/ft. The critical section is at $h/2$ = 30 in. from the support, and the design shears and moments at the section are

$$V_d = 1.4 \times 0.625 \times 6 = 5.25 \text{ kips}$$
$$M_d = 3V_d = 15.75 \text{ ft-kips}$$
$$V = 1.7 \times 160 = 272 \text{ kips}$$
$$M = 1.7 \times 160 \times 6 = 1632 \text{ ft-kips}$$

The section modulus I/c_t = 10 × 60²/6 = 6000 in.³ and the cross-sectional area = 60 × 10 = 600 in.² Then

$$f_d = \frac{15.75 \times 12}{6000} = 0.0315 \text{ ksi}$$

$$f_{pc} = \frac{390}{600} = 0.65 \text{ ksi}$$

The distance from c.g.c. to c.g.s. is 12 in. (Fig. 21). Therefore,

$$f_{pe} = f_{pc} + \frac{390 \times 12}{6000} = 1.43 \text{ ksi}$$

With $\sqrt{f_c'} = \sqrt{5000}$ = 70.7 psi = 0.0707 ksi, Eq. (43) gives M_{cr} = 6000 (6 × 0.0707 + 1.43 − 0.0315) = 10,940 in.-kips = 911 ft-kips

The vertical component V_p of the tendon tension is 390 sin 10° = 67 kips. Since d = 42 in. and 0.8h = 48 in., use d = 48 in. From Eqs. (41) and (42),

$$V_{ci} = \frac{911}{1632/272} + 5.25 + 0.6 \times 10 \times 48 \times 0.0707 = 177 \text{ kips}$$

$$V_{cw} = (3.5 \times 0.0707 + 0.3 \times 0.65)\, 10 \times 48 + 67 = 279 \text{ kips}$$

With the required capacity-reduction factor $\phi = 0.85$, Eq. (7) of Sec. 1 gives

$$\frac{\phi A_v f_y d}{s} = V_u - \phi V_{ci} = 5.25 + 272 - 0.85 \times 177 = 127 \text{ kips}$$

Using ⅝-in. U stirrups, $A_v = 0.62$ in.² and

$$s = \frac{0.85 \times 0.62 \times 40 \times 48}{127} = 8.0 \text{ in.}$$

27. Prestress Transfer Bond Pretensioned tendons usually transfer their stress to the surrounding concrete through bond. The length of transfer varies with many factors. For plain wires, it averages about 100 diameters; for seven-wire strands it is often taken as 50 to 75 diameters.

If a flexural crack should occur near the end of a pretensioned member, there is a tendency for the tendons to be pulled out as a result of bond slippage. For seven-wire strands without mechanical anchorage, the following formula gives the minimum length of embedment L in inches, required to prevent such slippage:

$$L = (f_{ps} - \tfrac{2}{3} f_{se})D$$

where f_{ps} = stress in steel to be developed at ultimate moment, ksi
f_{se} = effective prestress, ksi
D = diameter of strand, in.

28. Anchorage For tendons with end anchorages, where the prestress is transferred to the concrete by direct bearing, the stress may be transmitted by steel plates, steel blocks, or reinforced-concrete blocks. Stress analysis for anchorages is complicated, and as a result they are often designed by experience, tests, and usage. Since they are usually supplied by the prestressing companies, the engineer does not ordinarily have to design them.

The allowable bearing stress depends on several factors, such as the amount of reinforcement at the anchorage, the ratio of bearing area to the total area, and the assumptions made in computing the stress. The value $0.60f'_c$ is commonly used, assuming uniform bearing over the contact area. Many codes and bridge design specifications prescribe the allowable bearing stress as $0.6f'_{ci}\sqrt[3]{A'/A_b} < f'_{ci}$, where A_b is the bearing area of the anchor plate and A' is the area of the maximum portion of the anchorage surface that is geometrically similar to and concentric with the area A_b.

Because of strict economy in the design of end anchorages, it has not been unusual for poor concrete to fail under application of prestress. Therefore, concrete must be of high quality and must be carefully placed at the anchorages.

End Block. The portion of a prestressed member surrounding the anchorages of the tendons is often called the end block. The theoretical length of the block, sometimes called the lead length, is the distance required to transfer the prestress and distribute it throughout the entire beam cross section. End blocks are required for beams with posttensioning tendons but are not needed where all tendons are pretensioned wires or seven-wire strand.

In posttensioned members a closely spaced grid of vertical and horizontal bars must be placed near the end face of the end block to resist bursting, and closely spaced vertical and horizontal reinforcement is required throughout the length of the block.

In pretensioned beams, vertical stirrups acting at a unit stress of 20,000 psi to resist at least 4 percent of the total prestressing force should be placed within the distance $d/4$ of the end of the beam, with the end stirrup as close to the end of the beam as is practicable. The Portland Cement Association Laboratories developed an empirical equation for the design of stirrups to control horizontal cracking in the ends of pretensioned I-girders,

$$A_t = 0.021 \frac{T}{f_s} \frac{h}{l_t}$$

where A_t = required total cross-sectional area of stirrups at the end of girder, to be uniformly distributed over a length equal to one-fifth of the girder depth

T = total effective prestress force, lb
f_s = allowable stress for the stirrups, psi
h = depth of girder, in.
l_t = length of transfer, assumed to be 50 times the strand diameter, in.

TYPICAL SECTIONS

29. Beam Sections Cross sections commonly used for prestressed-concrete beams are the rectangle, the symmetrical I, the unsymmetrical I, the T, the inverted T, and the box. The suitability of these shapes will depend on the simplicity, availability, and reusability of formwork, ease in placing concrete, functional and aesthetic requirements, and theoretical considerations. The rectangular shape is easiest to form but uneconomical in material. The T is suitable for high ratios of M_G/M_T, where there is little danger of overstressing at transfer and where the concrete is effectively concentrated at the compressive flange. The inverted T is good for low ratios of M_G/M_T (to avoid overstressing at transfer) but does not have a high ultimate moment. The I and box have more concrete near the extreme fibers and are efficient both at transfer and under ultimate loads, but have weaker webs and require more complicated forming.

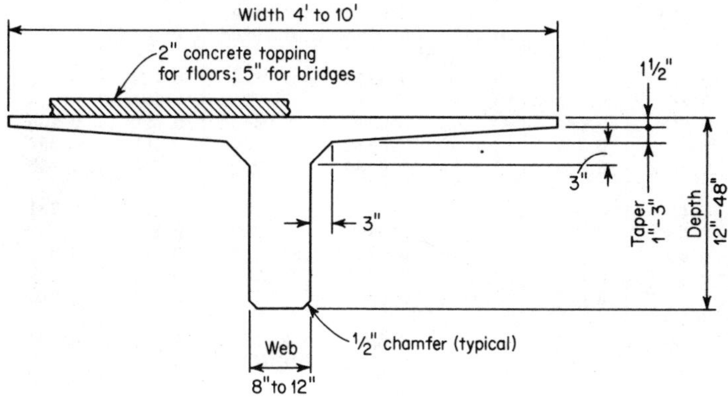

Table of properties
Width = 6'; web = 8"; taper = 2"; no topping

Depth, in.	Area, in.^{2}	I, in.^{4}	c_b, in.
12	265	2,360	9.2
16	297	6,170	11.8
20	329	11,730	14.5
24	361	19,700	17.0
28	393	30,400	19.5
32	425	44,060	21.9
36	457	61,000	24.2

Fig. 22 Single T-section.

Typical T-sections are shown in Figs. 22 and 23. Properties of T-, I-, and box sections are given in Table 12.

30. Span-Depth Ratios For reasons of economy and aesthetics, higher span-depth ratios are almost always used for prestressed concrete than for reinforced concrete. Higher ratios are possible because deflection can be much better controlled. On the other hand, when these ratios get too high, camber and deflection become quite sensitive to variations in

TABLE 12 Properties of Sections

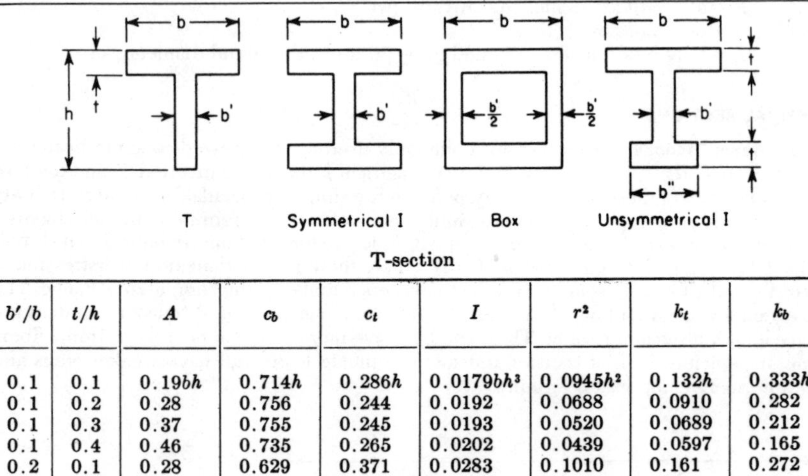

T-section

b'/b	t/h	A	c_b	c_t	I	r^2	k_t	k_b
0.1	0.1	$0.19bh$	$0.714h$	$0.286h$	$0.0179bh^3$	$0.0945h^2$	$0.132h$	$0.333h$
0.1	0.2	0.28	0.756	0.244	0.0192	0.0688	0.0910	0.282
0.1	0.3	0.37	0.755	0.245	0.0193	0.0520	0.0689	0.212
0.1	0.4	0.46	0.735	0.265	0.0202	0.0439	0.0597	0.165
0.2	0.1	0.28	0.629	0.371	0.0283	0.1010	0.161	0.272
0.2	0.2	0.36	0.678	0.322	0.0315	0.0875	0.129	0.272
0.2	0.3	0.44	0.691	0.309	0.0319	0.0725	0.105	0.234
0.2	0.4	0.52	0.684	0.316	0.0320	0.0616	0.090	0.195
0.3	0.1	0.37	0.585	0.415	0.0365	0.0985	0.169	0.237
0.3	0.2	0.44	0.626	0.374	0.0408	0.0928	0.148	0.248
0.3	0.3	0.51	0.645	0.355	0.0417	0.0819	0.127	0.231
0.3	0.4	0.58	0.645	0.355	0.0417	0.0720	0.112	0.203
0.4	0.1	0.46	0.559	0.441	0.0440	0.0954	0.171	0.216
0.4	0.2	0.52	0.592	0.408	0.0486	0.0935	0.158	0.229
0.4	0.3	0.58	0.609	0.391	0.0499	0.0860	0.141	0.220
0.4	0.4	0.64	0.612	0.388	0.0502	0.0785	0.128	0.205
1.0	1.0	1.00	0.500	0.500	0.0833	0.0833	0.167	0.167

Symmetrical I- and Box Sections

b'/b	t/h	A	c_b	c_t	I	r^2	k_t	k_b
0.1	0.1	$0.28bh$	$0.500h$	$0.500h$	$0.0449bh^3$	$0.160h^2$	$0.320h$	$0.320h$
0.1	0.2	0.46	0.500	0.500	0.0671	0.146	0.292	0.292
0.1	0.3	0.64	0.500	0.500	0.0785	0.123	0.246	0.246
0.2	0.1	0.36	0.500	0.500	0.0492	0.137	0.274	0.274
0.2	0.2	0.52	0.500	0.500	0.0689	0.132	0.264	0.264
0.2	0.3	0.68	0.500	0.500	0.0791	0.117	0.234	0.234
0.3	0.1	0.44	0.500	0.500	0.0535	0.121	0.243	0.243
0.3	0.2	0.58	0.500	0.500	0.0707	0.122	0.244	0.244
0.3	0.3	0.72	0.500	0.500	0.0796	0.111	0.222	0.222
0.4	0.1	0.52	0.500	0.500	0.0577	0.111	0.222	0.222
0.4	0.2	0.64	0.500	0.500	0.0725	0.113	0.226	0.226
0.4	0.3	0.76	0.500	0.500	0.0801	0.105	0.211	0.211

TABLE 12 Properties of Sections *(Continued)*

Unsymmetrical I-sections

b''/b	b'/b	t/h	A	c_b	c_t	I	r^2	k_t	k_b
0.3	0.1	0.1	0.21bh	0.650h	0.350h	0.0260bh^3	0.1236h^2	0.190h	0.354h
0.3	0.1	0.2	0.32	0.675	0.325	0.0345	0.1080	0.160	0.332
0.3	0.1	0.3	0.43	0.672	0.328	0.0387	0.0900	0.134	0.274
0.3	0.2	0.1	0.29	0.610	0.390	0.0316	0.1090	0.179	0.280
0.3	0.2	0.2	0.38	0.647	0.353	0.0378	0.0994	0.153	0.282
0.3	0.2	0.3	0.47	0.655	0.345	0.0402	0.0856	0.131	0.248
0.5	0.1	0.1	0.23	0.597	0.403	0.0326	0.1420	0.238	0.352
0.5	0.1	0.2	0.36	0.611	0.389	0.0464	0.1288	0.210	0.331
0.5	0.1	0.3	0.49	0.606	0.394	0.0535	0.1090	0.180	0.274
0.5	0.2	0.1	0.31	0.572	0.428	0.0373	0.1204	0.210	0.282
0.5	0.2	0.2	0.42	0.595	0.405	0.0488	0.1160	0.195	0.286
0.5	0.2	0.3	0.53	0.599	0.401	0.0540	0.1020	0.170	0.254
0.5	0.3	0.1	0.39	0.557	0.443	0.0430	0.1103	0.198	0.250
0.5	0.3	0.2	0.48	0.582	0.418	0.0510	0.1065	0.183	0.255
0.5	0.3	0.3	0.57	0.592	0.408	0.0553	0.0970	0.164	0.238
0.7	0.1	0.1	0.25	0.554	0.446	0.0381	0.1525	0.276	0.342
0.7	0.1	0.2	0.40	0.560	0.440	0.0560	0.1391	0.248	0.316
0.7	0.1	0.3	0.55	0.557	0.443	0.0651	0.1182	0.212	0.267
0.7	0.2	0.1	0.33	0.540	0.460	0.0425	0.1290	0.239	0.280
0.7	0.2	0.2	0.46	0.552	0.448	0.0578	0.1258	0.228	0.281
0.7	0.2	0.3	0.59	0.553	0.447	0.0657	0.1113	0.202	0.249
0.7	0.3	0.1	0.41	0.534	0.466	0.0467	0.1140	0.214	0.244
0.7	0.3	0.2	0.52	0.546	0.454	0.0598	0.1150	0.210	0.254
0.7	0.3	0.3	0.63	0.550	0.450	0.0663	0.1051	0.191	0.234

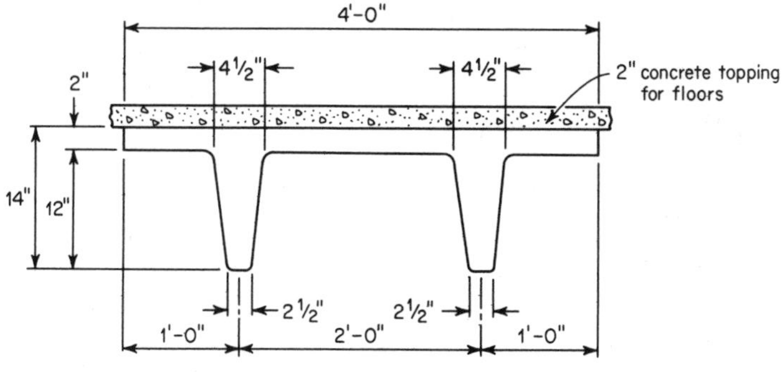

Table of properties

Topping	Area, in.^{2}	I, in.^{4}	c_b, in
With 2" topping	276	4,456	11.74
Without topping	180	2,862	10.00

Fig. 23 Double T-section.

loadings, in properties of materials, in magnitude and location of prestress, and in temperature. Furthermore, the effects of vibration become more pronounced. Care should be taken with cantilever beams, since they are particularly sensitive to deflection and vibration.

Span-depth ratio limitations should vary with the nature and magnitude of the live load, the damping characteristics, the boundary conditions, the shape and variations of the section, the modulus of elasticity, and the span. If the structure is carefully investigated for camber, deflection, and vibration, there is no reason to adhere to any given ratio.

The limiting values in Table 13 may be used as a preliminary guide for building design. In general, with span-depth ratios some 10 percent below the tabulated values, problems of camber, deflection, and vibration are not likely to develop unless the loadings are extremely heavy and vibratory in nature. On the other hand, these ratios can be exceeded by 10 percent or more if careful study ensures acceptable behavior. The ratios are intended for both hard-rock concrete and lightweight concrete but should be reduced by about 5 percent for lightweight concrete having E_c less than 3,000,000 psi. For long spans (say, in excess of about 70 ft) and for heavy loads (say, live loads over 100 psf) the values should be reduced by 5 to 10 percent. For in-place concrete in composite action with precast elements, the total depth may be considered in computing span-depth ratios.

Experience with prestressed-concrete railway bridges is not sufficient to establish span-depth ratios. Usual ratios have been in the range of 10 to 14 for box sections up to 100 ft or more. For simple-span highway bridges of the I-beam type, up to about 200 ft, a span-depth ratio of 20 is considered conservative, 22 to 24 is normal, while 26 to 28 would be the critical limit. Box sections can have ratios about 5 to 10 percent higher than I-beams, while T-sections spaced far apart should have ratios about 5 to 10 percent lower than I-beams. Again, there is no reason to believe that a fixed span-depth ratio will apply to all cases.

31. Cable Layouts Typical cable layouts for pretensioned and posttensioned simple spans are shown in Figs. 24 and 25, respectively. Layouts for single and double cantilevers are shown in Figs. 26 and 27.

32. Tendon Protection and Spacing Minimum concrete protection for tendons is governed by requirements for fire resistance and for corrosion protection. ACI318-77 specifies the following minimum thickness (inches) of concrete cover for prestressing steel, ducts, and nonprestressed steel:

	Min cover, in.
Cast against and permanently exposed to earth	3
Exposed to earth or weather:	
Wall panels, slabs, and joists	1
Other members	1½
Not exposed to weather or in contact with the ground:	
Slabs, walls, joists	¾
Beams, girders, columns:	
Principal reinforcement	1½
Ties, stirrups, or spirals	1
Shells and folded-plate members:	
Reinforcement ⅝ in. and smaller	⅜
Other reinforcement	Nominal diameter but not less than ¾

Minimum spacing of tendons is governed by several factors. First, the clear spacing between tendons, or between tendons and side forms, must be sufficient to permit easy placing of concrete. A minimum of 1⅓ times the size of the maximum aggregate is recommended. Second, to develop the bond between steel and concrete properly, the clear distance between bars should be at least the diameter of the bars for special anchorage and 1½ times the diameter for ordinary anchorage, with a minimum of 1 in. These limitations may not be necessary for small wires and strands, which are often bundled together.

ACI318 calls for a minimum clear spacing at each end of the member of four times the diameter of individual wires or three times the diameter of strands, in order to develop the transfer bond properly.

33. Partial Prestress Partial prestress in prestressed concrete may mean one or more of the following:

1. Tensile stresses are permitted in concrete under working loads.
2. Nonprestressed reinforcements, whether of mild steel or high-tensile steel, are employed in addition to tendons.
3. Tendons are stressed to a lower level than usual.

TABLE 13 Approximate Limits for Span-Depth Ratios

	Continuous spans		Simple spans	
	Roof	Floor	Roof	Floor
One-way solid slabs	52	48	48	44
Two-way solid slabs (supported on columns only)	48	44	44	40
Two-way waffle slabs (3-ft waffles)	40	36	36	32
Two-way waffle slabs (12-ft waffles)	36	32	32	28
One-way slabs with small cores	50	46	46	42
One-way slabs with large cores	48	44	44	40
Double tees and single tees (side by side)	44	40	40	36
Single tees (spaced 20-ft centers)	36	32	32	28
Cantilever solid slab: roof 20, floor 18				

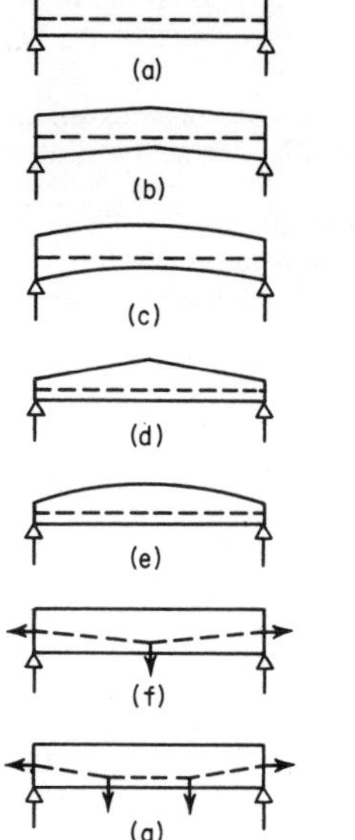

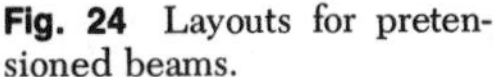

Fig. 24 Layouts for pretensioned beams.

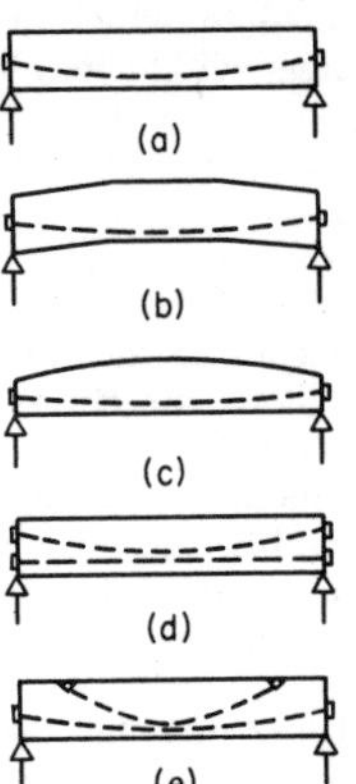

Fig. 25 Layouts for posttensioned beams.

Figure 28 shows load-deflection curves for a bonded beam with its concrete subjected to varying degrees of prestress. The case of full prestress permitting no tension under working load is represented by curve *b*. Partial prestressing permitting tension up to the modulus of rupture is shown by curve *c*. A nonprestressed concrete beam is shown in curve *d*. An overprestressed region is indicated between curves *a* and *b*.

The advantages of partial prestress compared with full prestress are (1) better control of camber, (2) saving in prestressing steel, (3) saving in end anchorages, (4) possible greater resilience in the structure, and (5) economical utilization of mild steel.

The disadvantages of partial prestress are (1) earlier appearance of cracks, (2) greater deflection under overload, (3) higher principal tensile stress under working load, and (4) slight decrease in ultimate flexural strength for the same amount of steel.

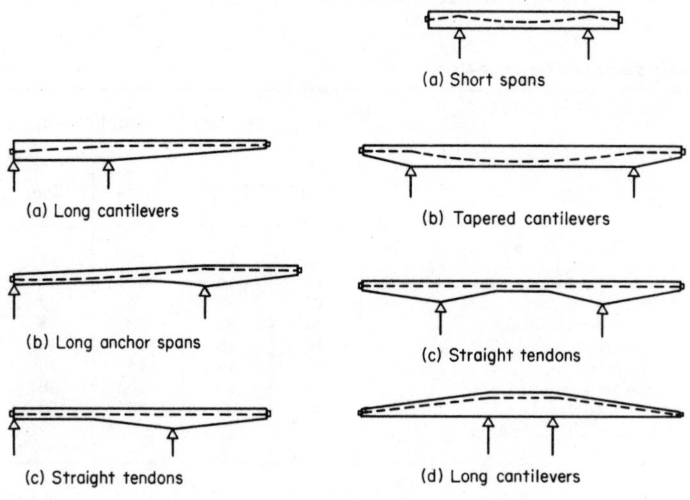

Fig. 26 Typical layouts for single cantilevers.

Fig. 27 Typical layouts for double cantilevers.

Nonprestressed reinforcements can be placed at various positions in a prestressed beam to improve its behavior and strength at different stages. Frequently, one set of reinforcement can serve to strengthen the beam in several ways:

1. To provide strength immediately after transfer of prestress:
   a. Along the compression flange, which may be under tension at transfer
   b. Along the tension flange, which may be under high compressive stress at transfer
2. To reinforce certain portions of the beam for special or unexpected loads during handling, transportation, and erection
3. To distribute cracks under working loads
4. To increase ultimate capacity of the beam
5. To help carry high compression in the concrete
6. To reinforce the concrete along directions which are not prestressed: web, end block, and flange slab reinforcements

When nonprestressed reinforcements are used to carry compression, the compressive stress in the steel is generally quite high because of shrinkage and creep in the concrete. When they are used to carry tension, the reinforcements cannot function effectively until the concrete has cracked. However, the design of tension reinforcements is usually made on the assumption that they will be stressed to the usual allowable values (e.g., 20,000 psi for intermediate-grade steel) and that their total tension will replace the tension in the portion of concrete which might be lost as a result of cracking.

Figure 29 shows the stresses and strains produced in various reinforcements under different stages of loading. It is to be noted that nonprestressed reinforcements will be stressed very little under working loads but will be effective at ultimate load, especially for underreinforced beams.

34. Combination of Prestressed and Reinforced Concrete While a combination of prestressed and reinforced concrete is represented in the use of nonprestressed reinforcement, the flexural strength is essentially supplied by the tendons, with the nonprestressed steel playing a minor role. For certain types of construction, a full combination of prestressed and reinforced concrete may be the best design, making use of the advantages

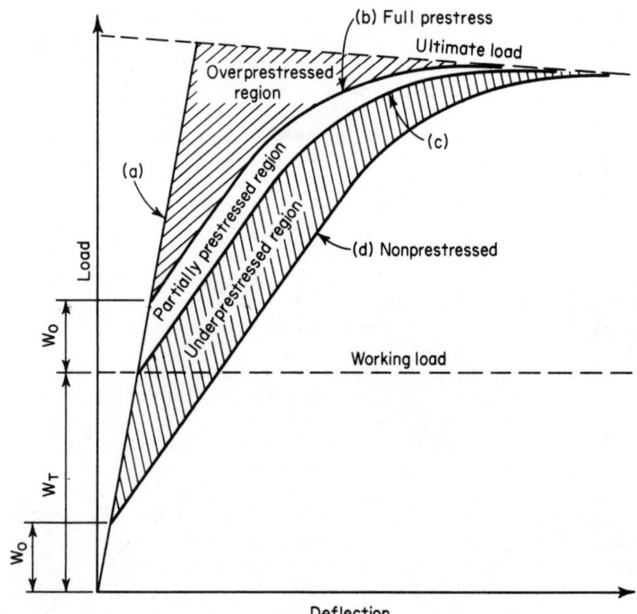

Fig. 28 Load-deflection curves for varying degrees of prestress.

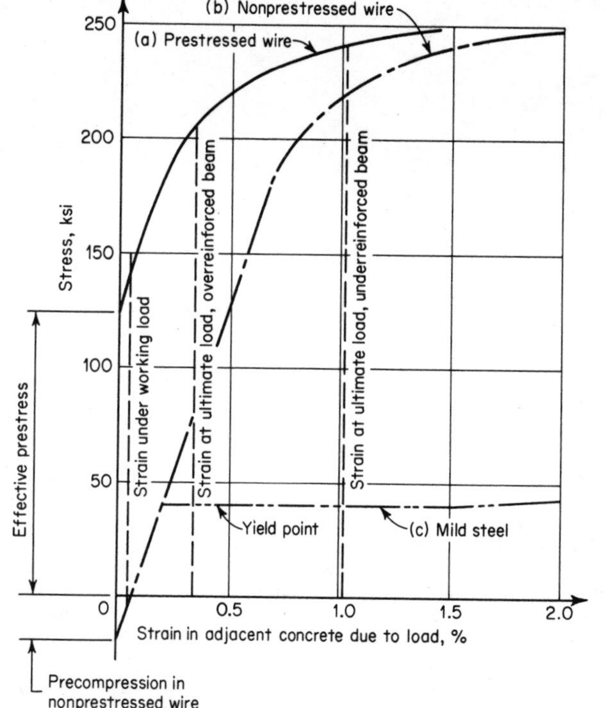

Fig. 29 Stress-strain diagrams.

of both. Reinforced concrete has the advantage of simplicity in construction, monolithic behavior, no camber, less creep, and reasonably high ultimate strength. Prestressed concrete utilizes high-strength steel economically, produces a favorable distribution of stress under certain conditions of loading, and controls deflection and cracking.

Certain structural elements and systems favor reinforced concrete, others prestressed concrete, still others partially prestressed concrete. Some will be best designed with a combination of reinforced and prestressed concrete having the nonprestressed steel carrying perhaps 50 percent or more of the total ultimate load.

One occasion for the use of this combination is the case of high live-load to dead-load ratio, when prestressing alone may produce excessive camber. Another is the case of high added dead load requiring prestressing in stages, which may be cumbersome. A third case is the requirement of high ultimate strength or resilience to resist dynamic loadings. There is also reason to believe that a heavy amount of nonprestressed steel used in conjunction with unbonded tendons will result in economy and in developing a high ultimate stress in the tendons.

For precast columns, prestressing will help control cracking during transportation and erection and will contribute to the bending strength. Nonprestressed steel will increase both the axial load and the flexural capacity. Hence a combination may be the best solution for certain cases. The use of nonprestressed reinforcement for joineries and continuity is, of course, often a simple and economical solution.

Nonprestressed steel does not act until the concrete cracks, and does not contribute toward the precracking strength. Hence if cracking could result in a primary or a secondary failure, nonprestressed steel may be of no help. The possibility of corrosion of the prestressing steel if the member cracks too early or too often should also be investigated.

CONTINUOUS BEAMS

Continuous beams may be fully cast in place with tendons continuous from one end to the other (Fig. 30). They may also be precast in smaller elements which are made continuous by special posttensioning arrangements (Fig. 31).

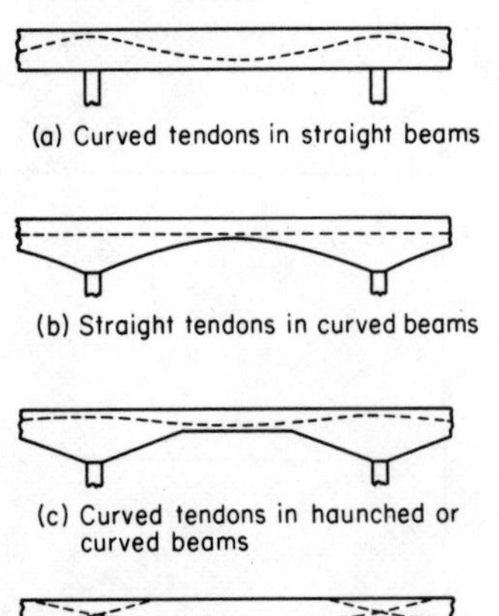

(a) Curved tendons in straight beams
(b) Straight tendons in curved beams
(c) Curved tendons in haunched or curved beams
(d) Overlapping tendons

Fig. 30 Layouts for fully continuous beams.

No support reactions are induced by prestressing a statically determinate system. In a continuous beam, or any statically indeterminate system, support reactions are generally induced by the application of prestress, because the bending of the beam due to prestress may tend to deflect the beam at its supports. These reactions produce secondary moment in the beam.

35. Continuous-Beam C Lines Under the action of prestress alone, neglecting the weight of the beam and all other external loads, the *C* line (the line of pressure in the concrete) in a simple beam coincides with the c.g.s. line. In a continuous beam, the *C* line departs from the c.g.s. line in amounts required to resist the secondary moments. When

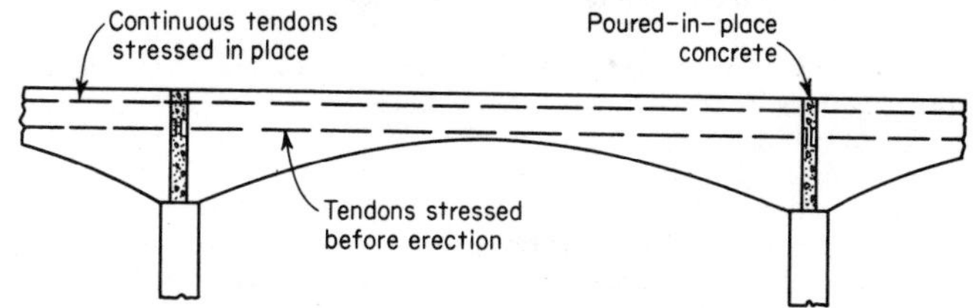

(a) Continuous tendons stressed after erection

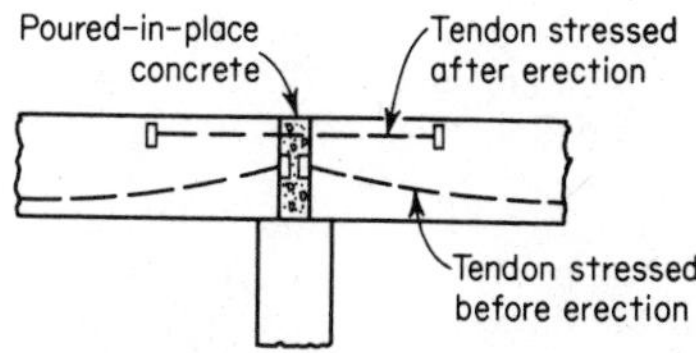

(b) Short tendons stressed over supports

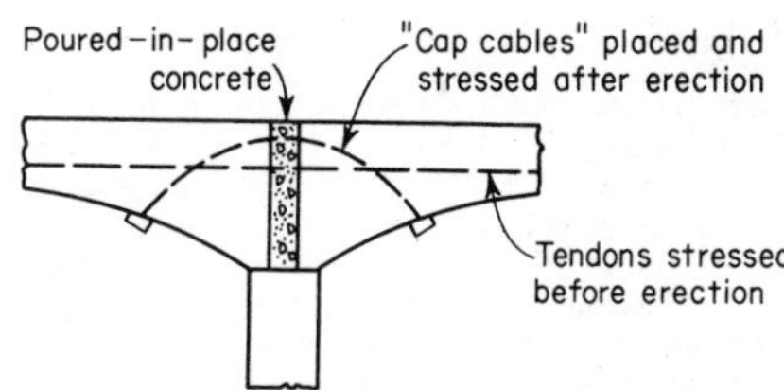

(c) Cap cables over supports

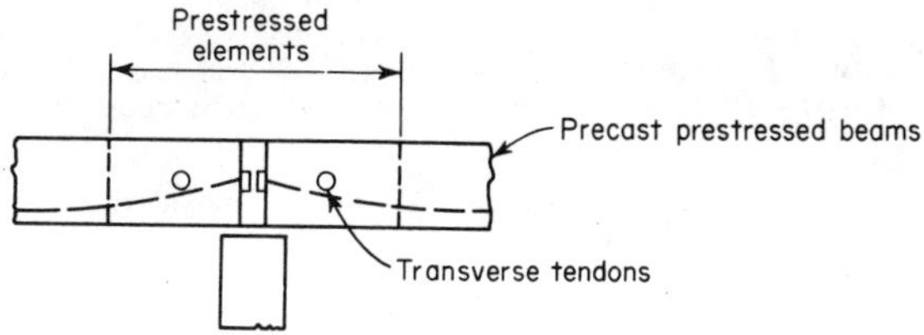

(d) Continuous elements over supports transversely prestressed

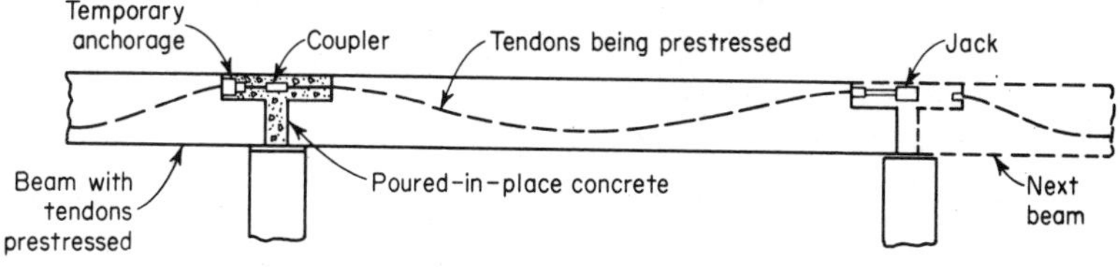

(e) Couplers over supports

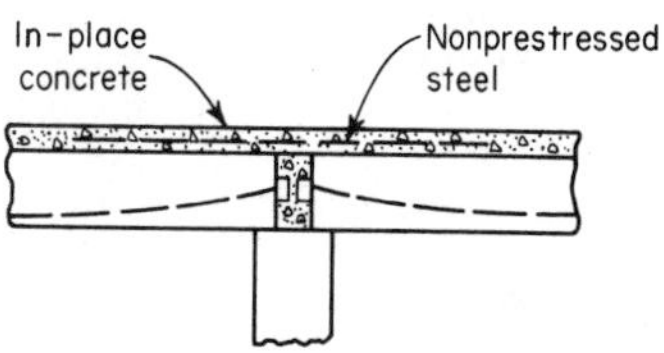

(f) Nonprestressed steel over supports

Fig. 31 Layouts for partially continuous beams.

the c.g.s. line is coincident with the C line, it is termed a "concordant cable." All simple beams and some continuous beams have concordant cables, while most continuous beams have nonconcordant cables.

While a nonconcordant cable usually gives a more economical solution, the concordant cable is sometimes preferred because it induces no external reactions. The concordant cable works better in a precast continuous beam. Several methods have been proposed for obtaining concordant cables. A simple rule is the following: Every real moment diagram for a continuous beam on nonsettling supports, produced by any combination of external loadings, whether transverse loads or moments, plotted to any scale, is one location for a concordant cable in that beam. This theorem is easily proved. Since the moments in a continuous beam are computed on the basis of no deflection over the supports, and since any c.g.s. line following the corresponding moment diagram will produce a similar moment diagram, that c.g.s. line will also produce no deflection over the supports; hence it will induce no reactions and is a concordant cable.

The C line can be located from the c.g.s. line by a linear transformation. Linear transformation is defined as the location of the C line from the c.g.s. line by displacements at the interior supports without changing its intrinsic shape within each span. Since the C line deviates from the c.g.s. line on account of the moments produced by the induced reactions, and since such moments vary linearly within the span, it follows that the C line can be linearly transformed.

A simple method of determining the C line is shown in Examples 8 and 9.

Example 8 A continuous prestressed-concrete beam with bonded tendons is shown in Fig. 32*a*. The c.g.s. is eccentric at *A*, bent sharply at *D* and *B*, and has a parabolic curve for the span *BC*. Locate the line of pressure (the C line) in the concrete due to prestress alone. Consider a prestress of 250 kips.

SOLUTION. The primary moment diagram due to prestress is shown in *b*. The corresponding shear diagram is shown in *c*, from which the loading diagram is drawn in *d*. The fixed-end moments are computed for this loading. At *A* there is the additional moment $0.2 \times 250 = 50$ kip-ft resulting from the prestress eccentricity. Moment distribution is performed in *e*.

The eccentricity of the line of pressure at *B* is $246/250 = 0.98$ ft. The line of pressure for the beam can be computed by plotting its moment diagram and dividing the ordinates by the value of the prestress. But this is not necessary; since the line of pressure deviates linearly from the c.g.s. line, it is only necessary to move the c.g.s. line so that it passes through the points located over the supports, as shown in *f*. The line of pressure at *D* is translated upward by the amount $(0.98 - 0.4)\ 30/50 = 0.35$ ft and is now located $0.80 - 0.35 = 0.45$ ft below the c.g.c. line. At midspan of *BC*, the line of pressure is translated upward by the amount $(0.98 - 0.4)\ 25/50 = 0.29$ ft and is now located 0.61 ft below the c.g.c. line.

Example 9 A uniform load of 1.2 kips/ft is applied to the beam of Example 8. Compute the stresses in the concrete at section *B*, where $I = 39{,}700$ in.^{4} and $A_c = 288$ in.^{2} (Fig. 33).

SOLUTION. The moment diagram for distributed load is plotted in *c*. Dividing these moments by the prestress of 250 kips gives the C line in *d*. Adding *d* to *f* of Example 8, the pressure line for both prestress and external load is given in *e*.

The resulting moment at section *B* is $250 \times 0.52 = 130$ kip-ft, from which

$$f = \frac{-250}{288} \pm \frac{130 \times 12 \times 18}{39{,}700} = -0.867 \pm 0.707 \text{ ksi}$$
$$= -160 \text{ psi top fiber}$$
$$= -1574 \text{ psi bottom fiber}$$

36. Load-Balancing Method This method is convenient for both the analysis and the design of prestressed continuous beams. When the external load is exactly balanced by the transverse component of the prestress, the beam is under a uniform stress $f = F/A_c$ across any section. For any change in load from that balanced condition, only the effects of change need be computed. For example, if the additional moment is M, the additional stresses are given by $f = My/I$. Thus, after load balancing, the analysis of prestressed continuous beams is reduced to the analysis of nonprestressed continuous beams. Since such analysis will be applied to only the unbalanced portion of the load, approximate methods may often prove sufficient.

Design by the load-balancing method gives a different visualization of the problem. It becomes a simple matter to lay out the cables in an economical manner and to compute the required prestress and the corresponding fiber stresses in concrete. This is illustrated in Example 10.

In the load-balancing method it is often assumed that the dead load of the structure is balanced by the effective prestress. Therefore, a slight amount of camber may exist under the initial prestress. It is not always necessary to balance all the dead load, since such balancing may require too much prestress, and a limited amount of deflection may not be objectionable. On the other hand, when the live load is large compared with the dead load, it may be necessary to balance some of the live load in addition to the dead load.

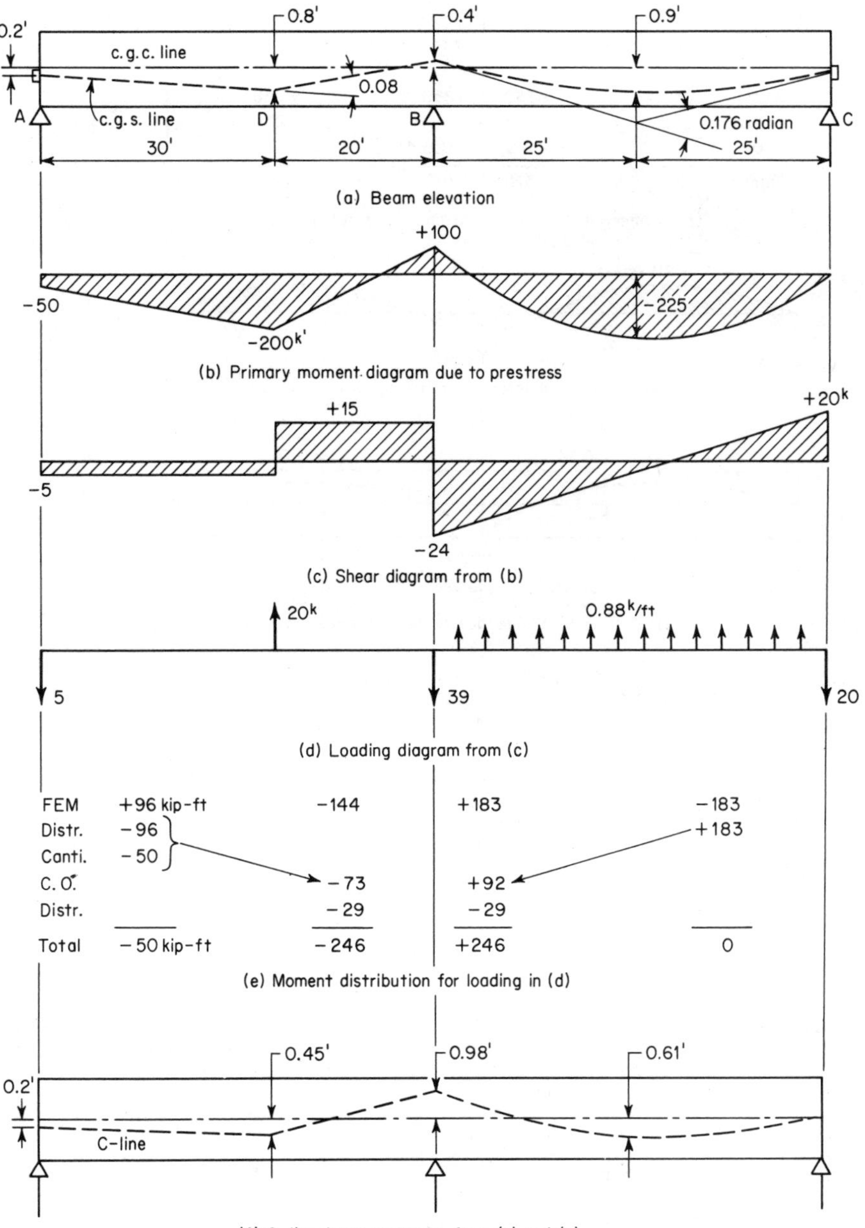

Fig. 32 Example 8.

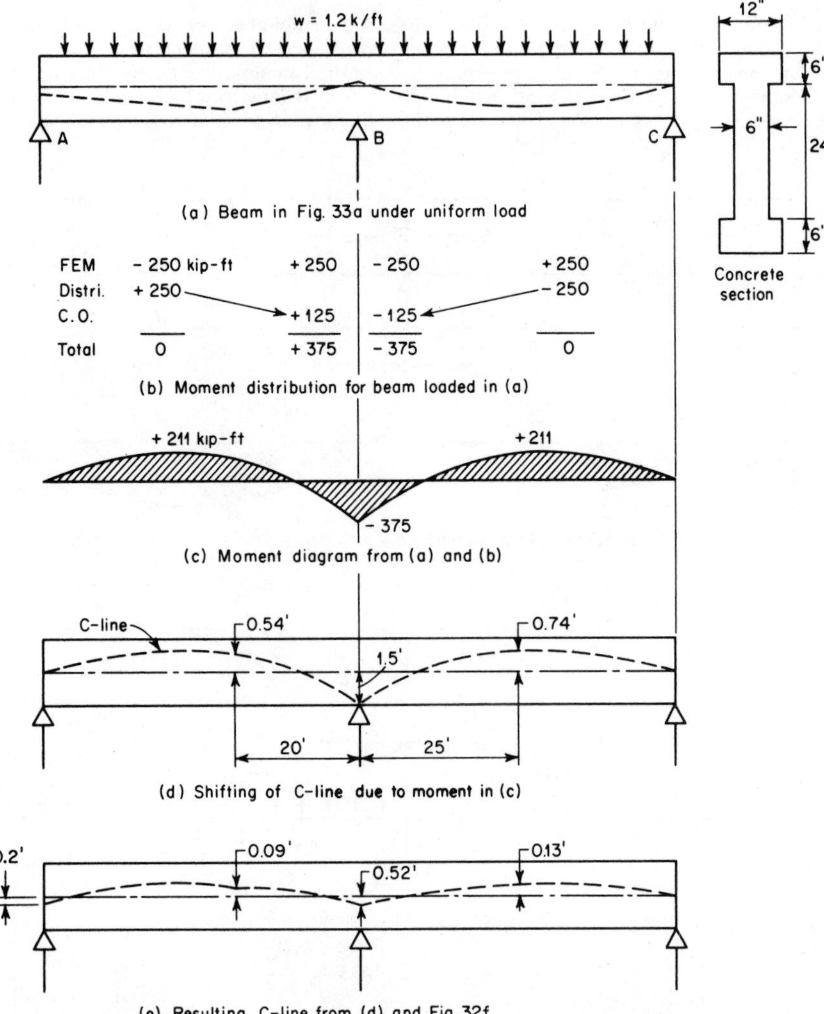

Fig. 33 Example 9.

Example 10 For the continuous beam in Fig. 34, determine the prestress F required to balance a uniform load of 1.03 kips/ft, using the most economical location of cable. Assume a concrete protection of at least 3 in. for the c.g.s. Compute the midspan section stresses and the reactions for the effect of prestress and an external load of 1.6 kips/ft.

SOLUTION. The most economical cable location is one with the maximum sag so that the least amount of prestress will be required to balance the load. A 3-in. protection is given to the c.g.s. over the center support and at midspan (a theoretical parabola based on these clearances will have slightly less than 3 in. at a point about 20 ft in from the exterior support). The c.g.s at the beam ends should coincide with the c.g.c. and cannot be raised, not only because such raising will destroy the load balancing but because it will not help to increase the efficiency of the cable, since unfavorable end moments will be introduced.

The cable now has a sag of 18 in., and the prestress F required to balance the load of 1.03 kips/ft is

$$F = \frac{wL^2}{8h} = \frac{1.03 \times 50^2}{8 \times 1.5} = 214 \text{ kips}$$

The fiber stress under this balanced-load condition is

$$f = \frac{F}{A_c} = \frac{214{,}000}{360} = -593 \text{ psi}$$

Owing to the additional load of 1.6 − 1.03 = 0.57 kip/ft the negative moment at the center support is

$$M = \frac{wL^2}{8} = 0.57 \times \frac{50^2}{8} = -178 \text{ kip-ft}$$

and $$f = \frac{Mc}{I} = \frac{6 \times 178 \times 12{,}000}{12 \times 30^2} = \pm 989 \text{ psi}$$

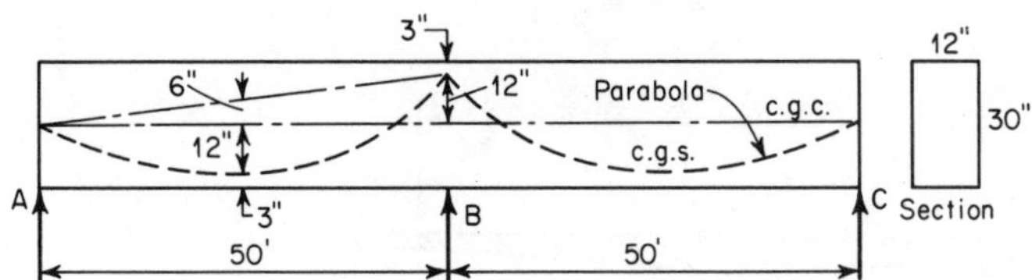

Fig. 34 Example 10.

The resulting stresses at the center support are

$$f = -593 + 989 = +396 \text{ psi tension top}$$
$$f = -593 - 989 = -1582 \text{ psi compression bottom}$$

The reactions due to 1.03 kips/ft can be computed from the vertical components of the cable and are, very closely,

Exterior support: $R_A = 1.03 \times 25 - \frac{214}{50} = 25.8 - 4.3 = 21.5$ kips

Interior support: $R_B = 51.6 + 2 \times 4.3 = 60.2$ kips

Under the action of 0.57 kip/ft load, the reactions are, by the elastic theory,

Exterior support: $R_A = 10.6$ kips
Interior support: $R_B = 35.6$ kips

Hence the total reactions due to 1.6 kips/ft load and the effect of F = 214 kips are

Exterior support: $R_A = 21.5 + 10.6 = 32.1$ kips
Interior support: $R_B = 60.2 + 35.6 = 95.8$ kips

37. Ultimate Strength of Continuous Beams The ultimate strength of prestressed continuous beams can be estimated by limit analysis. Plastic hinges form at points of maximum moment in underreinforced beams. Complete plastic hinges may not develop at cross sections where shear is large and in overreinforced beams, in which case the action will be only partly plastic.

Cracking in prestressed continuous beams can be computed by the elastic theory; cracking begins when the tensile fiber stress reaches the modulus of rupture of the concrete.

DESIGN EXAMPLES

Example 11 Design a precast, prestressed roof T panel with the cross section shown in Fig. 35. Given: simple span 100 ft, roofing and piping 6 psf, live load 20 psf, and with f'_{ci} = 3500 psi, f'_c = 5000 psi, E_c = 4,000,000 psi.

Allowable tension: $f_b = 6\sqrt{f'_c} = 424$ psi at full dead load plus live load
Allowable compression: $= 0.45f'_c = 2250$ psi at full dead load plus live load
$= 0.60f'_{ci} = 2100$ psi at transfer

Pretensioning tendon ½-in. 7-wire strand (Table 5). Posttensioning ¼-in. buttonhead wires (Table 4).

$f_{pu} = 250{,}000$ psi, $f_0 = 175{,}000$ psi, $f_{se} = 145{,}000$ psi, $f'_v = 40{,}000$ psi. Properties of T (Fig. 22): $A_c = 457$ in.^{2}, $I_c = 61{,}000$ in.^{4}, $c_b = 24.2$ in.

SOLUTION.

$$w_G = 457 \times \frac{150}{144} = 476 \text{ plf} \qquad w_s = 6 \times 6 = 36 \text{ plf}$$

$$w_L = 6 \times 20 = 120 \text{ plf} \qquad w_T = 632 \text{ plf}$$

$$M_G = 476 \times \frac{100^2}{8} = 595$$

$$M_S = 36 \times \frac{100^2}{8} = 45$$

$$M_L = 120 \times \frac{100^2}{8} = \underline{150}$$

$$M_T = 790 \text{ kip-ft}$$

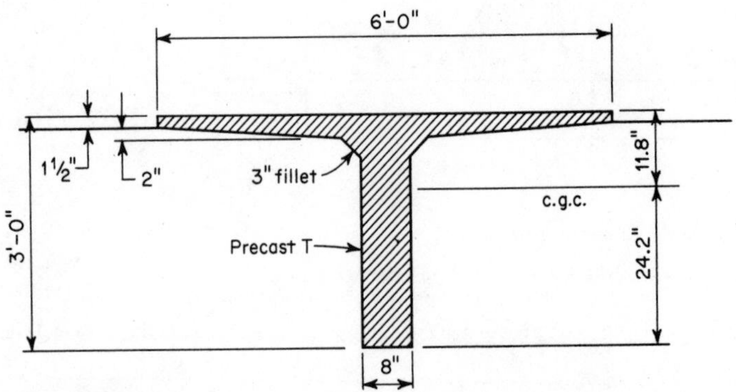

Fig. 35 Example 11.

AMOUNT OF PRESTRESS. Since M_G is large, we can probably locate c.g.s. at the lowest position assuming concrete protection $d' = 4$ in. to the c.g.s. at midspan (Fig. 36). From Eq. (32).

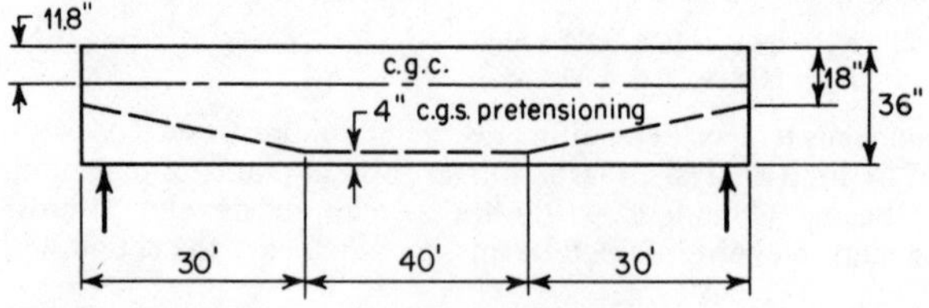

Fig. 36 Location of c.g.s. for girder pretensioning.

$$M_T = F(k_t + c_b - d') + \frac{f'_b I}{c_b}$$

$$k_t = \frac{61{,}000}{457 \times 24.2} = 5.52 \text{ in.}$$

$$790 \times 12{,}000 = F(5.5 + 24.2 - 4) + 424 \times \frac{61{,}000}{24.2}$$

$$F = 327{,}000 \text{ lb}$$

The corresponding top fiber stress is [Eq. (12)]

$$f_t = \frac{F}{A} - \frac{Fec}{I} + \frac{Mc}{I}$$

$$= \frac{-327{,}000}{457} + \frac{327{,}000 \times 20.2 \times 11.8}{61{,}000} - \frac{790 \times 12{,}000 \times 11.8}{61{,}000}$$
$$= -716 + 1276 - 1832 = -1272 < 2250 \text{ psi}$$

The stress in the steel at transfer is somewhat lower than f_0 and may be taken as 165,000 psi. Thus

$$F_0 = \frac{165}{145} \times 327{,}000 = 372{,}000 \text{ lb}$$

Then, with M_G = 595 kip-ft, we have

$$f_t = \frac{-372{,}000}{457} + \frac{372{,}000 \times 20.2 \times 11.8}{61{,}000} - \frac{595 \times 12{,}000 \times 11.8}{61{,}000}$$
$$= -814 + 1455 - 1380 = -739 \text{ psi}$$
$$f_b = -814 - \frac{24.2}{11.8}(1455 - 1380)$$
$$= -814 - 154 = -968 \text{ psi}$$

These fiber stresses at transfer indicate a fairly uniform stress distribution at the midspan section, which is desirable.

PRETENSIONING TENDONS. To supply 327 kips, using ½-in. 7-wire strands with A_{ps} = 0.1438 in.^{2} and F = 145 × 0.1438 = 20.8 kips requires 327/20.8 = 15.7 strands. Use 16 strands arranged as in Fig. 37.

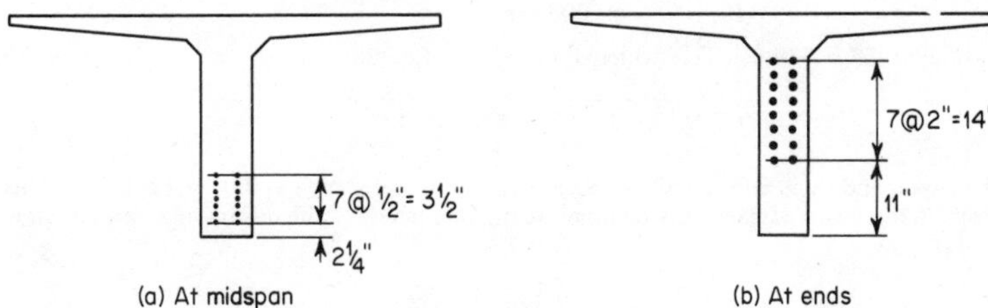

Fig. 37 Arrangement of pretensioning tendons.

CAMBER AT TRANSFER. The camber at transfer can be computed from Eq. (e) of Fig. 19, to which is added the deflection due to the weight of the beam

$$\Delta = \frac{L^2}{8EI}\left[M_2 + M_1 - \frac{M_1}{3}\left(\frac{2a}{L}\right)^2\right] - \frac{5}{48}\frac{M_G L^2}{EI}$$
$$= \frac{100^2 \times 144}{8 \times 4000 \times 61{,}000}\left[372 \times 6.2 + 372 \times 14 - \frac{372 \times 14}{3}\left(\frac{2 \times 30}{100}\right)^2 - \frac{5}{6} \times 595 \times 12\right]$$
$$= 0.69 \text{ in. upward}$$

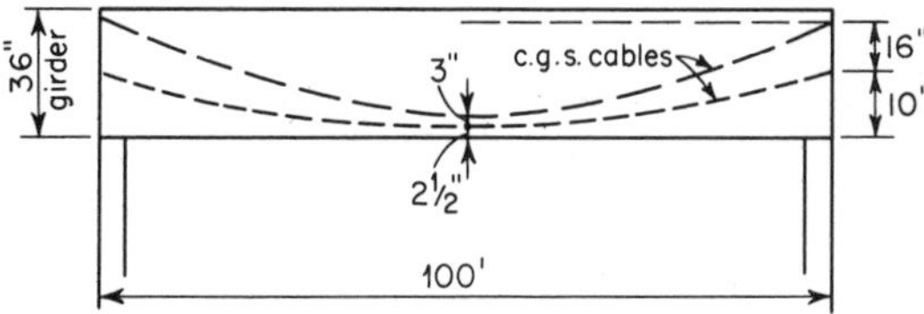

Fig. 38 Location of c.g.s. for girder posttensioning.

POSTTENSIONING. If the girder is posttensioned (rather than pretensioned) with ¼-in. wires with A_{ps} = 0.049 in.^{2}, the effective prestress is 0.049 × 145 = 7.1 kips per wire. The number of wires required is 327/7.1 = 46. Two parabolic tendons of 23 or 24 wires each can be used (Fig. 38).

CRACKING. Cracking strength can be determined from Eq. (16b), assuming the modulus of rupture $f_r = 7.5\sqrt{f_c'}$ = 530 psi.

$$M = F(e + k_t) + \frac{f_r I}{c}$$

$$= \frac{327(20.2 + 5.52)}{12} + \frac{530 \times 61{,}000}{24.2 \times 1000 \times 12}$$
$$= 812 \text{ kip-ft}$$

This corresponds to a uniform load $w = 8 \times 812/100^2 = 0.650$ klf, or a live load of $650 - 476 - 36 = 138$ plf $= 138/6 = 23$ psf.

ULTIMATE STRENGTH (ART. 18). Assuming that the ultimate strength of the steel is fully developed, and with $A_{ps} = 16 \times 0.1438 = 2.3$ in.^{2},

$$T' = 2.3 \times 250 = 575 \text{ kips}$$
$$A' = \frac{575}{0.85 \times 5} = 135 \text{ in.}^2$$

The area A' is supplied by the 6-ft flange with $k'd$ about 2 in. The moment arm is $36 - 4 - 1 = 31$ in. and the ultimate moment

$$M_u = 2.3 \times 250 \times 31/12 = 1486 \text{ kip-ft}$$

This corresponds to a uniform load $w' = 8 \times 1486/100^2 = 1.19$ klf. The ACI required ultimate load capacity is $U = 1.4D + 1.7L = 1.4\,(476 + 36) + 1.7 \times 120 = 0.921 \text{ klf} < 1.19$ klf.

STIRRUPS. The maximum shear at design load is $632 \times 50 = 31.6$ kips. Thus

$$v = \frac{VQ}{Ib} = \frac{31{,}600 \times 8 \times 24.2^2/2}{61{,}000 \times 8} = 152 \text{ psi}$$

At c.g.c., $f = 327{,}000/457 = 716$ psi. The principal tension is [Eq. (38)]

$$S_t = \sqrt{152^2 + 358^2} - 358 = 32 \text{ psi}$$

which is very low.

Combined moment and shear failure will not occur in a simple beam such as this, where the tendons are relatively low in the beam. Hence, only nominal stirrups are needed. The minimum requirement is given by Eq. (44).

$$A_v = \frac{A_{ps}}{80}\frac{f_{pu}}{f_y}\frac{s}{d}\sqrt{\frac{d}{b_w}} = \frac{2.3}{80} \times \frac{250}{40} \times \frac{s}{32}\sqrt{\frac{32}{8}} = 0.0112s \text{ in.}^2$$

For single No. 4 bars, $A_v = 0.20$ in.^{2} and $s = 0.20/0.0112 = 17.9$ in. According to the ACI Code, s cannot exceed 24 in. or $\frac{3}{4}d$, whichever is smaller. Therefore, use No. 4 single stirrups at 18 in.

Example 12 The top flange of a composite section 4 in. thick by 60 in. wide is to be cast in place. Design a precast section with a total depth of 36 in. (including the thickness of the slab) for the following moments:

$M_t = 320$ kip-ft = total moment on section
$M_c = 220$ kip-ft = moment in composite section
$M_p = 100$ kip-ft = moment on precast portion
$M_g = 40$ kip-ft = girder load moment

The allowable stresses are

$f_b = 1.80$ ksi = compression in precast section at transfer
$f_t' = 0.30$ ksi = tension in precast section at transfer
$f_t = 1.60$ ksi = compression in composite section at working load
$f_b = 0.16$ ksi = tension in composite section at working load

Initial prestress = 150 ksi; effective prestress = 125 ksi.

SOLUTION. Assume the lever arm = $0.65h$ for M_T.

$$F = \frac{M_T}{0.65h} = \frac{320 \times 12}{0.65 \times 36} = 164 \text{ kips}$$
$$F_0 = 164 \times 150/125 = 197 \text{ kips}$$

The concrete area required for an inverted T can be approximated by

$$A_c = \frac{1.5F_0}{f_b} = \frac{1.5 \times 197}{1.8} = 164 \text{ in.}^2$$

The resulting trial section is shown in Fig. 39. Section properties for the precast portion are

$$\begin{aligned} 4 \times 14 &= 56 \times 2 = 112 \\ 28 \times 4 &= 112 \times 18 = 2016 \\ A_c &= 168 \qquad 2128 \div 168 = 12.7 \text{ in.} = c_b \end{aligned}$$

$$56\left(\frac{4^2}{12} + 10.7^2\right) = 6{,}500$$
$$112\left(\frac{28^2}{12} + 5.3^2\right) = 10{,}450$$
$$I_c = \overline{16{,}950} \text{ in.}^4$$

Section properties for the composite section are

$$4 \times 60 = 240 \times 2 = 480$$
$$\frac{168}{408} \times 23.3 = \frac{3920}{4400} \div 408 = 10.8 \text{ in.} = c_t$$
$$240\left(\frac{4^2}{12} + 8.8^2\right) = 18{,}800$$
$$168 \times 12.5^2 = 26{,}250$$
$$I \text{ of precast portion} = 16{,}950$$
$$I' = \overline{62{,}000} \text{ in.}^4$$

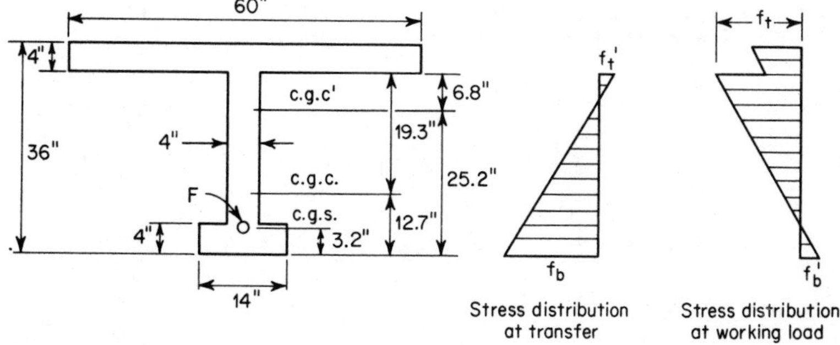

Fig. 39 Example 12.

1. The c.g.s. of the composite section must be located for optimum capacity to resist moment, but not so low as to overstress the precast portion at transfer. Equating the tensile stress in the top fiber to the allowable value gives

$$f_t' = -\frac{F_0}{A_c} + \frac{F_0ec}{I_c} - \frac{M_Gc}{I_c}$$
$$0.30 = -\frac{197}{168} + \frac{197 \times e \times 19.3}{16{,}950} - \frac{40 \times 12 \times 19.3}{16{,}950}$$
$$= -1.17 + 0.224e - 0.545 \qquad (a)$$
$$e = 9.0 \text{ in.}$$

Thus, the c.g.s. can be located 12.7 − 9 = 3.7 in. above the bottom fiber.

2. The required value of F is determined by equating the tensile stress in the bottom fiber of the composite section at working load to the allowable value

$$f_b' = -\frac{F}{A_c} - \frac{Fec}{I_c} + M_p\frac{c}{I_c} + M_c\frac{c'}{I'}$$
$$0.16 = -\frac{F}{168} - \frac{F \times 9 \times 12.7}{16{,}950} + \frac{100 \times 12 \times 12.7}{16{,}950} + \frac{220 \times 12 \times 25.2}{62{,}000}$$
$$= -F(0.00595 + 0.00675) + 0.90 + 1.07 \qquad (b)$$
$$F = 143 \text{ kips}$$
$$F_0 = 143 \times 150/125 = 172 \text{ kips}$$

Since the eccentricity was computed for F_0 = 197 kips in Eq. (a), it must be revised for F_0 = 172 kips. Thus,

$$0.30 = -1.02 + 0.196e - 0.545$$
$$e = 9.5 \text{ in.}$$

which locates the c.g.s. 12.7 − 9.5 = 3.2 in. above the bottom fiber. Correcting Eq. (b) for e = 9.5 in. gives

$$0.16 = -F(0.00595 + 0.00712) + 0.90 + 1.07$$
$$F = 139 \text{ kips}$$
$$F_0 = 139 \times 150/125 = 167 \text{ kips}$$

3. With this value of F_0 and the assumed trial cross section, the compressive stress (bottom fiber) in the precast section at transfer is

$$f_b = -\frac{167}{168} - \frac{167 \times 9.5 \times 12.7}{16{,}950} + \frac{40 \times 12 \times 12.7}{16{,}950}$$
$$= -1.0 - 1.19 + 0.36$$
$$= -1.83 \text{ ksi}$$

which is close enough to the allowable value 1.80 ksi.

4. The compressive stress in the top fiber of the precast portion at working load

$$f_t = -\frac{139}{168} + \frac{139 \times 9.5 \times 19.3}{16{,}950} - \frac{100 \times 12 \times 19.3}{16{,}950} - \frac{220 \times 12 \times 6.8}{62{,}000}$$
$$= -0.83 + 1.51 - 1.37 - 0.29$$
$$= -0.98 \text{ ksi}$$

which is less than the allowable value 1.60 ksi.

The stress in the top fiber of the cast-in-place flange can be computed by using the appropriate values in $f = Mc/I$. However, this stress will not be critical (Fig. 39).

Section 3

Concrete Construction Methods

FRANCIS A. VITOLO

Former President, Corbetta Construction Company, Inc., White Plains, N.Y.

The principal goal of a designer is to obtain the most economical structure while maintaining its basic utilitarian functions and its architectural integrity, all without sacrificing quality. The owner's choice of the numerous forms of contract will vary both the designer's and the contractor's roles in attempting to achieve this goal. When the designer and the contractor are part of the same organization or when they are part of a team, the open line of communications permits them to share the effort.

The actual construction contract can assume one of three basic forms.

With a *guaranteed maximum price* contract, contractors agree to perform the work at a guaranteed maximum price for a fixed fee. As an inducement in keeping costs down, they share in any savings. Under this arrangement, the contractor will naturally strongly influence the designer. On a *cost plus fixed fee* contract, the contractor agrees to perform the work at a fixed fee, without guaranteeing the maximum price. The major incentive to the contractor is to maintain low costs, in order to establish, or perpetuate, a reputation that will promote future contracts. The owner and the designer must share the desire for economy. These forms of contract permit work to start before final drawings are prepared, offering completion months sooner.

The *lump sum* form of contract is the most common because it offers a high degree of competition. This gives contractors the incentive to devote their skills, ingenuity, and inventiveness to reducing costs in order to maintain a fair margin of profit. By law, except in rare instances, all public works contracts must be lump sum.

Contractors are constantly planning ahead, weighing which forthcoming projects will best fit their bidding schedules. Therefore, prospective bidders should be made aware of imminent projects as early as possible. When a firm bid date has been established, the availability of bid documents should be advertised at least a month in advance. When feasible, bids should be solicited from recognized qualified contractors. Since many contractors are preparing bids for other contracts, conflicting bid dates should be avoided. (Public works agencies receive bids on known specific days of the week or month.) During the last hours, indeed minutes, before bid time the general contractor is swamped with late proposals from prospective subcontractors and material suppliers, with comparing qualified bids, with bargaining, etc. Therefore, bid openings should not be scheduled on Mondays or on the day following a holiday; afternoon is preferable to morning.

The time allotted for bid preparations should include a reasonable allowance for obtaining and distributing bid documents, for quantity take-offs and analyses of the results, for the planning, devising, and evaluation of construction methods and of proper sequencing, and finally for estimating the costs of the various elements and assembling them into a bid package. Some additional time should be allowed for the issuance of addenda or of clarifications so as not to face the necessity for granting a postponement.

The designer should be responsive to the questions raised by prospective bidders. All too often designers create an unhealthy climate by treating contractors as adversaries. The bid documents should contain all pertinent information in clear, concise form, without resort to exculpatory phrases. A properly conducted prebid conference, where contractors' questions are given straightforward answers, will often eliminate misunderstandings, promote good relationships, and benefit all parties.

Bid documents should be made as simple as possible. Alternate and unit prices should be solicited only if they will be of significant value and are explicit for all trades involved.

The specification writer should refrain from naming proprietary articles, which inhibit competitive pricing and later may cause the contractor to relinquish control of deliveries, etc.

A reasonable time should be allowed for completion of the project, taking into consideration such factors as urgency, local climatic conditions, accessibility, and availability of materials, equipment, and manpower (especially if they are unusual).

Finally, the designer should encourage rather than inhibit contractor resourcefulness. Specific procedures, sequences, etc., should be given only if they are a function of the design.

1. General Considerations The period of original layout and preliminary general design establishes the basic pattern for the structure from which the design details are developed. Even at this early stage, the designer should be mindful of the construction point of view and guide the design accordingly.

It is not suggested that the engineer deviate from improved or advanced ideas of design merely for the purpose of facilitating construction. It is suggested, however, regardless of the design or the use for which the structure is to be constructed, that various principles and practices can be incorporated which will reduce construction costs.

Reductions in costs or time are inherent in mass production, which requires repetition. A detail may be complicated, or of an unusual configuration, and still be economically mass-produced, as long as sufficient repetition is involved to warrant the planning and equipment required. In the project layout, this will involve a repeated, geometric pattern, possibly of column, beam, or girder spacings, which in turn will produce slabs of similar sizes and dimensions. In order to meet architectural, design, or other standards of the owner, the repetition may consist of sections of the structure rather than individual units. While many variations might occur in individual elements within the section, the repeated sections would be similar in every respect.

Architectural or mechanical features which cause variations throughout an otherwise repetitive series of elements or sections may, in many instances, be congregated in one area to take advantage of the repetition in the remainder of the series.

2. Formwork Formwork, a major item of cost in concrete construction, accounts for 35 to 60 percent of the cost of the concrete work.

The cost of form fabrication, per unit of concrete surface formed, is inversely proportional to the number of reuses of the form, i.e., the number of times the form can be used without refabrication or changes between uses. With repetitive use, the form is fabricated only once. As the number of reuses is increased, the quality of the form can be economically increased. Sufficient reuse will justify plastic, steel, or concrete forms for areas which normally are economically restricted to wood forms.

Forms are preferably fabricated in a central location with stationary, powered equipment, with all the advantages of assembly-line mass production. However, mass production requires a quantity of identical units.

Columns. Increasing the size of several smaller odd-size columns to the dimensions of a greater number of larger columns will increase the quantity and, therefore, the cost of the concrete. However, the savings in form costs will, in many instances, exceed the cost of additional concrete and result in a decrease in the overall cost of the structure. This cost reduction will also apply to spread footings, pile caps, buttresses, beams, girders, and, in some circumstances, slabs.

In multistory structures consideration should be given to repetition of column sizes from floor to floor. A minor reduction in column dimensions from floor to floor, due to smaller loadings, not only causes refabrication of the column form panels but also requires changes to the top of the form, since the slabs, girders, and beams will normally retain the same dimensions from floor to floor. Minor reduction in column dimensions from floor to floor should be accumulated, and one comparatively large reduction made after several floors. When making the accumulated reduction, consideration should be given to a reduction in only one dimension of the column, thereby requiring the refabrication of only two sides of the column forms.

From the above, it will be noted that when it is undesirable to use the same column cross section throughout one floor or sectional area, it will be advantageous to use rectangular columns with one dimension equal. This will vary only two of the four column side panels.

Beams and Girders. The same general principles apply with respect to beam and girder design. These require form sides, bottoms, and shores. Variation in beam widths affects the slab forms and the beam bottom form. A variation in the height affects two beam side forms and the shores (as well as reshores). Obviously, where a change from a standard is required, changing only the width involves dimensional variations in a much smaller percentage of the formwork and shores.

When selecting the original dimensions of a series of beams or girders a width equal to that of the column face into which they frame will reduce the form details required at the head of the column. Similarly, selection of a beam depth equal to that of intersecting beams will reduce form details at beam intersections and will produce greater quantities of shoring of equal lengths.

Slabs. The length and width of a slab are not particularly significant, except for their effect on the lengths of beams and girders into which the slab frames. Variation of the slab depth in small areas of lighter loadings, however, may seriously affect the forms of the adjacent framing. This requires beams with special sides, special column top details (unless the beams are equal to or wider than the column width), and special-length shores and reshores for this slab area.

Walls. For wall forms the side-panel dimensions become important only with reference to the length of adjacent framing members. Variation in the thickness of the wall, however, does affect considerably the fabrication and installation of box-outs for openings and sleeves and the side forms of the adjacent framing members.

Foundation walls which require forms sloped at the bottom are undesirable. They are more economically placed with stepped forms.

Tie Beams. For tie beams placed monolithically with slabs on ground, side forms are not used for depths below slab up to 2 or 3 ft, depending on soil conditions. The cost of sloping trench sides and placing additional concrete is less than the cost of the forms. It will be noted that when forms are required, they cannot be recovered. This means not only a form cost based upon no reuse, but lost material that cannot be refabricated for use elsewhere. Such lost forms are expensive and should be avoided whenever possible.

Accessories. Form costs include the cost of form accessories, such as ties, and in the case of fireproofing of structural steel, form hangers. These accessories are wholly, or in part, used only once as they become embedded in the concrete. The standardization of girder, beam, and column dimensions, and wall and slab thicknesses, all helps to reduce the number of sizes and lengths of form accessories to be stored, handled, and placed.

3. Reinforcing Steel The contractor uses assembly-line mass-production methods for the cutting, bending, and handling of reinforcing steel. This is done in a central yard, equipped with heavy-duty powered equipment. When possible, jigs and templates are established, near the cutting yard, for the assembly-line fabrication of column and beam cages, footing and pile-cap mats, and other assemblies. As in the form yard, unit costs are lowered with a reduction in the numbers of bar sizes, lengths of bars of the same size, types of bend to be processed, and the number of cage or mat variations to be assembled.

Columns. In many cases, simplifying the formwork (for example, increasing smaller odd-size columns to the size of larger columns) automatically changes the column reinforcing to a standard size. A small increase in the bar size of isolated columns will often eliminate a quantity of various cage sizes and result in only a negligible increase in the weight of the reinforcing steel.

The use of overall rectangular column ties and beam stirrups combined with interior C-type ties, in lieu of double interlocking rectangular ties, will facilitate the assembly of column and beam cages.

Long Bars. Except for special designs, bar lengths which exceed standard mill lengths should be avoided. Bars exceeding approximately 20 ft in length which require bends on both ends are awkward to handle in the bending yard. After one end is bent, they must be turned around to bend the other end. Furthermore, yards are not normally set up with sufficient vacant area to turn the bars in the vicinity of the bending machine. In this situation, the contractor will prefer to bend one end and splice a short, bent bar to the other. Long bars requiring a sharp bend toward the center are awkward to bend, handle, store, and place. Again, it is preferable to splice two bars, one straight and the other with a bend on one end.

Obviously, the use of spliced bars to eliminate bending does increase the weight and cost of the reinforcing material. However, a review of bending costs, and the higher handling and placing costs of some unusual bent bars, will usually show a saving in labor costs which exceeds the cost of the additional steel used for splicing. This cost differential can be considerable in areas of high labor rates.

Concentrations of Steel. The problem of concentrations of reinforcing steel which interfere with the concrete placement, or with the placement of the reinforcement itself, should be a matter of routine checking during the design. When corrective measures must be taken during the preparation of shop drawings, delays result during the period required by the designer to investigate them. Further delays may be encountered in changing the drawings to conform to the designer's corrective measures. When the situation is encountered during placement of the steel, the resultant delays and/or additional labor can be very costly.

Concentrations occur mainly in heavily reinforced members or at the framing connections of these members. However, they may also occur in framing systems which are considered as lightly reinforced. They are not readily apparent in the latter systems and, not being suspected, are often overlooked during the design and preparation of shop drawings. These concentrations may be the result of the reinforcing from several beams or girders intersecting over a column which, when added to the column reinforcement, may restrict or prohibit the concrete from passing into the column forms. The intersection of beam and girder reinforcing, combined with considerable slab reinforcing which is continuous over the beams, can also create the same problem. The latter situation becomes more serious when the slab reinforcing is fanned, with the vertex passing over the beam or intersection.

It should be remembered that dowels which are to be placed prior to concreting may have been noted but not pictured on the drawings, and are easily overlooked. These dowels may introduce concentrations.

Lap splices double the steel area within the splice length. Although not shown or indicated on the contract drawings, additional splices may be required in continuous bars which exceed stock lengths, or in the case of long bent bars which cannot be placed in one piece.

Occasionally there are concentrations known to the engineer which, for design reasons, are not readily avoidable by changing bar sizes or arrangements or by increasing the size of the member. In these cases, difficulties in concrete placement may be overcome by using a smaller aggregate. Care should be exercised in using this solution, however, because of possible difficulties in vibrating the concrete, possible segregation of the concrete, and the added cost of using an additional concrete mix.

Thin concrete members, or members of special design, may require closer placement tolerances to ensure compliance with design computations. Any such requirement should be specifically noted, and detailed if necessary, in the specifications or on the drawings.

Inclusion on the drawings of specific information covering dowel lengths and locations, end-anchorage requirements, and required and permissible splice locations and details will avoid misunderstandings and delays in the preparation of shop drawings and in the field work.

4. Concrete The designer specifies the strength and other qualities which the concrete should have to serve its purpose. The experienced concrete contractor can produce the specified concrete and will accept this responsibility when given control of the entire operation, from the design of mix through the curing operations. In other words, the specification should give, in as much detail as is necessary, the strength and other qualities required. However, the means and methods used to achieve these results, including materials, design mix, mixing, transporting, placing and curing, should be left to the contractor. Of course, the engineer must retain overall control. This is best accomplished simply by requiring compliance with recognized standards such as the municipal building code and the various standards of the industry. The reliable contractor will use these standards as minimum requirements, supplementing them when necessary to produce the desired results. Any additional requirements, limitations, and restrictions serve only to increase the cost of the work and to transfer responsibility from the contractor to the engineer or owner.

The selection of the raw materials, and the combining of them into a concrete mix, are done by the contractor on the basis of knowledge of the materials themselves, and of available facilities. A rigidly specified mix may be difficult to place in restricted form openings caused by thin sections, concentrations of reinforcing, or the presence of embedded items. Under these conditions segregation and excessive honeycombing are likely to occur. With control of the mix design, the contractor can, without difficulty or loss of time, readily adjust the concrete supply to overcome these problems. The answer may be merely a matter of using smaller aggregate, or increasing the slump (with the other adjustments necessary to ensure strength, wearability, and durability) above that normally required.

The use of high-early-strength cement should, in most cases, be permitted at the contractor's option. When it is used by those who understand it, its performance and results cannot be questioned. Although more expensive than regular cement, it is often the secret of success in cutting costs when it is used to accelerate the release of formwork, to bring a project back to schedule, and to anticipate delays from inclement (or freezing) weather.

Some projects are adapted to the use of paving machines, some to transit-mixed concrete, and others centrally mixed concrete, or a combination of mixing and transportation methods. The contractor can best judge the method which will prove most economical and, at the same time, produce the required results.

Testing. The testing of concrete is basically an inspection control. Therefore, the testing laboratory should be an independent agency retained by the owner. The contractor's reputation can only be adversely affected by problems resulting from the placement of substandard concrete, and he must have the fair and honest tests which can be assured by a reliable, experienced laboratory.

Removing Forms. It is generally specified that forms be left in place for various lengths of time depending on their location, i.e., columns, walls, beam bottoms, beam sides, slabs, etc. These requirements are for the purpose of ensuring that the concrete has attained sufficient strength to warrant removal of the forms, and are based upon the theoretical strengths of concrete that should have developed in the specified period. These strengths are not confirmed until cylinders have been tested, usually not sooner than 28 days after the concrete is placed. Ordinarily, this is a considerable time after the forms have been removed.

Stripping on strength requires that the concrete strength be supported by test cylinders prior to form removal in lieu of the theoretical strengths as noted above. Specifications which permit stripping of forms on the basis of attained strength, rather than time, may result in savings from quicker turnover of formwork and a reduction in completion time of the structure. Although this requires additional test cylinders, since the 28-day tests must be made to meet building-code requirements, the additional cost of testing is negligible compared with the saving that can result. However, when stripping on the basis of proved strength is permitted by the specifications, stripping on the basis of time should be optional in the event of damaged test cylinders or questionable test results.

Curing. Curing affects not only the strength and quality of the concrete but also, depending on the method used, the properties of the surface. Some commercial curing agents will adversely affect the bonding of paint and other surface finishes and the color stability of paints. Curing agents may leave surfaces not scheduled to receive further treatment with an undesirable color. Here, again, it is suggested that the desired results be explicitly noted, and the means and methods of accomplishment be left to the contractor.

5. Embedded Items Items to be embedded in the concrete include inserts, hangers, nailers, reglets, anchors, sleeves, frames, and structural steel. Too often, the importance of studying the details of embedded items is overlooked, resulting in a missed opportunity to reduce costs or avoidable difficulties in the field.

Hangers, anchors, and similar embedded items which protrude from the face of the concrete must necessarily pierce the forms. Unless these items occur in the same form location in succeeding repetitive sections, forms must be either patched or refabricated. Furthermore, they impede stripping of the form and may, depending upon their size and number, interfere with placement of the form supports and shoring. In the case of hooked anchors, the form must be fabricated with a separate waste panel for each use in order to minimize form damage upon stripping. The waste panel represents an additional operation and increases both forming and stripping costs. Embedded items which are attached to the form face, without penetrating the form itself, are preferred.

Architectural metal and delicate steel items with attached anchors are placed in one operation prior to placement of the concrete. When using embedded inserts in lieu of attached anchors, this placement requires two operations. However, the cost of supporting these items with attached anchors in the forms, and their protection during concreting and subsequent construction, may more than offset the cost of the additional operation when embedded inserts are used. When contractors are given the option of placement by either method, their experience will indicate the most economical method, and the possibility of damage in the event the items are installed in the formwork. It also helps to avoid concreting delays, in the event of late delivery of the item.

The question of whether to embed heavy castings, structural-steel members, and other assemblies in the formwork prior to concreting, or to "box out" for these items and place them later should also be left to the contractor's option.

Contract documents should show in exact detail the number, location, position, size, and type of each embedded item, with details of attached or detached anchorages where either is permissible.

Each embedded item should be studied with respect to its effect on the reinforcing steel, concrete cover required, and any changes resulting from its proximity to openings, columns, beams or girders, and mechanical-trade items.

6. Special Designs During the design stage, the engineer is minutely familiar with all phases of the project. During this state of the work, problems of a special or unique nature may arise which will be of vital concern to the contractor in bidding, planning, and executing the work. Their details should be properly documented and noted in the contract documents.

The following paragraphs indicate the type of information that should be noted but which is normally made available to the contractor only after a question is raised and the information requested. The latter practice may result in delays in planning and may on occasion require expensive changes in planning or in the field.

The contract drawing notation of the load used in the design of vehicle passageways such as bridges, elevated bypasses, ramps, and entrances informs contractors as to the extent they may use these facilities during construction.

Specific information should be noted concerning framing designed for composite action, deflection to be controlled through camber, and those special structures in which a definite sequence of form release is essential to avoid a reversal of stress for which the framing is not designed.

Definite and specific information on the sequence of concreting, shoring, and reshoring is essential where columns or beams are required to assume gradually increasing increments of load and deflection as the concreting progresses upward in the structure.

Exact elevations to which formwork is to be installed should be given for elevated highways, bridges, and other structures where deflection must be controlled so that finished elevations will conform to proper slope for drainage and to required grade for high-speed roadways. The formwork elevations are also required for concrete framing members which are cambered to provide for deflection that will result from hung framing or mechanical equipment.

Formwork for long-span arches is normally required to be released in predetermined sectional sequences to assure that all the arch ring remains in compression during the operation. Since this often establishes limitations in the design and operation of the falsework, form design and job planning will be facilitated if this information is made available with the contract documents. The contractor's attention should be drawn to those structures, or parts of structures, which depend upon other parts of the construction for their stability. These notes should indicate the degree of completion required to assure stability.

In the case of cantilevered or other similar unbalanced construction, the contractor should, if possible, be given the option to locate construction joints within permissible areas noted on the drawings.

The above indicates the type of information desired by the contractor, and it should be anticipated and furnished in the design of folded plates, domes, arches, thin shell, very long spans, or other structures in which construction details involve more than routine construction procedures.

7. Tolerances It has always been understood that concrete construction, being an on-site operation, cannot meet the close tolerances required of manufactured items of other materials. Prior to the advent of precast concrete, tolerances, as such, were not specified, but were matters of judgment between the engineer and contractor. Depending upon the member involved and its intended use, the degree of accuracy was a rule-of-thumb consideration of workmanship. Precast concrete has come to be recognized as a manufactured item, and tolerances are more frequently specified and adhered to.

Dimensional and strength tolerances in concrete work have been defined and standardized by the industry, particularly by the American Concrete Institute and the American Society for Testing and Materials. The adoption of these standards by the industry has minimized questions of acceptability of concrete work on the basis of dimensional and strength accuracy. However, this acceptability extends only to the finished concrete work and not the completed structure. Confusion still results from the engineer's acceptance of these tolerances, on the one hand, while on the other he details metal frames for openings, installation of masonry units, mechanical installations, etc., to tolerances which cannot be guaranteed with the accepted standards. Failure of the engineer to recognize and to provide for this situation creates unnecessary problems for the concrete contractor and others concerned.

Tolerances for warp, camber, and concrete finishes have not reached the standardization and recognition of other tolerances and should be more closely checked and provided for in the specifications.

8. Shop Drawings Upon the execution of the contract between the owner and the contractor, the owner's engineer assumes two obligations: (1) the approval of shop, or working, drawings and material samples submitted by the contractor, and (2) the inspection and supervision of the actual field work.

Shop drawings are prepared by the contractor for the purpose of expediting construction. They are, with few exceptions, generally an amplification of the information contained in the contract drawings and specifications. They clarify the work to be accomplished by means of (1) larger-scale details, (2) detailed dimensions computed from overall dimensions shown on the contract drawings, (3) combining details which may appear on several contract drawings, (4) combining on the shop drawing information from both the contract drawings and specifications, (5) transforming tables, schedules, and written information into a visual form of sketches and diagrams, and (6) giving the field personnel, on one or several drawings, all the information, but only that, required by them to accomplish the particular portion of the work with which they are immediately concerned. Shop drawings also serve to "proofread" the contract drawings, for it is during preparation of the shop drawings that most errors, inconsistencies, and missing information are discovered.

Shop drawings are submitted to the designer for approval. This provides a review of the work and an opportunity to correct errors and to make alterations prior to any expensive preparatory work in the field. They also show designers the contractor's interpretation of the work to be performed, and give them an early opportunity for clarification in the event that the contractor's interpretation differs from that intended.

Needless to say, designers should review the shop drawings carefully. If it is not their intention to do so, the requirement that they be submitted to them should be deleted from the contract for reasons of economy. This, in turn, will alert the contractor to the fact that the drawings are not being checked.

9. Material Samples Material samples are submitted to the designer for approval, for the purpose of ensuring a mutual understanding and interpretation of the material-requirement sections of the specifications. These sample-submission requirements of the specifications should be limited to those materials which the engineer plans to test, or about which a question might arise as to the expected color shade, finish, or quality of fabrication. The submission of samples of trade-name materials which are specifically identified in the specifications is a waste of time and money for both the engineer and the contractor.

INSPECTION

10. The Resident Engineer The engineer's contact with the actual construction is through the resident engineer. The resident engineer was once viewed as a policeman, hired by the engineer, whose main duties were to see that the contractor complied with the terms of the contract. Through a broader understanding by both engineer and contractor this viewpoint has changed, and the resident engineer's position has assumed its proper perspective.

The resident engineer is the field supervisor and coordinator who expedites the completion of the construction work. With the understanding that teamwork between engineer and contractor is essential to both parties, resident engineers are impartial, even though employed by, and reporting to, the engineer. They are impartial in that they protect the owner by making sure that the contractor fulfills the contractual obligations and at the same time protects the contractor from improper requests or orders which are beyond the scope of the contract. Their experience and authority permit them to make on-site decisions, and corrections and changes which may be needed to expedite the work. They should be aware of the limits of their authority and instructed as to the nature or extent of such corrections or changes as are beyond that authority and which should be referred to the engineer. Only by familiarity with the contract documents, and a knowledge of the intent of the designer are they able to interpret the documents and make those decisions required to produce a more satisfactory project. They should be sufficiently experienced to realize that there are often several ways of accomplishing a result, and that the method should be the option of the contractor provided the proposed method will produce the desired result.

CONTRACT DOCUMENTS

The contract documents, other than the proposal form, consist of the contract drawings and specifications. Both are interdependent and inseparable, and together they inform the contractor exactly what the owner and engineer require of the contractor. It is upon these documents that the contractor has based the proposal to construct the project.

11. Preparation Contractors cannot anticipate the intentions of the engineer or owner, and must assume that the contract documents represent what is required. Under competitive bidding, they cannot do otherwise and expect to receive the contract award. At the same time, where ambiguities are found during the short time prior to bidding and sufficient time is not available for clarification, contractors must increase their proposals to cover any possible additional expenses which may be involved.

A review of the contract documents gives the contractor a fairly good indication of the problems and progress to be expected during the planning and preparatory stages and during the period of actual construction. Well-prepared documents which appear to be accurate, complete, and concise, while fully covering essential details, indicate that the design was carefully executed and that a minimum of delay and problems can be expected to develop during the contract. Incomplete or carelessly prepared documents suggest a long construction period with many delays in the preparation of shop drawings, and in the actual construction work, pending detailed information, corrections, and clarification. They also suggest the probability of discussions and possible disagreements as to the extent of the work included in the contract price. This applies to the contract specifications as well as to the drawings, for since they complement each other, the sum is only as good as the least effective part. A well-prepared set of drawings can come to naught when accompanied by hastily or carelessly prepared specifications.

12. Specifications The specifications inform the contractor of those requirements and obligations which are more effectively given in words, or which cannot be satisfactorily transmitted by diagrams or notes on the drawings. As instructions or information, they, like the drawings, should be exact, complete, and yet concise. They should be prepared by competent personnel, and completely checked to avoid conflicts between specifications and drawings.

The specifications should be prepared specifically for the individual project. The inclusion of extraneous material, copied from specifications for other projects, may cause confusion in an otherwise clear presentation. In an attempt to cover the engineer and owner from every conceivable angle, specification writers sometimes create more problems than they had hoped to avoid.

13. Intent The construction contract implies to the contractor that the owner knows what is wanted, is willing to pay for what is ordered, and has retained a competent designer who has fulfilled his obligations and will accept full responsibility for the design. The contractor is concerned when items appear in the specifications or on the contract drawings which attempt to transfer the designer's responsibility to the contractor. The following are examples of this transfer of responsibility which, in every case, requires the requested work to be done at no cost to the engineer or owner:

1. The contractor agrees that all foundation work required for mechanical equipment, located but not detailed, shall be included in the contract price.
2. The intent of the contract is to construct facilities complete for their intended use. The contractor agrees to furnish and install any and all items required to complete these facilities, whether or not shown or indicated by the contract documents.
3. The contractor agrees that the contract documents have been reviewed and found to be complete and correct in every respect, including but not limited to the structural adequacy.

The contractor, upon viewing items such as these in the contract, can only conclude that those responsible for the preparation of the documents are, at best, uncertain as to their completeness or accuracy, and must decide whether to increase his proposal to cover eventualities, or to forgo submitting a proposal.

14. Scope of Work The "scope of work" section of the contract specifications establishes the extent of the work covered by the contract. It is that part of the contract documents which enumerates the items desired by the owner and upon which the contractor's proposal is based.

As this section is the nucleus of the entire contractual agreement, there should be no question or doubt, on the part of anyone concerned, as to the meaning or intent of the section or any of its parts. The drawings and remainder of the specifications furnish details and further information on the items included in the scope of work section but are not construed as changing the limitations of the contract work as set forth in that section.

In contracts wherein one contractor is to construct the entire project, this section may be very short and general in nature. In contracts involving several contractors working concurrently under separate contracts, and in contracts involving additions or alterations to existing structures, the construction of only a portion of a project, a portion of the work covered by the contract drawings, etc., the section on scope of work must be more explicit and may of necessity be voluminous by comparison.

15. Drawings The contract drawings are the picture half of the contract documents and usually show the work required by the written specifications. They should show all the information, in sufficient detail, which, when combined with the specifications, will enable the contractor to construct the project as it is intended. They should be legible, well correlated as to sections and details, and accurately drawn to scale.

On larger projects, the architectural, structural, and mechanical drawings may be independent sets prepared by separate design firms. The better the correlation of the various sets of drawings prior to their being issued for proposals, the less the likelihood of confusion, delay, and expense during construction. When conflicts are found during construction, work stops and labor is idled until such time as the inconsistency is clarified or corrected. Checking drawings in the field is very expensive for the contractor, designer, and owner.

Contract drawings inevitably contain errors. Revised drawings are required to correct these errors and to incorporate changes required for various other reasons. Once a change is made, the revised drawing should be issued promptly. While the revision may not affect work currently being done in the field, it may seriously affect planning, scheduling, and other items preparatory to future work. In many instances, it is advisable to notify the contractor verbally of a pending revision to avoid unnecessary preparatory work.

Of immediate concern to the contractor upon receipt of a revised drawing is the date and details of the revision. A description of the revision in the drawing title box is seldom sufficient. A more satisfactory procedure is to outline the revised dimension, detail, or section with an irregular, heavy pencil line on the reverse side of the tracing. On the front side, in heavy print, the irregular line is identified with an arrow and a revision number. The revision is then readily apparent, and will not require a detailed comparison of the entire drawing with the previous one to determine the extent of the revision. This procedure not only facilitates finding the revision, but also helps to guard against the overlooking of parts of a revision when several details on various sections of the drawing are revised under one date and revision number.

Whenever dimensions are changed, the drawing should be corrected to scale to avoid a distorted picture. Where this is not practical because of the size or complexity of the drawing, a note under the revision number should state "not to scale." In either case, revisions should be thoroughly checked to ensure correlation with details not changed, and with the architectural and mechanical drawings, to ensure that the correction of one error has not introduced another.

References to drawings in correspondence should always include the drawing title, number, and date, to avoid any possible confusion as to the exact issue of the drawing involved.

Section 4

Masonry Construction

WALTER L. DICKEY

Consulting Civil and Structural Engineer, Los Angeles, Calif.

NOTATION

A_g = gross area of masonry cross section
A_s = area of tension reinforcement
A_v = area of stirrup
b = width of beam
d = depth of beam to reinforcement
D = diameter of bar
E_m = modulus of elasticity of masonry
E_s = modulus of elasticity of steel
f_m = compressive stress in extreme fiber
f'_m = approved ultimate compressive stress of masonry
f_s = tensile stress in reinforcement
f_v = tensile stress in stirrup
h = vertical or horizontal distance between supports, also width of bond beams in plane of wall
h' = effective height of wall or column
M_m = resisting moment as limited by masonry
M_s = resisting moment as limited by steel
n = E_s/E_m
Σ_0 = sum of perimeters of reinforcing bars
p = A_s/bd
t = overall dimension of column, also wall thickness
v_m = shear stress in masonry
v' = shear carried by stirrups

MATERIALS

1. Burned-Clay Units Burned-clay units include common and face brick, hollow clay tile, terra-cotta, and ceramic tile. The latter two are not considered to be structural materials.

Solid units are those whose net cross-sectional area in any plane parallel to the bearing surface is not less than 75 percent of the gross area. Units whose net area is less than 75 percent of the gross area are called *hollow units.*

2. Brick Building brick is available in three grades. Grade SW is for use where the brick may freeze when permeated with water, and where a uniform, high degree of resistance to weathering is desired. Grade MW may be used for exposure to temperatures below freezing, where the brick is not likely to be permeated. It is suitable for the face of a wall above grade. Grade NW is intended for backup or interior masonry. Physical requirements are given in Table 1 (ASTM C62).

TABLE 1 Physical Requirements for Building Brick and Facing Brick

Grade	Compressive strength, flat, min, psi		Water absorption, 5-hr boil, max, %		Saturation* coefficient, max, %	
	Avg of 5	Individual	Avg of 5	Individual	Avg of 5	Individual
SW, severe weathering	3,000	2,500	17.0	20.0	0.78	0.80
MW, moderate weathering	2,500	2,200	22.0	25.0	0.88	0.90
NW, no exposure	1,500	1,250	No limit	No limit	No limit	No limit

* Ratio of 24-hr cold absorption to 5-hr boil absorption.

Facing brick is available in grades SW for high resistance to frost action and MW for moderate resistance to frost action (ASTM C216). Type FBX is intended for use where a high degree of mechanical perfection, narrow color range, and minimum variation in size are required. Type FBS is used where wider variation in color and size is acceptable. Type FBA is manufactured and selected to produce characteristic architectural effects from nonuniformity in size, color, and texture.

3. Structural Clay Tile There are two grades of load-bearing wall tile. Grade LBX is suitable for general use in masonry construction and for use in masonry exposed to weathering. This grade is suitable for direct application of stucco. Grade LB is intended for masonry not exposed to frost action and for exposed masonry protected with a facing of 3 in. or more of other masonry.

Non-load-bearing tile (Grade NB) includes partition tile, furring tile, and fireproofing tile.

Facing tile is intended for general use in interior and exterior walls and partitions. Type FTX is a smooth-faced tile low in absorption, easily cleaned, and resistant to staining. Type FTS is a smooth or rough tile of moderate absorption, moderate variation in face dimensions, and medium range in color. These are available in two classes: standard and special-duty, the latter having superior resistance to impact and transmission of moisture and greater lateral and compressive load resistance.

Floor tile is available in grades FT1 and FT2, both of which are suitable for use in flat or segmented arches or in combination tile and concrete ribbed-slab construction.

Other types of hollow units, such as brick-block, may be used satisfactorily and should not be ruled out simply because they are not included in ASTM specifications for tile.

Ceramic glazed facing tile, facing brick, and solid masonry units are available in ASTM C126 Grade S for comparatively narrow mortar joints and Grade G where variation in face dimension must be very small.

Physical requirements for structural clay tile are given in Table 2.

4. Concrete Units Concrete building brick (ASTM C55), solid load-bearing units (ASTM C145), and hollow load-bearing units (ASTM C90) are available in Types I and II. The moisture content of Type I units is controlled while that of Type II is not. The moisture limits are related to the shrinkage characteristics of the units and to the relative humidity of the jobsite (Table 3). This is to control shrinkage and the consequent hazard of cracking due to structural restraint. These limits may be waived if special precaution is taken to prevent such stress, as by control joints, etc. These units are of two grades. Grade N is for general use, as in exterior walls that may or may not be exposed to water penetration, and for interior or back-up use. Grade S is limited to use where not exposed to weather.

Hollow non-load-bearing units (ASTM C129) are available in Types I and II.

Physical requirements for concrete masonry units are given in Table 4.

5. Mortar Bond is more important to the proper functioning of masonry than is the strength of the mortar itself. Mortars for unit masonry may be specified as to proportions, or on the basis of property specifications. Proportions for five different types of mortar are given in Table 5. No strength requirements are stipulated. However, for reinforced masonry, or other masonry where assurance of strength is important, field sampling and testing should be done. The alternate method of proportioning requires mixing of the ingredients to a required flow and meeting the 2-in. cube strengths given in Table 6.

TABLE 2 Physical Requirements for Structural Clay Tile

Type and grade	Absorption, % (1 hr boiling)		Compressive strength, psi (based on gross area)			
			End-construction tile		Side-construction tile	
	Avg of 5 tests	Individual	Min avg of 5 tests	Individual	Min avg of 5 tests	Individual
Load-bearing (ASTM C34):						
LBX	16	19	1,400	1,000	700	500
LB	25	28	1,000	700	700	500
Non-load-bearing (ASTM C56), NB	...	28				
Floor tile (ASTM C57):						
FT1	...	25	3,200	2,250	1,600	1,100
FT2	...	25	2,000	1,400	1,200	850
Facing tile (ASTM C212):						
Standard	...	...	1,400	1,000	700	500
Special-duty	...	...	2,500	2,000	1,200	1,000
Glazed units (ASTM C126)	...	...	3,000	2,500	2,000	1,500

TABLE 3 Maximum Moisture Content for Type I Units
(Percent Absorption, Average of Three Units)

Linear shrinkage, %	Conditions at job site		
	Humid*	Intermediate†	Arid‡
0.03 or less	45	40	35
0.03–0.045	40	35	30
0.045–0.065 max	35	30	25

*Average annual humidity above 75 percent.
†Average annual humidity 50 to 75 percent.
‡Average annual humidity less than 50 percent.

REINFORCED MASONRY

6. Materials Mortar and grout for reinforced masonry are covered by ASTM C476, which is a proportion specification. The mortar may consist of 1 part portland cement, ¼ to ½ part lime, and fine aggregate 2¼ to 3 times the sum of the volumes of cement and lime. The ½ part lime (i.e., 1:½:4½) generally gives better bond and water retention as well as better workability. The mortar may also consist of 1 part portland cement and 1 part Type II masonry cement with fine aggregate 2¼ to 3 times the combined volumes of cement. Masonry cement is favored in some areas, but many agencies prefer the lime, in which all

TABLE 4 Requirements for Concrete Masonry Units

Product	Compressive strength, min, psi on average gross area		Water absorption, max, pcf, average of 5 units			
			Oven-dry weight of concrete, pcf			
	Average of 5 units	Individual unit	Over 125	105–125	105 or less	85 or less
Concrete building brick (ASTM C55):						
Grades NI, NII	3500*	3000*	13	15	18	
Grades SI, SII	2500*	2000*				20
Solid load-bearing concrete masonry units (ASTM C145):						
Grades NI, NII	1800	1500	13	15	18	
Grades SI, SII	1800	1500				20
Hollow load-bearing concrete masonry units (ASTM C90):						
Grades NI, NII	1000	800	13	15	18	
Grades SI, SII	1000	800				20
Hollow non-load-bearing concrete masonry units (ASTM C129), Types I, II	800	500				

*Brick flatwise.

TABLE 5 Unit-Masonry Mortar. Proportioning by Volume

Mortar type	Parts by volume			Aggregate measured in damp, loose condition
	Portland cement	Masonry cement	Hydrated lime or lime putty	
M	1	1 (Type II)		Not less than 2¼ and not more than 3 times the sum of volumes of cement and lime
S	½	I (Type II)		
N		I (Type II)		
O		1 (Type I or II)		
M	1		¼	
S	1		Over ¼ to ½	
N	1		Over ½ to 1¼	
O	1		Over 1¼ to 2½	
K	1		Over 2½ to 4	

TABLE 6 Unit-Masonry Mortar. Proportioning by Strength

Mortar type	Average 28-day compressive strength, psi
M	2500
S	1800
N	750
O	350
K	75

the ingredients can be known and tested to give predictably better results. Two types of grout are specified (Table 7). A high water-cement ratio is used to obtain fluidity, but it is reduced rapidly by absorption of the masonry units. Recommended size of aggregate is ⅛ in. maximum for grout spaces to 2 in., ⅜ in. maximum for spaces to 3 in., and ¾ in. for wider spaces.

Cold-drawn wire complying with ASTM A82 and reinforcing bars recognized by ACI for reinforced concrete are used. In general, bar or wire under ¼ in. is not deformed. For the few situations in which placement is critical, positive positioners must be used.

7. Design Tests indicate that the structural performance of reinforced masonry is analogous to that of reinforced concrete within the extremely low limits of stress that are permitted. The bond between the units is such that masonry can be assumed to act as a homogeneous material within the range of working stresses. The assumption that masonry carries no tension is ultraconservative in most cases and may lead to erroneous conclusions.

TABLE 7 Grout Proportions by Volume

	Parts by volume of portland cement or portland blast-furnace slag cement	Parts by volume of hydrated lime or lime putty	Aggregate, measured in a damp, loose condition: Fine	Aggregate, measured in a damp, loose condition: Coarse
Fine grout....	1	0–1/10	2¼–3 times the sum of the volumes of the cementitious materials	
Coarse grout..	1	0–1/10	2¼–3 times the sum of the volumes of the cementitious materials	1–2 times the sum of the volumes of the cementitious materials

The net section, particularly in hollow units, is an important consideration. Mortar joints are sometimes raked for appearance, which results in a greatly reduced effective section. This is especially critical for forces perpendicular to the wall. Design for shear is similar to that for reinforced concrete.

8. Allowable Stresses Two levels of stress are permitted by the Uniform Building Code, depending upon inspection (Table 8). A special inspector employed by the owner or the owner's agent must be present at all times during construction of the masonry if the allowable stresses requiring special inspection are used.

If the value of f'_m is determined by tests (Art. 13), the allowable stresses given in Table 8 may be used.

9. Beams The following procedure is suggested for the design of reinforced-masonry beams. For the allowable stresses *fim* and f_s and the corresponding values of E_m and E_s, determine

$$k = \frac{1}{1 + f_s/nf_m} \qquad j = 1 - \frac{k}{3} \qquad K = \frac{1}{2} f_m jk$$

For preliminary design, k and j may be assumed to be 0.30 and 0.90, respectively. Determine b, d, and A_s from

$$M = Kbd^2 \qquad M = A_s f_s jd$$

Values of K as a function of np and f_m are given in Fig. 1.

Bending stress can be checked by

$$f_m = \frac{M}{bd^2} \frac{2}{jk}$$

Shearing stress is checked by $v_m = V/bjd$. If the allowable stress for no web reinforcement is exceeded, web reinforcement must be provided. Stirrup spacing is given by $s = f_v A_v / bv'$.

TABLE 8 Maximum Working Stresses, psi, for Reinforced Solid and Hollow Unit Masonry

Type of stress[a]	Hollow clay units Grade LB or hollow concrete units[b] Grade A		Grouted solid hollow units, concrete Grade A, clay Grade B, or solid units 2500 psi on gross area		Solid units 3000 psi on gross area		Special testing[f] f'_m established by prism tests				
Ultimate compressive strength f'_m	675	1350	750	1500	900	1800	2000	2700	3000	3500	4000
Special inspection required	No	Yes	No	Yes	No	Yes	Yes	Yes	Yes	Yes	Yes
Compression—axial, walls,[h] $0.2\ f'_m$	135	270	150	300	180	360	400	540	600	700	800
Compression—axial, columns,[i] $0.18 f'_m$	122	244	135	270	162	324	360	486	540	630	720
Compression—flexural, $0.33 f'_m$	225	450	250	500	300	600	667	900			
Shear:											
No shear reinforcement,											
Flexural,[c] $1.1\sqrt{f'_m}$	25	40	25	42	25	47	49	50			
Shear walls[d]											
$M/Vd \leq 1$,[g] $0.9\sqrt{f'_m}$	17	33	17	34	17	34					
$M/Vd = 0$, $2\sqrt{f'_m}$	25	50	25	50	25	50					
Reinforcing taking all shear:											
Flexural, $3\sqrt{f'_m}$	75	110	75	115	75	127	134	150			
Shear walls[d]											
$M/Vd \leq 1$,[g] $1.5\sqrt{f'_m}$	35	55	35	58	35	64	67	75			
$M/Vd = 0$, $2\sqrt{f'_m}$	60	73	60	77	60	85	89	104	110	118	120
Modular ratio n, $30{,}000/f'_m$	44	22	40	20	33	17	15	11	10		

Bearing:											
Full area, $0.25f'_m$	170	340	187	375	225	450	500	675	750	875	1000
⅓ or less of area,[e] $0.3f'_m$	200	400	225	450	270	540	600	810	900	1050	1200
Bond—plain bars	30	60	30	60	30	60	60				
Bond—deformed bars	100	140	100	140	100	140	140				

[a]Allowable values according to Uniform Building Code, 1976.

[b]Stresses for hollow unit masonry are based on net section.

[c]Web reinforcement shall be provided to carry the entire shear in excess of 20 psi whenever there is required negative reinforcement for a distance of one-sixteenth the clear span beyond the point of inflection.

[d]Where determinations involve rigidity considerations in combination with other materials or where deflections are involved, the moduli of elasticity and rigidity under columns entitled "Yes" for special inspection shall be used.

[e]This increase shall be permitted only when the least distance between the edges of the loaded and unloaded areas is a minimum of one-fourth of the parallel side dimensions of the loaded area. The allowable bearing stress on a reasonably concentric area greater than one-third, but less than the full area, shall be interpolated between the values given.

[f]Special testing shall include preliminary tests conducted to establish f'_m and at least one field test during construction of walls per each 5000 sq ft of wall but not less than three such field tests for any building.

[g]Use straight-line interpolation for M/Vd values between 0 and 1.

[h]See Eq. (2).

[i]See Eq. (3).

Bond stress is given by $u = V/\Sigma_0 jd$ and development length L of reinforcing bars by $L = f_s D/4u$.

Values of k, j, and $2/kj$ are given in Table 9.

Lintels are designed as beams supporting the triangular portion of the wall bounded by lines at 45° from each support. Concentrated loads from beams framing into the wall above the opening may be assumed to be distributed over a length equal to the base of the trapezoid formed by drawing, from the edges of the beam or bearing plate, lines at 60° with the horizontal. Alternatively, the wall spanning an opening may be designed as a deep beam.

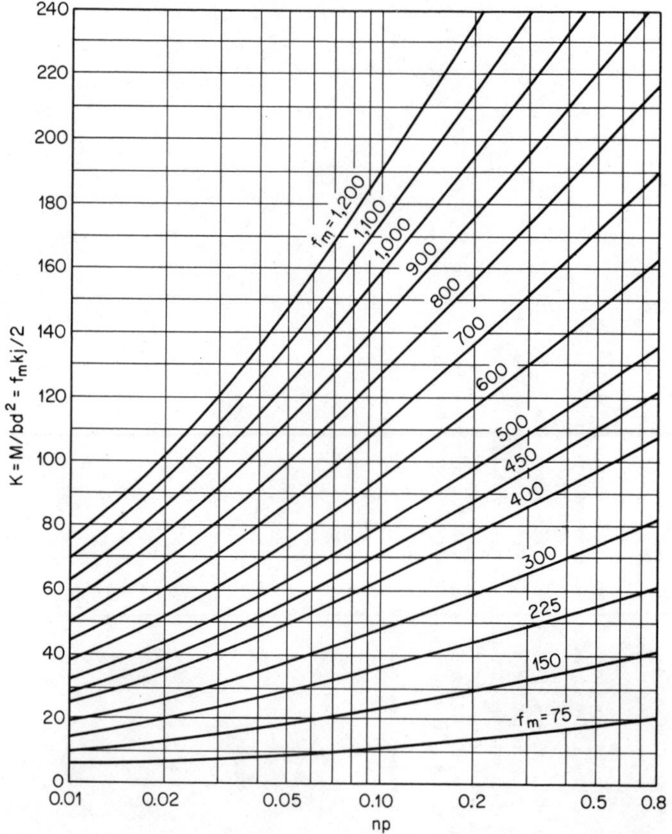

Fig. 1 Values of K for reinforced masonry. *(From Concrete Masonry Association of California.)*

Deep Beams. Walls are often designed to span from caisson to caisson to eliminate grade beams. Methods of design have been developed by the Portland Cement Association and others. Tests have shown that these give large factors of safety; so simplified methods may be used up to ratios of $h/t = 48$. In one approximate method a strip at the bottom of the wall is designed to carry the total load. Another, which is simple and quite conservative, is to proportion the tension steel as in a beam or tied arch; the reactions are assumed to be carried by a vertical strip at each end, which is the width of the end bearing plus twice the wall thickness, with the stress limited by

$$f_m = 0.20 f'_m \left[1 - \left(\frac{h/t}{48}\right)^3\right] \tag{1}$$

10. Walls The ratio of height or length of a reinforced-masonry bearing wall to its thickness is limited by the Uniform Building Code to 25. The corresponding ratios for nonbearing walls are 30 and 48 for exterior and interior walls, respectively. The allowable axial stress f_m is given by

$$f_m = 0.20f'_m \left[1 - \left(\frac{h}{40t}\right)^3\right] \tag{2}$$

where f'_m = approved ultimate compressive stress, not to exceed 6000 psi. This formula does not take end conditions into consideration. The following effective heights h' are suggested: $h' = 2h$ for cantilever walls, $h' = 1.8h$ for cantilever guided at top, $h' = h$ for pin-ended wall, $h' = 0.75h$ for wall pinned at one end and fixed at the other, $h' = 0.5h$ for

TABLE 9

np	k	j	$2/kj$
0.010	0.131	0.956	15.93
0.020	0.181	0.939	11.76
0.030	0.216	0.927	9.95
0.040	0.245	0.918	8.89
0.050	0.270	0.909	8.14
0.055	0.281	0.906	7.87
0.060	0.291	0.902	7.60
0.065	0.301	0.899	7.38
0.070	0.310	0.896	7.19
0.075	0.319	0.893	7.01
0.080	0.327	0.890	6.85
0.085	0.336	0.888	6.71
0.090	0.343	0.885	6.58
0.095	0.351	0.883	6.45
0.100	0.358	0.880	6.34
0.105	0.365	0.878	6.24
0.110	0.371	0.876	6.14
0.115	0.378	0.873	6.04
0.120	0.384	0.871	5.97
0.125	0.390	0.869	5.89
0.130	0.396	0.867	5.82
0.135	0.401	0.866	5.75
0.140	0.407	0.864	5.68
0.145	0.412	0.862	5.62
0.150	0.417	0.860	5.56
0.155	0.422	0.859	5.51
0.160	0.427	0.857	5.46
0.165	0.432	0.855	5.41
0.170	0.437	0.854	5.36
0.175	0.441	0.852	5.31
0.180	0.446	0.851	5.26
0.185	0.450	0.849	5.22
0.190	0.455	0.848	5.18
0.195	0.459	0.846	5.14
0.200	0.463	0.845	5.11
0.250	0.500	0.833	4.80
0.300	0.530	0.823	4.58
0.350	0.556	0.814	4.41
0.400	0.579	0.806	4.27
0.450	0.600	0.800	4.17
0.500	0.618	0.794	4.07

wall fixed at both ends. However, judgment must be used in evaluation of the degree of fixity or lateral support.

Minimum reinforcement is 0.2 percent of the gross cross-sectional area of the wall, with at least one-third in either direction.

Lateral forces will generally govern the design of a wall. If there are pilasters or intersecting walls, moments can be based on end fixity.

Shear walls must be checked for tie-down and for the condition of no live load as well as full live load.

Typical wall elevations are shown in Fig. 2. The spacing H should not exceed 12 ft for low-shrinkage units and 8 ft for other units. The spacing S should not exceed 8 ft, except in seismic areas for which $S_{max} = 4$ ft. Where reinforcement is continuous $c \geq 2b$.

Arrangement of horizontal reinforcement is usually determined by the position of openings and by code requirements. According to the UBC, horizontal reinforcement is required in the top of footings, at the top of wall openings, at roof and floor levels, and at the top of parapet walls. Only horizontal reinforcement which is continuous in the wall is considered in determining the minimum area of reinforcement.

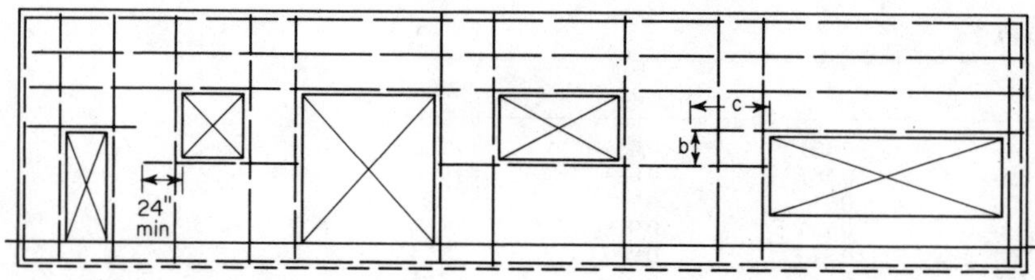

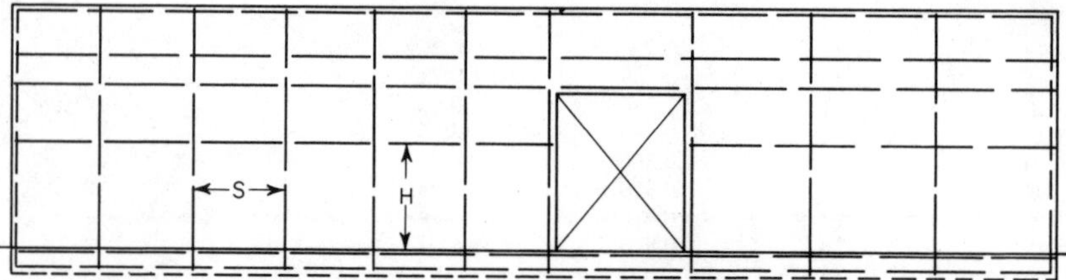

Fig. 2 Wall reinforcing. *(From Concrete Masonry Association of California.)*

Vertical reinforcement is generally determined by design for wind, seismic, and vertical loads. A recommended reinforcement for 8-in. masonry walls is No. 5 at 4 ft.

11. Columns The allowable load (Uniform Building Code) on reinforced-masonry columns is

$$P = A_g(0.18f'_m + 0.65p_gf_s)\left[1 - \left(\frac{h}{40t}\right)^3\right] \tag{3}$$

The least dimension must be 12 in., except that it may be 8 in. if the column is designed for one-half the allowable stress. The unsupported length must not exceed 20 times the least dimension.

The ratio p_g may not be less than 0.5 percent or more than 4 percent. The minimum number of reinforcing bars is four and the smallest diameter ⅜ in. Lateral ties at least ¼ in. in diameter must be spaced not over 16 bar diameters, 48 tie diameters, or the least dimension of the column. Ties may be in the bed joints or tight around the vertical bars. Additional ties to enclose anchor bolts are advisable at tops of columns. This is especially important if the element supported may be subjected to lateral loads, as from earthquake or wind.

Combined vertical load and bending is evaluated by

$$\frac{f_a}{F_b} + \frac{f_b}{F_b} \leq 1$$

12. Diaphragms Horizontal diaphragms are often used to distribute lateral forces to shear walls or other bracing systems. In the design of diaphragms in reinforced-masonry buildings, the walls are usually considered as flanges of deep plate girders whose webs are the diaphragms. Longitudinal shear between the web and the flange must be considered in determining the connection to the wall. Horizontal truss systems usually consist of struts and tension-rod diagonals. This type of bracing is not generally recommended for masonry buildings because deflection is usually excessive compared with other systems.

TESTS AND INSPECTION

13. Compressive Strength of Masonry Some codes permit tests to determine the approved ultimate compressive stress f'_m. Two methods are used.

The strength may be established by preliminary tests of prisms built of the same materials under the same conditions and, insofar as is possible, with the same bonding and workmanship as will be used. Prisms for walls are 16 × 16 in. and of the thickness and type of construction of the prototype. Shorter prisms may be needed to fit within the test depth of available machines. However, if the length is less than the width, the results may be inaccurate because the direction of splitting may be changed. Prisms for columns are 8 × 8 in. in plan and 16 in. high. All specimens should have a height-to-thickness ratio not less than 2. If other ratios are used, the value of f'_m is taken as the compression strength of the specimens multiplied by the following correction factor:

h/d	1.5	2.0	2.5	3.0
Factor	0.86	1.00	1.11	1.20

Prisms must be tested under conditions specified by the governing code.

The approved value of f'_m may be established by tests of individual units (Table 10).

Table 10

Compressive Strength of Units, psi	Assumed f'_m, psi
1,000	900
1,500	1,150
2,500	1,550
4,000	2,000
6,000	2,400
8,000	2,700
10,000	2,900
12,000	3,000

DETAILING AND CONSTRUCTION

14. Detailing The cost of masonry can be greatly influenced by the detailing. It is important to detail the units and the reinforcement so that they can be placed with the rhythmic procedure of good laying. Care must be given to modular layouts to minimize field cutting of units. Wide spacing of steel will usually result in economy. Steel in grout spaces must have at least ¼ in. cover for embedment. If spacer bars are used on joint reinforcement, they can serve as chairs to keep the reinforcement clear so that mortar may flow beneath. In general, the thickness of joints should be at least twice the diameter of the embedded bar. However, adequate bond might also be achieved by the use of welded spacers to serve as anchors in transferring stress.

Steel details are similar to those for reinforced concrete, but with the additional limitations of modular spacing and emphasis on wide spacing of reinforcement.

15. Concrete Foundations Horizontal concrete surfaces that are to receive masonry should be clean and preferably slightly damp, with the aggregate exposed on the surface roughened to assure good bond. Grout spaces should be kept clear of mortar and the bottom course grouted solid before additional courses are placed.

Foundation dowels should not be bent to a slope of more than 1 in 6. Incorrectly positioned dowels should be grouted into a cell adjacent to the cell containing a vertical wall-reinforcing bar if necessary. Wall steel need not be tied to the dowels, although this is a good way to keep it in place.

16. Workmanship Since masonry construction consists of assemblages of small units, workmanship is of utmost importance. For grouted masonry, bed joints should be beveled from the inside, rather than furrowed. This may leave an open space on the inside face of the bed joint, but this will be filled with grout and will serve as a mechanical key. Units must be pressed down firmly while the mortar is moist and plastic. Head joints should be shoved tight and full, although in grouted work the back of the joint can be open. Joints in hollow masonry should be full for the thickness of the face shell. One of the advantages of this is that it breaks capillary action. Units must not be tapped to relocate them after the mortar has lost its initial plasticity. Such tapping will break the bond and cause weakness and leakage. If adjustments are necessary, the old mortar should be removed. Mortar fins protruding more than the thickness of the joint into the grout space, from either head or bed joints, should be removed but not allowed to fall into the grout space.

Racking (stepping back successive courses) is acceptable, but toothing is not. Toothed joints cannot be pressed together to make a tight bond. If toothing is permitted, caulking material may be used to seal the joints, or the joints made by pressing mortar onto the surface and tucking mortar tightly into the joint.

Joints should be tooled to compress the mortar against the edges of the masonry to make the joint watertight as well as neat.

Wetting. To assure proper bonding, clay units should be wetted so that the surface is slightly damp. The absorption rate should not exceed 0.025 oz/sq in./min when placed in water to a depth of ⅛ in. Concrete units should not be wetted before laying except in very hot, dry weather, when the bearing surfaces to receive mortar may be slightly moistened immediately before laying. Wetting should be minimized since it causes slight expansion and subsequent shrinkage which may cause cracking.

Admixtures must be used sparingly and carefully. Generally they are used to reduce the water content. They affect the flow after suction of the mortar and should be checked for their effect on bond.

Section 5

Thin-Shell Concrete Structures

DAVID P. BILLINGTON
Professor of Civil Engineering, Princeton University

Thin shells are defined as curved or folded slabs whose thicknesses are small compared with their other dimensions. They may be classified in a number of ways, two of which are type of curvature and method of generation.

Singly curved shells are developable surfaces (Fig. 1*a*). They are usually cylindrical or conical. Shells of positive double curvature (two curvatures in the same direction) usually have a domelike appearance (Fig. 1*b*). They are nondevelopable and, therefore, are stiffer than singly curved shells if they are properly supported. Shells of negative double curvature (two curvatures in opposite directions) usually have the form of a saddle or warped plate (Fig. 1*c*), and are also nondevelopable.

Thin-shell surfaces may be generated by rotation or by translation of a plane curve. Surfaces of rotation are formed by revolving a plane curve about an axis in its plane. Such surfaces may all be analyzed in a similar manner, but this does not necessarily imply similarity of behavior.

Surfaces of translation are formed by moving a plane curve along some other plane curve. Again, there is no necessary similarity of behavior within this class. For example, elliptic paraboloids and spherical domes can be built of nearly the same shape, except at the edges, and their behavior is nearly the same although their methods of generation are quite different. On the other hand, elliptic paraboloids and hyperbolic paraboloids covering the same areas can behave quite differently even though their methods of generation are the same.

Many shell structures which have been built do not fit easily into the categories defined above. Some examples are *wave-form translation shells* (Fig. 2), which have alternating positive and negative curvature, and *free-form shells*. The overall behavior pattern of such shells may not be too difficult to establish, but a detailed analysis, particularly for bending moments, may require substantial idealization of the structure which, in the end, may best be confirmed by physical-model analyses.

1. Thin-Shell Concrete Roofs Thin-shell concrete structures have been used for roofs because of their possibilities for wide spans, for impressive appearance, and for economical construction. In some cases all three factors are important, but more commonly one or two predominate.

Thin shells for wide spans are limited by high cost of the extensive centering or scaffolding required. Where the roof can be cast in a number of similar and self-supporting parts, scaffolding can be reused and its cost per square foot of covered area substantially

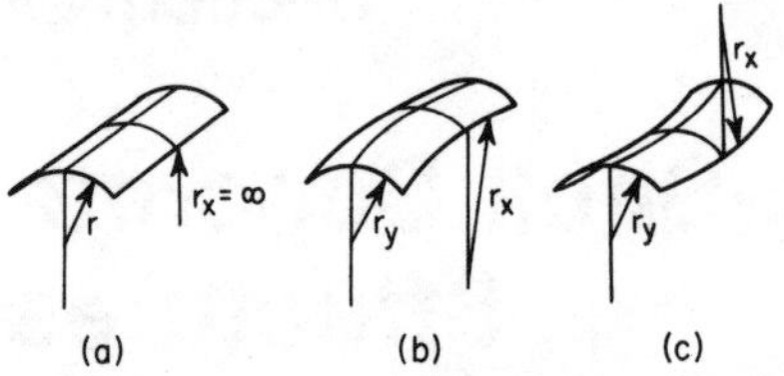

Fig. 1 *(a)* Singly curved shell; *(b)* shell of positive double curvature; *(c)* shell of negative double curvature.

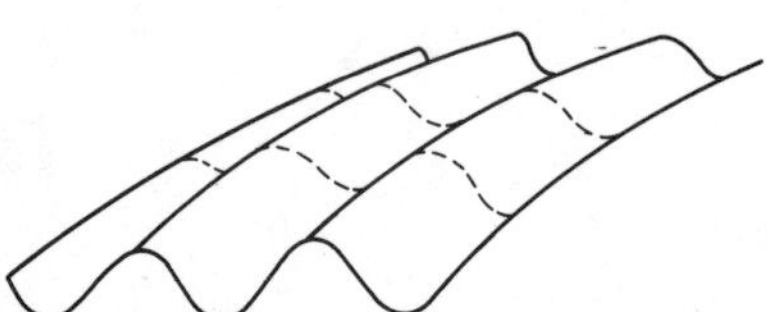

Fig. 2 Waveform translation shell.

reduced. A further limitation is the lack of engineering experience, particularly with regard to analysis for creep, large deflections, and buckling. Thin-shell structures which have performed well on short spans may exhibit difficulties when scaled up to much larger dimensions, because commonly accepted analyses and tolerances in construction may fail to account for some critical factor on very large spans.

The architect's desire for roofs of impressive appearance may lead to complex shapes which, even for small spans where scaffolding costs are not excessive, tend to increase analytic and construction difficulties. Precasting techniques, such as those of Nervi, can be exploited to produce attractive structures with a reduction in field-built formwork.

Architectural exuberance may lead to another difficulty, that of enforcing a shape and a type of support which are not well suited to each other. For example, a roof composed of intersecting circular semicylinders requires diagonal arches of elliptical profile, which is a relatively inefficient arch shape, whereas the doubly curved, hyperbolic-paraboloid cross vault leads to diagonal arches of parabolic profile. Although the parabola is a better arch shape, form costs for the doubly curved thin shell may be greater.

The use of thin shells for reasons of construction economy is limited by the complexity of the shape and the form reuse possible. It is generally true that the carrying action of doubly curved shells is more efficient than that of singly curved ones, but the formwork is more complex. In addition, the unit cost of materials in place will often be higher for the thinner sections of the more complex structure. Furthermore, since minimum concrete thickness is often controlled by construction practice or by buckling, and minimum reinforcement by temperature and shrinkage requirements, increases in the complexity of shape often cannot be offset by materials saving. This is particularly true in structures of smaller span.

2. Behavior of Roof Structures A roof structure is given curvature so that the loading is carried predominantly by in-plane forces. Ideal moment-free behavior cannot be attained in practice, but in well-designed thin shells such behavior should predominate. Ideal

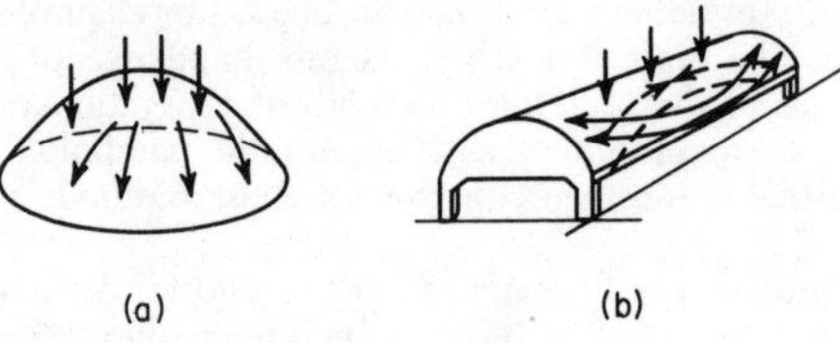

Fig. 3

behavior can be roughly explained in terms either of arch action, where normal loads are carried directly to supports by axial forces (Fig. 3*a*), or of beam action, where loads are carried by compression, tension, and shear, all in the plane of the middle surface (Fig. 3*b*). In many thin shells both actions are present.

In addition to in-plane forces (Fig. 4*a*), there will be moments and radial shears (Fig. 4*b*) whose importance generally depends upon conditions at the boundaries where the shell is supported, joined to an edge member, or connected to another shell. Generally speaking, if displacements of the edges are not excessive, the bending forces can be kept small and the behavior explained by considering only the in-plane forces.

Most thin-shell structures can resist a variety of loading patterns by in-plane forces and, therefore, usually need be analyzed only for uniformly distributed loadings. Except for the supporting system, it is usually not necessary to consider partial live loadings, e.g., snow load on one side of a roof.

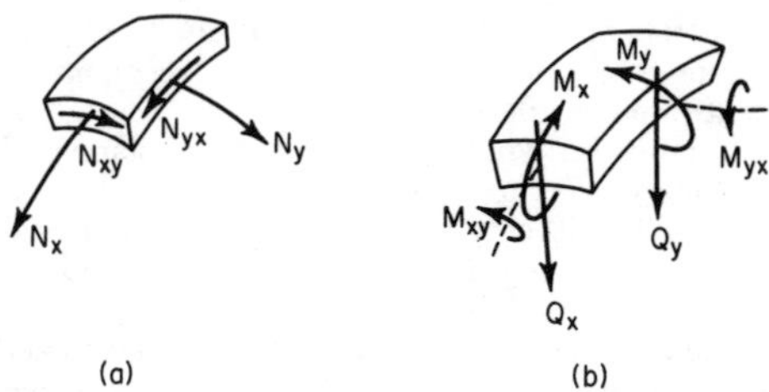

Fig. 4 *(a)* In-plane forces; *(b)* bending forces.

These comments refer to thin shells under distributed loads. Where heavy concentrated loads occur, large local bending moments may arise even where the boundaries are carefully designed. However, the local failure which may result does not necessarily cause collapse of the roof; there are many examples of locally damaged thin shells which adjusted to their defects.[1] If heavy load concentrations are anticipated, local stiffening of the shell should be provided.

3. Thin-Shell Concrete Walls Thin shells are frequently used for ground-storage reservoirs, elevated water tanks, and natural-draft cooling towers because they are relatively easy to analyze, inexpensive to build, and have served well over long periods. In such structures the dead load causes predominantly in-plane compression and the live load (internal pressure) causes uniform horizontal in-plane tension which can be carried by reinforcement or, for vertical walls, by prestressing. Where the predominant live load is wind, as for high cooling towers, the resulting in-plane tension (uplift forces) on the windward side of the tower is partly balanced by dead-load compressions and partly taken by reinforcement. Even though the maximum tension is normally below the cracking strength of the concrete, reinforcement should be designed to take all of the uplift tension.

STRUCTURAL ANALYSIS

4. Thin-Shell Theory A summary of thin-shell theory is given in Ref. 2. For the shells considered here, two simplifications are often admissible: (1) the membrane theory and (2) a simplified bending theory. The principal guide for the choice of a theory is the successful long-term structural behavior of the type of thin shell and its scale. Very simplified theories have proved successful for dome roofs under gravity loading and circular cylindrical walls under internal pressure.

Analysis is commonly based on the following assumptions:

1. The material is homogeneous, isotropic, and linearly elastic. Although none of these assumptions is correct for concrete, tests have indicated that, under working loads, the concrete thin shell behaves very nearly as if they were.[3]

2. The system behaves according to the *small-deflection theory,* which essentially requires that deflections under load be small enough so that changes in geometry do not alter the static equilibrium. The usual measure of validity for this theory is that the radial displacements of the shell be small compared with its thickness.

3. The thickness of a thin shell is denoted by h and is always considered small in comparison with its radii of curvature r. Thin-shell theory has been used where r/h is as low as 30.[4] The surface bisecting the thickness is called the *middle surface,* and by specifying its form, and h at every point, the shell is defined geometrically.

4. The in-plane forces (Fig. 4*a*) are distributed uniformly over the thickness. They are often expressed as *stress resultants,* defined as forces per unit length on the middle surface. A stress resultant divided by h yields a stress. Membrane stress resultants are denoted by N'_x, N'_y, and $N'_{xy} = N'_{yx}$ and can be obtained solely from equations of equilibrium. Unprimed resultants signify the more accurate values, which include the effects of bending. An extensive compilation of membrane-theory formulas is given in Ref. 5.

The membrane theory is based upon the assumption of no bending or transverse shear in the shell; only in-plane forces are considered. In many thin shells it provides a reasonable basis for design except at the boundaries where the shell is supported or stiffened. This is because local restraints usually exist at boundaries, or because the reactions required by the membrane theory cannot be supplied by the edge members. The substantial bending that can occur at the boundaries is usually evaluated by an approximate bending theory in which the effects of edge loads and edge displacements on both stress resultants and bending moments are considered.

Bending theory in general implies a formulation which includes both in-plane and bending forces. For most thin shells this theory is complicated and difficult to use as a basis for design. For surfaces of rotation a simplified theory works well for many practical cases; this is given in Art. 11. Where this simplification is used, the following general method of analysis is recommended:

1. The loading is assumed to be resisted solely by the membrane stress resultants. Thus the shell is made statically determinate by releasing its internal moment resistance and its external edge restraints. This corresponds to the primary system (reduced structure) in the analysis of a statically indeterminate structure.

2. The forces and displacements at the boundaries of the primary system will, in general, be incompatible with the actual boundary conditions. These discrepancies are errors which must be corrected.

3. Corrections corresponding to unit edge effects are determined.

4. Compatibility is obtained by determining the magnitudes of the corrections which will eliminate the errors.

This approach is illustrated for domes in Art. 10 and cylindrical tanks in Art. 11. For some other types of thin shells the membrane theory provides a useful approximate analysis, but for structures of large scale or for shell types without a record of successful long-term behavior it is essential to use a more complete analysis procedure, usually either a numerical solution or an empirical solution on a well-scaled model. Reference 6 gives a summary of general numerical programs applied to thin-shell concrete structures.

5. Stability *Shell Roofs.* Buckling can be initiated by relatively high permanent-load compressive stresses, which with time can lead to creep and eventually large displacements. Displacements of a shell normal to its middle surface imply a change in curvature, so that thin shells, since they rely on their curvature for resistance to load, are particularly susceptible to large displacements and thus to buckling. Although much has been written on the stability of thin shells,[7,8,9] there is little experimental work on thin-shell concrete structures of the shape and boundary conditions usually found in roofs.

The buckling pressure q_{cr} on a spherical thin shell[8] is given by

$$q_{cr} = CE\left(\frac{h}{a}\right)^2 \tag{1}$$

where $C = 2/\sqrt{3(1-\nu^2)}$
ν = Poisson's ratio
h = thickness of shell
a = radius of shell

Experiments show that C is much smaller than $2/\sqrt{3(1-\nu^2)}$. Schmidt[10] shows values as low as 0.06 with none above 0.32. For translational shells, he suggests that Eq. (1) be written in the form

$$q_{cr} = CE\frac{h^2}{r_x r_y} \tag{2}$$

where r_x and r_y are the two principal radii of curvature.

These equations show that instability becomes a problem for shells which are very thin, flat, or of low-modulus concrete. Concrete creep can contribute to large deflections; it may be reduced by providing reinforcing steel in both faces of the shell. The effect of creep can be estimated by assuming a reduced value of E or, if the principal membrane stresses at any point are known, by determining the tangent modulus of elasticity and dividing it by a factor for long-term deflections. The factor should not be less than 2.

Shell Walls. Permanent load does not play as important a role as transient loadings such as wind in the buckling of shell walls. For a cylinder, fixed at its base, free at its top, and under wind pressure, a bifurcation analysis gives the critical pressure as

$$q_{cr} = KE\left(\frac{h}{r}\right)^3 \tag{3}$$

Values of K for a cylinder with $r/h = 100$ can be found from Fig. 5 for the typical wind-pressure distribution shown in the figure.[11] Comparisons between these results and wind-tunnel test results show the former to be about 36 percent high.[12] Critical pressures with K from Fig. 5 and using $c = 1$ are 1860 and 430 psf for $H/r = 1$ and 6, respectively. Since r/h seldom exceeds 100 for thin concrete cylindrical shells, this shows that such shells will not buckle under wind pressure unless there is a high thermal gradient, as in cooling-tower shells, and insufficient circumferential steel to control cracking.

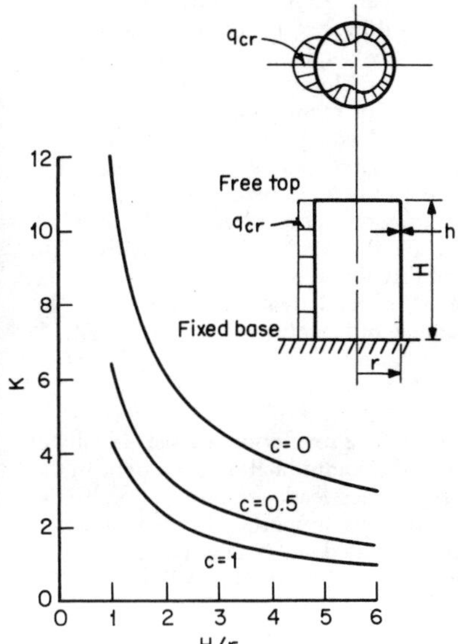

Fig. 5 Values of K in Eq. (3) for $r/h = 100$.

For hyperboloids under wind loading, simplified bifurcation results are from two to three times higher than wind-tunnel results.[13] The latter lead to an estimate of buckling pressure as[14]

$$q_{cr} = CE\left(\frac{h}{r}\right)^{\alpha} \tag{4}$$

where q_{cr} = buckling pressure, psi, along windward meridian
C = empirical coefficient = 0.052
h = thickness at throat, i.e., at point on meridian where horizontal radius is minimum
r = radius of shell parallel circle at throat
α = empirical coefficient = 2.3

The value of α has been derived from theoretical considerations as $7/3$ (Ref. 15). The coefficient C shows wide scatter[14] and in Europe has been taken as 0.077 ± 0.009 (Ref.

16). A restudy of the results reported in Ref. 14 shows 0.052 to be a more reasonable lower-bound estimate.[13] The value of q_{cr} by Eq. (4) should be compared with the design wind pressure at the top of the tower to ensure an adequate factor of safety against buckling.

For cooling towers, q_{cr} should be at least twice q_z, as defined in Art. 11, when dead load is included and cracking considered. This is because thermal gradients across cooling-tower shells, and possibly other types, are large enough to cause vertical cracks and hence to reduce substantially the circumferential stiffness of the shell.[17] Since this bending stiffness plays a major role in the buckling capacity, its reduction can reduce the buckling pressure drastically. The solution is to provide enough circumferential reinforcement to control the cracking and to use a thickness sufficient to keep $q_{cr} > 2q$.

Imperfections in the geometry of a completed shell can lead to overstressing or to buckling or both. In the failure of the Ardeer cooling tower in 1973 imperfections were recorded of over 18 in. in a 6-in.-thick shell, and these directly contributed to the failure.[18] Nevertheless, field tolerances must be permitted on very large structures, but always subject to specification by an experienced designer of thin-shell concrete structures.

6. Dynamic Behavior All the analyses presented in this section are for static behavior. For concrete-shell roofs wind is not usually a dynamic problem because of the weight of the concrete; a more usual problem is the tearing off of roofing because of the wind suction that arises over much of a shell roof. Seismic loading on shell roofs can be a difficult problem if the roof is supported at a few isolated points, in which case the horizontal shears from seismic motions can be large and need to be considered.

For shell walls the principal design loading can frequently be either seismic or wind. Quasistatic approximations, such as those in Ref. 42, to either loading have normally led to successful structures. If there appears to be a possibility that the natural frequencies of a shell are close to important wind-gust frequencies, special dynamic studies may be warranted. Reference 18 gives a discussion of dynamic wind loading on cooling towers.

DOME ROOFS

7. Behavior of Domes The behavior of domelike shells of revolution consists ideally of meridional (latitudinal) arch action and circumferential (longitudinal) hoop action. In the hemispherical dome (Fig. 6*a*), uniformly distributed surface loads are transmitted by the arch action of meridional forces directly to the continuous support. The hoop action is compressive near the crown and tensile near the edge. If the dome is a partial hemisphere supported by uniformly distributed forces tangent to the edge, the hoop forces can be restricted to compression only (Fig. 6*b*).

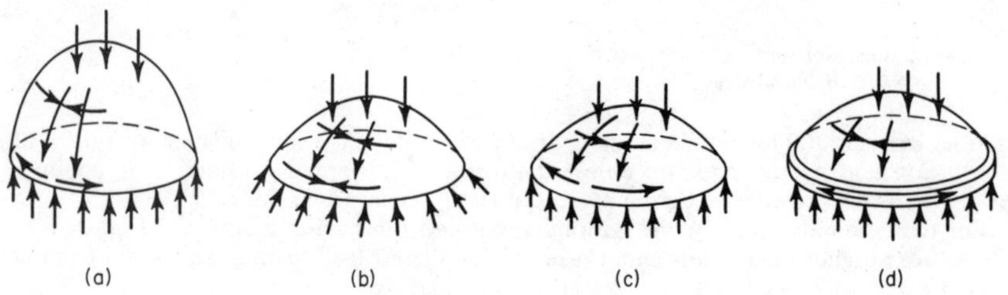

Fig. 6

In the dome of Fig. 6*c*, since the support provides only vertical reactions, the edge moves out and appreciable meridional bending is developed. This can be reduced by an edge ring (Fig. 6*d*). Dome rings are often prestressed so that tension is eliminated and meridional bending reduced. A monolithically connected cylindrical wall support acts somewhat like an edge ring, but tensile forces are distributed over a wider edge region and meridional bending can be larger than with the ring.

Bending can be greatly reduced by using a meridional curve such as a cycloid or an ellipse, which permits shallow-rise domes having a free edge with vertical tangent. The difficult formwork and casting problems tend to make such domes impractical.

The behavior of a surface of revolution is strongly affected by boundary conditions. For example, where only column supports are provided along a circular base, that part of the shell edge region between columns acts as a deep beam to carry the meridional forces to the columns. Figure 7 shows a dome built on a triangular plan with the boundaries formed by passing planes or surfaces through the sphere. Here the behavior departs radically

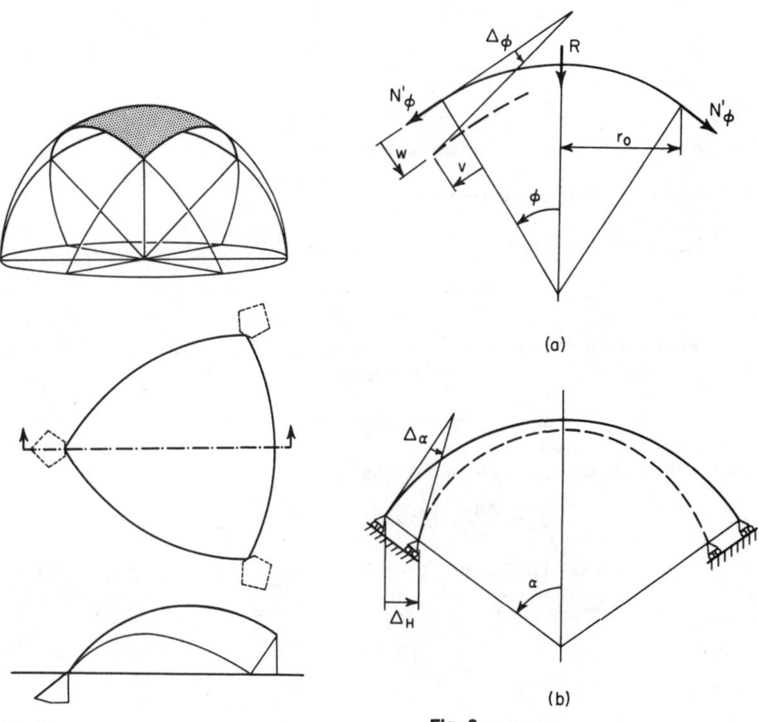

Fig. 7 **Fig. 8**

from the ideal behavior of arch action with hoop restraints, because there is support under only a few meridians and only the hoops near the crown are continuous. The portion at the crown tends to act like a dome, but the arch forces must be carried to the corners by the edge regions. The vertical reactions can be provided by slender columns and the horizontal reactions by prestressing within the shell edge region.

8. Membrane Theory *Surfaces of Revolution.* The equation for vertical equilibrium (Fig. 8*a*) gives directly an expression for the meridional stress resultant:

$$N'_\phi = -\frac{R}{2\pi r_0 \sin\phi} \tag{5}$$

where R = resultant of the vertical components of all loads above the angle ϕ.
A second equilibrium equation for the forces acting on a differential element (Fig. 9) in the direction normal to the surface gives

$$\frac{N'_\phi}{r_1} + \frac{N'_\theta}{r_2} + p_z = 0 \tag{6}$$

where r_1 = radius of curvature of meridian
r_2 = length of shell normal from surface to shell axis

These are the principal radii of curvature of the shell at the point under consideration.

Using Eq. (5) and the relation $r_0 = r_2 \sin\phi$, Eq. (6) gives

$$N'_\theta = \frac{R}{2\pi r_1 \sin^2\phi} - p_z \frac{r_0}{\sin\phi} \tag{7}$$

The displacements v and w (Fig. 8*a*) are given by

$$v = \sin\phi \left[\int \frac{f(\phi)}{\sin\phi}\, d\phi + C\right] \tag{8a}$$

$$w = v\cot\phi - \frac{r_2}{Eh}(N'_\theta - \nu N'_\phi) \tag{8b}$$

where ν = Poisson's ratio
C = constant determined by support conditions

$$f(\phi) = \frac{1}{Eh}[N'_\phi(r_1 + \nu r_2) - N'_\theta(r_2 + \nu r_1)] \tag{8c}$$

The rotation Δ_ϕ of the tangent to the meridian at any point is

$$\Delta_\phi = \frac{v}{r_1} + \frac{dw}{r_1 d\phi} \tag{8d}$$

Values of the displacements are usually needed only at the supports. For supports as shown in Fig. 8*b*, $v_\alpha = 0$, and the horizontal displacement is given by

$$\Delta_H = -\frac{r_2 \sin\alpha}{Eh}(N'_\theta - \nu N'_\phi)_{\phi=\alpha} \tag{9a}$$

The rotation of the meridional tangent at the edge is

$$\Delta_\alpha = \frac{\cot\alpha}{r_1} f(\alpha) + \frac{1}{r_1}\frac{d}{d\alpha}\frac{\Delta_H}{\sin\alpha} \tag{9b}$$

where f is the function defined in Eq. (8*c*). The positive directions of Δ_H and Δ_α are shown in Fig. 8*b*.

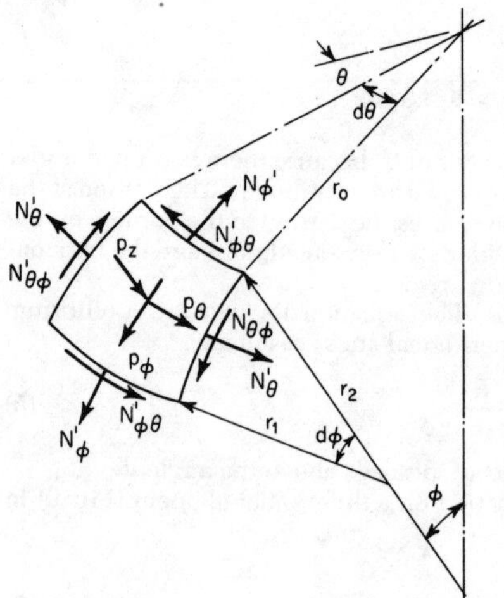

Fig. 9

Spherical Dome of Constant Thickness. With q = weight per sq ft of dome surface, $p_z = q\cos\phi$ (Fig. 9). Then, with $r_1 = r_2 = a$, Eqs. (5) and (7) give

$$N'_\phi = -\frac{aq}{1 + \cos\phi} \tag{10a}$$

$$N'_\theta = aq\left(\frac{1}{1+\cos\phi} - \cos\phi\right) \tag{10b}$$

The distribution of these resultants is shown in Fig. 10. If the dome is a spherical segment, there is no hoop tension at the edge if $\phi \lessapprox 51°50'$ and there is continuous support by forces tangent to the edge.

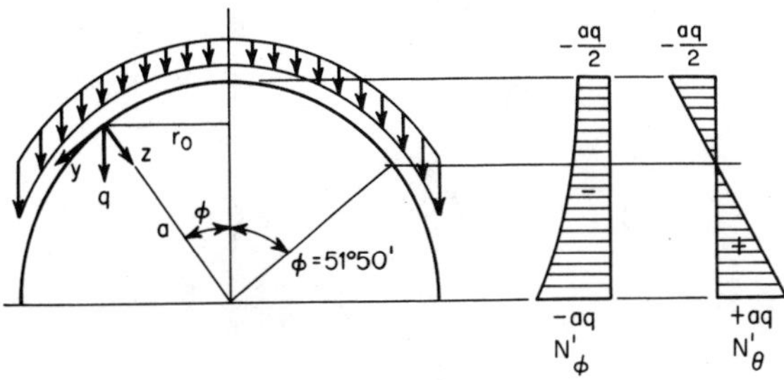

Fig. 10

With load p per sq ft of horizontal projection of the surface, $p_z = p\cos^2\phi$, and Eqs. (5) and (7) give

$$N'_\phi = -\frac{ap}{2} \tag{11a}$$

$$N'_\theta = -\frac{ap}{2}\cos 2\phi \tag{11b}$$

Stress resultants for other loadings are found similarly. Results for several cases are given in Table 1.

With N'_ϕ and N'_θ known, edge displacements for domes supported as in Fig. 8*b* are found from Eq. (9). Corresponding to Eq. (10),

$$\Delta_H = -\frac{a^2q}{Eh}\left(\frac{1+\nu}{1+\cos\alpha} - \cos\alpha\right)\sin\alpha \tag{12a}$$

$$\Delta_\alpha = -\frac{aq}{Eh}(2+\nu)\sin\alpha \tag{12b}$$

and, corresponding to Eq. (11),

$$\Delta_H = -\frac{a^2p}{2Eh}(-\cos 2\alpha + \nu)\sin\alpha \tag{13a}$$

$$\Delta_\alpha = -\frac{ap}{2Eh}(3+\nu)\sin 2\alpha \tag{13b}$$

Spherical Dome of Variable Thickness. A good approximation of the edge forces and displacements in a spherical dome of variable thickness is obtained by assuming a uniform thickness equal to that at the arc distance S_a from the edge,[19]

$$S_a = 0.5\sqrt{ah_a} \tag{14a}$$

where a = radius of dome

h_a = average thickness of thickened portion

This approximation involves the assumption that the thickened portion extends the distance S from the edge,

$$S = 2\sqrt{ah_a} \tag{14b}$$

Parabolic Dome of Constant Thickness. The membrane stress resultants for two cases of loading are given in Table 2. Displacements Δ_H and Δ_α can be determined by substituting these values in Eqs. (9).

TABLE 1 Membrane Forces in Spherical Domes

Loading	N_ϕ'	N_θ'	$N_{\phi\theta}'$
q per ft^2 of surface	$-aq\dfrac{\cos\phi_0-\cos\phi}{\sin^2\phi}$	$-N_\phi'-aq\cos\phi$	0
	For $\phi_0 = 0$ (no opening)		0
	$-aq\dfrac{1}{1+\cos\phi}$	$-N_\phi'-aq\cos\phi$	0
p per ft^2 projection	$-\dfrac{ap}{2}\left(1-\dfrac{\sin^2\phi_0}{\sin^2\phi}\right)$	$-N_\phi'-ap\cos^2\phi$	0
	For $\phi_0 = 0$ (no opening)		
	$-\dfrac{ap}{2}$	$-\dfrac{ap}{2}\cos 2\phi$	0
p_L per ft of edge	$-p_L\dfrac{\sin\phi_0}{\sin^2\phi}$	$-N_\phi'$	0
	For $\phi_0 = 0$ (load P_L at vertex)		
	$-P_L\dfrac{1}{2\pi a\sin^2\phi}$	$-N_\phi'$	0
p_w per ft^2 projection	$-p_w\dfrac{a}{3}\dfrac{\cos\phi\cos\theta}{\sin^3\phi}[3(\cos\phi_0-\cos\phi)-(\cos^3\phi_0-\cos^3\phi)]$	$-N_\phi'-ap_w\sin\phi\cos\theta$	$N_\phi'\dfrac{\tan\theta}{\cos\phi}$
	For $\phi_0 = 0$ (no opening)		
	$-p_w\dfrac{a}{3}\dfrac{\cos\phi\cos\theta}{\sin^3\phi}(2-3\cos\phi+\cos^3\phi)$	$-N_\phi'-ap_w\sin\phi\cos\theta$	$N_\phi'\dfrac{\tan\theta}{\cos\phi}$

9. Bending Theory For the bending near the edges of domes it is usually sufficient to use the Geckeler approximation, which can be interpreted physically as the substitution of an equivalent cylinder for the dome. The radius of curvature $r_2 = r_0/\sin\phi$ (Fig. 9) at the edge of the dome is taken as the radius of the equivalent cylinder, and the solution for bending in a full cylinder is used for the dome edge. Table 3 gives values for domes loaded by edge forces and moments. Values of the coefficient C are given in Table 4. Values of N_ϕ, N_θ, and M_ϕ are given for any point in the shell defined by the angle ψ. The edge displacements Δ_H and Δ_α are positive in the positive directions of H and M_α, respectively.

TABLE 2 Membrane Forces in Parabolic Domes

Loading	N_ϕ'	N_θ'	$N_{\phi\theta}'$
q per ft² of surface, Normal, ϕ	$-\dfrac{qr_0}{3}\dfrac{1-\cos^3\phi}{\sin^2\phi\cos^2\phi}$	$-\dfrac{qr_0}{3}\dfrac{2-3\cos^2\phi+\cos^3\phi}{\sin^2\phi}$	0
p per ft² of projection, ϕ	$-\dfrac{pr_0}{2\cos\phi}$	$-\dfrac{pr_0}{2}\cos\phi$	0

NOTE: r_0 = radius of curvature at vertex.

TABLE 3 Forces and Displacements in Domes of Revolution Loaded by Edge Forces Uniform around a Parallel Circle

	(H, s, ψ, ϕ, α, a)	(M_α, s, ψ, ϕ, α, a)
N_ϕ	$C_1 H\sin\alpha\cot(\alpha-\psi)$	$-2C_3\beta M_\alpha\cot(\alpha-\psi)$
N_θ	$2C_2\beta a H\sin\alpha$	$2C_1\beta^2 a M_\alpha$
M_ϕ	$\dfrac{C_3}{\beta}H\sin\alpha$	$C_4 M_\alpha$
Δ_H	$2\beta\dfrac{a^2}{Eh}H\sin^2\alpha$	$\dfrac{2\beta^2a^2\sin\alpha}{Eh}M_\alpha$
Δ_α	$2\beta^2\dfrac{a^2}{Eh}H\sin\alpha$	$\dfrac{4\beta^3a^2}{Eh}M_\alpha$

NOTE: $\beta^4 = 3(1-\nu^2)/a^2h^2$. See Table 4 for C.

TABLE 4 Coefficients C for Tables 3 and 8*

βs or βx†	C_1	C_2	C_3	C_4
0	1.0000	1.0000	0	1.0000
0.1	0.8100	0.9003	0.0903	0.9907
0.2	0.6398	0.8024	0.1627	0.9651
0.3	0.4888	0.7077	0.2189	0.9267
0.4	0.3564	0.6174	0.2610	0.8784
0.5	0.2415	0.5323	0.2908	0.8231
0.6	0.1431	0.4530	0.3099	0.7628
0.7	0.0599	0.3798	0.3199	0.6997
0.8	−0.0093	0.3131	0.3223	0.6354
0.9	−0.0657	0.2527	0.3185	0.5712
1.0	−0.1108	0.1988	0.3096	0.5083
1.1	−0.1457	0.1510	0.2967	0.4476
1.2	−0.1716	0.1091	0.2807	0.3899
1.3	−0.1897	0.0729	0.2626	0.3355
1.4	−0.2011	0.0419	0.2430	0.2849
1.5	−0.2068	0.0158	0.2226	0.2384
1.6	−0.2077	−0.0059	0.2018	0.1959
1.7	−0.2047	−0.0235	0.1812	0.1576
1.8	−0.1985	−0.0376	0.1610	0.1234
1.9	−0.1899	−0.0484	0.1415	0.0932
2.0	−0.1794	−0.0563	0.1230	0.0667
2.1	−0.1675	−0.0618	0.1057	0.0439
2.2	−0.1548	−0.0652	0.0895	0.0244
2.3	−0.1416	−0.0668	0.0748	0.0080
2.4	−0.1282	−0.0669	0.0613	−0.0056
2.5	−0.1149	−0.0658	0.0492	−0.0166
2.6	−0.1019	−0.0636	0.0383	−0.0254
2.7	−0.0895	−0.0608	0.0287	−0.0320
2.8	−0.0777	−0.0573	0.0204	−0.0369
2.9	−0.0666	−0.0534	0.0132	−0.0403
3.0	−0.0563	−0.0493	0.0071	−0.0423
3.1	−0.0469	−0.0450	0.0019	−0.0431
3.2	−0.0383	−0.0407	−0.0024	−0.0431
3.3	−0.0306	−0.0364	−0.0058	−0.0422
3.4	−0.0237	−0.0323	−0.0085	−0.0408
3.5	−0.0177	−0.0283	−0.0106	−0.0389
3.6	−0.0124	−0.0245	−0.0121	−0.0366
3.7	−0.0079	−0.0210	−0.0131	−0.0341
3.8	−0.0040	−0.0177	−0.0137	−0.0314
3.9	−0.0008	−0.0147	−0.0140	−0.0286
4.0	0.0019	−0.0120	−0.0139	−0.0258
4.1	0.0040	−0.0095	−0.0136	−0.0231
4.2	0.0057	−0.0074	−0.0131	−0.0204
4.3	0.0070	−0.0054	−0.0125	−0.0179
4.4	0.0079	−0.0038	−0.0117	−0.0155

*From Ref. 4.
†βs for domes (Table 3), βx for circular cylinders (Table 8).

The formulas in Table 3 derive from a bending-theory differential equation of the fourth order. In order to simplify the solution, the radius r_2 is taken constant, which is correct only for the spherical dome. It is because edge effects damp out so rapidly that the formulas can be used for most shells of revolution. The formulas can be relied on only if the opening angle α is greater than 25 or 30°.

Typical dimensions of spherical domes are given in Table 5.

TABLE 5 Typical Dimensions for Spherical Domes*

Section	D, ft	h, in.†	ϕ, deg	R, ft	a, ft
	100	3	30	13.4	100
			45	20.7	70.7
	125	3	30	16.8	125
			45	25.9	88.4
	150	3.5	30	20.1	150
		(3)	45	31.0	106.0
	175	4	30	23.5	175
		(3.5)	45	36.2	123.7
	200	4.5	30	26.8	200
		(4)	45	41.4	141.4

* From Portland Cement Association.

† Thickness usually increased by 50 to 75 percent at periphery.

10. Examples. Rigidly Supported Spherical Dome The dome of Fig. 11 is 2½ in. thick, except that it is thickened at the edge to 6 in. with a uniform taper over a length of 8 ft. The average dead load for the shell is taken at 40 psf. Roofing plus live load is 50 psf. Edge displacements will be calculated for a uniform thickness of 4 in. (Fig. 12).

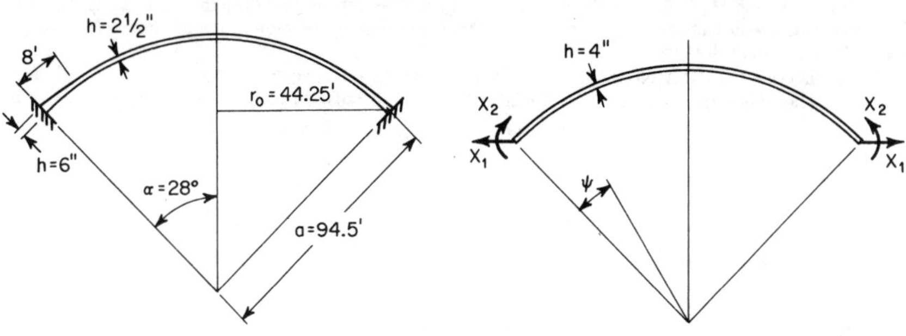

Fig. 11

Fig. 12

PRIMARY SYSTEM. The stress resultants are obtained from Eqs. (10):

$$N'_\phi = -\frac{94.5q}{1 + \cos\phi} \qquad N'_\theta = 94.5q\left(\frac{1}{1 + \cos\phi} - \cos\phi\right)$$

At the edge

$$N'_\alpha = -\frac{94.5q}{1 + 0.883} = -50.1q$$

and the horizontal and vertical components are

$$N'_{\alpha H} = N'_\alpha \cos\alpha = -50.1q \times 0.883 = -44.5q$$
$$N'_{\alpha V} = N'_\alpha \sin\alpha = -50.1q \times 0.469 = -23.8q$$

ERRORS. Edge displacements are considered positive in the directions of the redundant forces X_1 and X_2 (Fig. 12). Then from Eqs. (12)

$$\delta_{10} = \frac{94.5^2 q}{0.33E}\left(\frac{1.167}{1.883} - 0.883\right) 0.469 = -3340\frac{q}{E}$$

$$\delta_{20} = \frac{94.5q}{0.33E} \times 2.167 \times 0.469 = 291\frac{q}{E}$$

CORRECTIONS. The displacements of the edge resulting from unit values of the edge redundant forces X_1 and X_2 are found from the appropriate formulas in Table 3:

$$\beta^4 = \frac{3(1 - 0.167^2)}{(94.5 \times 0.33)^2} = 0.00300$$

$$\beta = 0.234 \qquad \beta^2 = 0.0548 \qquad \beta^3 = 0.0128$$

$$\delta_{11} = \frac{2 \times 0.234 \times 94.5^2 \times 0.469^2}{0.33E} = \frac{2780}{E}$$

$$\delta_{12} = \delta_{21} = \frac{2 \times 0.0548 \times 94.5^2 \times 0.469}{0.33E} = \frac{1390}{E}$$

$$\delta_{22} = \frac{4 \times 0.0128 \times 94.5^2}{0.33E} = \frac{1380}{E}$$

COMPATIBILITY. The equations of compatibility are

$$\delta_{11}X_1 + \delta_{12}X_2 + \delta_{10} = 0$$
$$\delta_{12}X_1 + \delta_{22}X_2 + \delta_{20} = 0$$

so that

$$2780X_1 + 1390X_2 = 3340q$$
$$1390X_1 + 1380X_2 = -291q$$

from which

$$X_1 = 2.63q = 0.237 \text{ kip/ft}$$
$$X_2 = -2.86q = -0.260 \text{ ft-kip/ft}$$

With $H = X_1$ and $M_\alpha = X_2$ known, N_ϕ, N_θ, and M_ϕ can be determined from the formulas in Table 3. The final stress resultants are obtained by adding these values to the membrane resultants N'_ϕ and N'_θ (see following example).

TEMPERATURE CHANGE. No forces result from expansion or contraction of the primary (membrane) structure. The only error in geometry is the lateral expansion caused by a temperature rise T, which is

$$\Delta_H = \delta_{10} = r_0 T\epsilon = 44.25 \times 6 \times 10^{-6}T = 265 \times 10^{-6}T$$

The equations of compatibility are

$$2780X_1 + 1390X_2 = -265 \times 10^{-6}TE$$
$$1390X_1 + 1380X_2 = 0$$

Example—Ring-Stiffened Spherical Dome The dome of this example is the same as that of the preceding example except that the edge is stiffened with a ring which is monolithic with the dome and which is free to slide and rotate on an immovable support (Fig. 13*a*).

PRIMARY SYSTEM. This consists of the shell and the ring as separate structures, subjected to the membrane stress resultant N'_α, which is assumed to act at the idealized junction of the shell and ring (Fig. 13*b*). Displacements are taken positive in the directions of the redundants X_1 and X_2.

The stress resultants and their horizontal and vertical components were determined in the preceding example. For the dead load $q = 40$ psf

$$N'_\phi = -\frac{3.78}{1 + \cos\phi} \qquad \text{kips/ft}$$

$$N'_\theta = 3.78\left(\frac{1}{1 + \cos\phi} - \cos\phi\right) \qquad \text{kips/ft}$$

$$N'_{\alpha H} = -1.78 \text{ kips/ft} \qquad N'_{\alpha V} = -0.95 \text{ kip/ft}$$

ERRORS. These consist of displacements at the junction of the shell and the ring. For the shell, using the results from the preceding example:

$$\delta^S_{10} = -3340 \times \frac{0.040}{E} = \frac{-133}{E} \qquad \text{ft}$$

$$\delta_{20}^{S} = 291 \times \frac{0.040}{E} = \frac{11.6}{E} \qquad \text{rad}$$

Figure 14*a* shows a portion of the ring under a horizontal force H and a moment M_α, each per unit length of arc. Neglecting the distance $b/2$, which is small compared with r, the horizontal displacement due to H is

$$\Delta_H = \frac{r^2 H}{EA} \tag{15}$$

where A is the area of the ring.

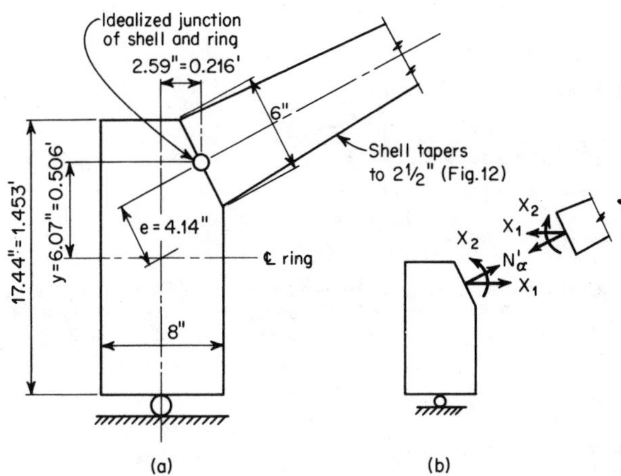

Fig. 13

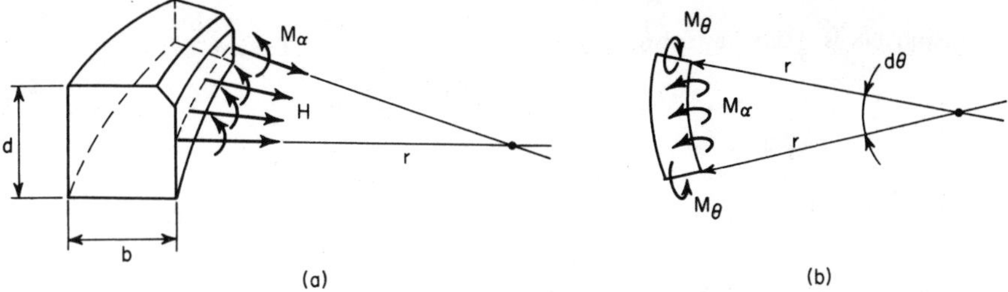

Fig. 14

From Fig. 14*b*, $M_\theta = rM_\alpha$. Assuming a linear stress distribution $f = M_\theta y/I$, the rotation Δ_α of the ring due to M_α is given by

$$\Delta_\alpha = \frac{\epsilon r}{y} = \frac{fr}{Ey} = \frac{M_\theta r}{EI} = \frac{r^2 M_\alpha}{EI} \tag{16}$$

The forces acting on the ring are

$$\begin{aligned} H &= N'_{\alpha H} = -1.78 \text{ kips/ft} \\ M_\alpha &= -0.506N'_{\alpha H} + 0.216N'_{\alpha V} \\ &= -0.506(-1.78) + 0.216(-0.95) = 0.695 \text{ ft-kip/ft} \end{aligned}$$

Then, from Eq. (16)

$$\delta_{20}^{R} = \frac{44.25^2 \times 0.695}{0.667(1.453^3/12)E} = \frac{7980}{E} \qquad \text{rad}$$

The horizontal displacement δ_{10} is, from Eq. (15),

$$\delta_{10}^R = \frac{44.25^2(-1.78)}{0.667 \times 1.453E} - 0.506\delta_{20}^R$$
$$= -\frac{3596}{E} - 0.506\frac{7980}{E} = -\frac{7634}{E} \quad \text{ft}$$

The last term in this equation is the horizontal displacement at the junction due to rotation of the ring. Thus, the δ_0 displacements are

$$\delta_{10} = \delta_{10}^S + \delta_{10}^R = -\frac{7767}{E}$$
$$\delta_{20} = \delta_{20}^S + \delta_{20}^R = \frac{7992}{E}$$

Corrections. With $X_2 = 1$ acting on the ring, Eq. (16) gives

$$\delta_{22}^R = \frac{44.25^2}{(0.667 \times 1.453^3/12)E} = \frac{11{,}490}{E} \quad \text{rad}$$

This rotation produces a displacement at the junction of the shell with the ring:

$$\delta_{12}^R = \delta_{21}^R = -0.506\frac{11{,}490}{E} = -\frac{5810}{E} \quad \text{ft}$$

With $X_1 = 1$ acting on the ring, the contribution to δ_{11} of the moment -1×0.506 resulting from the eccentricity of X_1 must be added to the value given by Eq. (15). This moment gives the rotation $-0.506 \times 11{,}490/E = -5810/E$. The resulting displacement at the junction is $0.506 \times 5810/E = 2940/E$. Thus, the displacement due to X_1 is

$$\delta_{11} = \frac{44.25^2}{0.667 \times 1.453E} + \frac{2940}{E} = \frac{4960}{E}$$

The displacements of the shell edge due to unit values of the redundants were determined in the preceding example: $\delta_{11}^S = 2780/E$, $\delta_{12}^S = \delta_{21}^S = 1390/E$, and $\delta_{22}^S = 1380/E$. Then, from $\delta = \delta^R + \delta^S$

$$\delta_{11} = \frac{7740}{E} \qquad \delta_{12} = \delta_{21} = -\frac{4420}{E} \qquad \delta_{22} = \frac{12{,}870}{E}$$

Compatibility is obtained by satisfying the equations

$$7740X_1 - 4420X_2 - 7767 = 0$$
$$-4420X_1 + 12{,}870X_2 + 7992 = 0$$

for which $X_1 = 0.81$ kip/ft and $X_2 = -0.34$ ft-kip/ft.

The stress resultants and moments throughout the dome can now be obtained by using X_1 and X_2 in the equations of Table 3 and combining the results with the membrane stress resultants. Table 6 gives these values for dead load plus live load (90 psf).

TABLE 6. Stress Resultants in Dome of Fig. 11
(Surface Load q = 90 psf)

ψ, deg	Dome with ring			Ring prestress = 3.24 kips		
	N_ϕ, kips/ft	N_θ, kips/ft	M_ϕ, ft-kips/ft	N_ϕ, kips/ft	N_θ, kips/ft	M_ϕ, ft-kips/ft
0	−2.92	26.64	−0.77	−0.94	−25.65	−0.37
1	−3.69	17.65	0.25	−0.28	−15.25	−0.85
2	−4.22	9.25	0.68	0.10	− 7.35	−0.93
3	−4.52	2.83	0.74	0.28	− 2.21	−0.77
5	−4.69	−3.82	0.44	0.28	1.80	−0.37
10	−4.38	−4.15	−0.03	0.00	0.37	+0.04
28	−4.25	−4.25	0.00	0.00	0.00	0.00

PRESTRESSED RING. The effect of prestressing the ring is easily determined. The radial pressure H_F is given by $H_F = F/r$, where F is the prestressing force. If F is not eccentric with respect to the center of gravity of the ring, Eqs. (15) and (16) give

$$\delta_{10} = \frac{44.25^2 H_F}{0.667 \times 1.453E} = \frac{2020H_F}{E} \text{ ft} \qquad \delta_{20} = 0$$

The equations of compatibility are

$$7740X_1 - 4420X_2 + 2020H_F = 0$$
$$-4420X_1 + 12{,}870X_2 = 0$$

from which $X_1 = -0.328H_F$ kip/ft and $X_2 = -0.114H_F$ ft-kip/ft.

The value of F required to counteract ring tension due to the dome dead load plus live load is obtained by combining the two solutions. For the dome load

$$T = r_0(N'_{\alpha H} + X_1) = 44.25(-1.78 + 0.81)90/40 = 96.5 \text{ kips}$$

and for the prestressing

$$C = r_0(H_F - 0.328H_F) = 29.7H_F \text{ kips}$$

With $C = T$, $H_F = 3.24$ kips/ft.

The stress resultants and moments caused by prestressing can be determined by using $H = X_1 = -0.328 \times 3.24 = -1.06$ kips/ft and $M_\alpha = X_2 = -0.114 \times 3.24 = -0.374$ ft-kip/ft in the equations of Table 3. The results are given in Table 6.

REINFORCEMENT

1. Dome hoop reinforcement is provided to take all the hoop tension. For the segment $\psi = 0°$ to $\psi = 1°$ (an arc length of 1.65 ft) the tensile force (Table 6) is

$$T_a = \frac{(26.64 - 25.65) + (17.65 - 15.25)}{2} \times 1.65 = 2.8 \text{ kips}$$

for which

$$A_s = \frac{2.8}{20} = 0.14 \text{ in.}^2$$
$$= \frac{0.14}{1.65} = 0.085 \text{ in.}^2\text{/ft, No. 3 at 15 in.}$$

Similarly, for the regions:

1 to 2°, $A_s = 0.18$ in.^{2} = 0.11 in.^{2}/ft, No. 3 at 12 in.

2 to 3°, $A_s = 0.10$ in.^{2} = 0.06 in.^{2}/ft, No. 3 at 22 in.

Beyond 3° there is no appreciable tension owing to edge effects.

2. Minimum dome reinforcement will be supplied throughout the shell equal to at least $0.0018bh$ for welded-wire fabric, as specified in ACI 318-77, Sec. 7.12.2, for slabs. For the 2½-in. shell, about 0.054 in.^{2}/ft in each direction is required. A welded-wire fabric 6 × 6 – 6/6 provides 0.058 in.^{2}/ft in each of two directions.

3. Meridional bending reinforcement will be provided to resist the combined effects of N_ϕ and M_ϕ. At the dome edge ($\psi = 0$) from Table 6, for total surface load plus prestressing:

$$M_\phi = -0.77 - 0.37 = -1.14 \text{ ft-kips/ft}$$
$$N_\phi = -2.92 - 0.94 = -3.86 \text{ kips/ft}$$

If the small compressive stress (3860/72 = 54 psi) is neglected, the resulting simplification is conservative, and with $d = 6 - 1.5 = 4.5$ in.,

$$A_s = \frac{M}{f_s jd} = \frac{1.14 \times 12}{20 \times 7/8 \times 4.5} = 0.16 \text{ in.}^2\text{/ft}$$

At $\psi = 1°$, $M_\phi = +0.25 - 0.85 = -0.60$ ft-kip/ft, from which, with $d = 6 - 3.5 \times 1.65/8 - 1.5 = 3.8$ in.,

$$A_s = \frac{0.60 \times 12}{20 \times 7/8 \times 3.8} = 0.11 \text{ in.}^2\text{/ft}$$

Since the moments from prestressing and surface loads are of opposite sign at this location, it is necessary to check for initial prestressing. Assuming initial prestress at a 25 percent increase and only dome dead load acting,

$$M_\phi = +0.25 \times 40/90 - 1.25 \times 0.85 = -0.95 \text{ ft-kip/ft}$$
$$A_s = 0.11 \times 95/65 = 0.16 \text{ in.}^2\text{/ft}$$

This moment is temporary and probably will not increase; so that f_s = 20,000 psi is conservative. It is more logical to use an ultimate-load analysis, where

$$M_u = -0.95 \times 1.5 = -1.42 \text{ ft-kips/ft}$$

From ACI 318, with f'_c = 4000 psi and f_y = 40,000 psi,

$$M_u = \phi A_s f_y \left(d - \frac{a}{2}\right) = 1.42$$

$$A_s f_y = 0.85 f'_c ab$$

from which a = 0.12 in. and A_s = 0.13 in.²/ft.

At $\psi = 2°$, $M_\phi = +0.68 - 0.93 = -0.25$ ft-kip/ft. With $d = 6 - 3.5 \times 3.30/8 - 1.5 = 3.0$ in.,

$$A_s = \frac{0.25 \times 0.12}{20 \times 7/8 \times 3.0} = 0.06 \text{ in.}^2/\text{ft}$$

Initially,

$$M_\phi = +0.68 \times 40/90 - 1.25 \times 0.93 = -0.86 \text{ ft-kip/ft}$$

$$M_u = \phi A_s f_y \left(d - \frac{a}{2}\right) = 1.5 \times 0.86 = 1.29 \text{ ft-kips/ft}$$

from which

$$A_s = 0.15 \text{ in.}^2/\text{ft}$$

Similar calculations at $\psi = 3°$, with h = 3.8 in., give A_s = 0.14 in.²/ft and at $\psi = 5°$, with $h = 2\frac{1}{2}$ in., A_s = 0.11 in.²/ft.

These computations show the need, largely due to initial prestressing, for top radial reinforcement of No. 3 at 8 in. (A_s = 0.17 in.²/ft) extending 10 ft from the edge (just beyond $\psi = 5°$).

The layout of reinforcement is shown in Fig. 15.

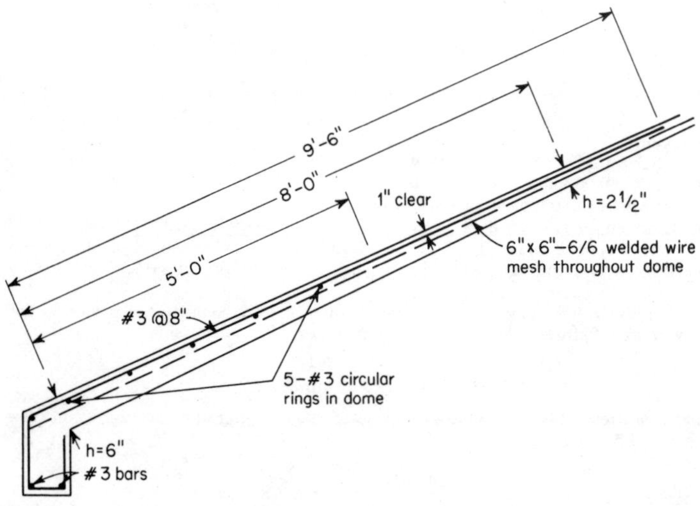

Fig. 15

RING PRESTRESSING. The final tension force required is

$$T = rH_F = 44.25 \times 3.24 = 143 \text{ kips}$$

Prestress will be furnished by circular rings of tensioned steel at an assumed final stress of 120,000 psi:

$$A_s = 143/120 = 1.2 \text{ in.}^2$$

This area may be supplied by 25 wires each 0.25 in. in diameter. An initial stress of 150,000 psi is required to compensate for losses assumed to be 20 percent.

Dome on Wall. If the dome is supported by a cylindrical wall, as in a tank, the analysis is the same as that for the ring-stiffened dome except that the displacements of the ring are replaced by the displacements of the wall (Art. 11). If the dome and wall are joined monolithically through a ring, four redundant forces are involved: two between the dome and the ring and two between the ring and the wall. An example is given in Ref. 20.

Where the dome is built integrally with a wall, or where the ring of a dome-ring structure is restrained either through friction or continuity with a wall, temperature effects will be important (as in the edge-fixed dome of Art. 10) and should be investigated.

SHELL WALLS

11. Cylindrical Tanks The membrane theory for cylindrical shells results in the following general equations (Fig. 16):

$$N'_\phi = -p_z r \tag{17a}$$

$$N'_{\phi x} = -\int \left(p_\phi + \frac{1}{r}\frac{\partial N'_\phi}{\partial \phi} \right) dx + f_1(\phi) \tag{17b}$$

$$N'_x = -\int \left(p_x + \frac{1}{r}\frac{\partial N'_{x\phi}}{\partial \phi} \right) dx + f_2(\phi) \tag{17c}$$

where f_1 and f_2 are determined from support conditions.

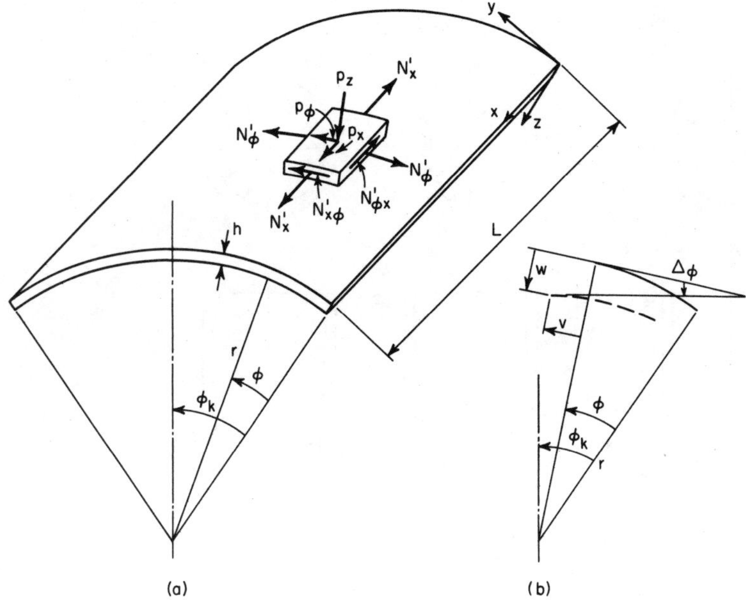

Fig. 16

The displacements u, v, w in the directions of the axes x, y, z are

$$u = \frac{1}{Eh}\int (N'_x - \nu N'_\phi)dx + f_3(\phi) \tag{17d}$$

$$v = -\int \frac{1}{r}\frac{\partial u}{\partial \phi}\, dx + \frac{2(1+\nu)}{Eh}\int N'_{\phi x} dx + f_4(\phi) \tag{17e}$$

$$w = \frac{\partial v}{\partial \phi} - \frac{r}{Eh}(N'_\phi - \nu N'_x) \tag{17f}$$

where f_3 and f_4 are determined from support conditions.

The rotation of the tangent to the circle at any point, positive as shown in Fig. 16*b*, is given by

$$\Delta_\phi = \frac{v}{r} + \frac{\partial w}{r\partial \phi} \tag{17g}$$

Internal Liquid Pressure (Fig. 17*a*). With γ = density of liquid, $p_\phi = p_x = 0$, and $p_z = -y(H - x)$, Eqs. (17) give (Fig. 18)

$$N'_\phi = \gamma(H - x)r \tag{18a}$$
$$N'_x = N'_{x\phi} = 0 \tag{18b}$$

The resulting displacements are

$$w = -\frac{\gamma r^2}{Eh}(H - x) \tag{19a}$$

$$\frac{dw}{dx} = \phi_x = \frac{\gamma r^2}{Eh} \tag{19b}$$

Variable Thickness. If the thickness of the tank wall varies linearly (Fig. 17*c*) the stress resultants for liquid pressure are given by Eq. (18). The displacements are

$$w = -\frac{\gamma r^2}{Eh_x}(H - x) \tag{20a}$$

$$\frac{dw}{dx} = \phi_x = \frac{\gamma r^2}{Eh_x}\frac{h_{\text{top}}}{h_x} \tag{20b}$$

where h_x = thickness at elevation x

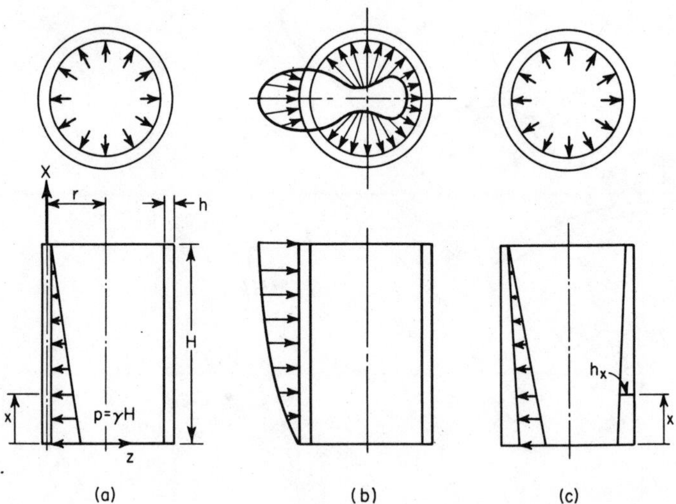

Fig. 17

Seismic Loading. If the horizontal acceleration can be considered as a percentage α of the dead weight q, so that $q_s = \alpha q$, the loading is $p_\phi = q_s \sin\theta$, $p_x = 0$, $p_z = q_s \cos\theta$, and Eqs. (17) give, for $N'_{\phi x} = N'_x = 0$ at $x = H$,

$$N'_\phi = -q_s r \cos\phi \tag{21a}$$
$$N'_{\phi x} = 2q_s(H - x)\sin\phi \tag{21b}$$
$$N'_x = \frac{q_s}{r}(H - x)^2 \cos\phi \tag{21c}$$

Wind Load. The wind pressure p_z on shell walls of rotation can be defined by (Fig. 17*b*).

$$p_z = K_z H_\phi p_{30} \tag{22}$$

where

$$K_z = 2.64\left(\frac{z}{z_g}\right)^{2/\alpha} \tag{23}$$

H_ϕ in Eq. (22) is a horizontal distribution factor. Coefficients H_ϕ from Ref. 21 for smooth cylinders with several ratios of L/D are given in Table 7*a*. Results of tests in Germany[22] on a full-scale hyperbolic cooling tower with small vertical ribs are also shown in this table. Measurements in the United States have given similar results.[23]

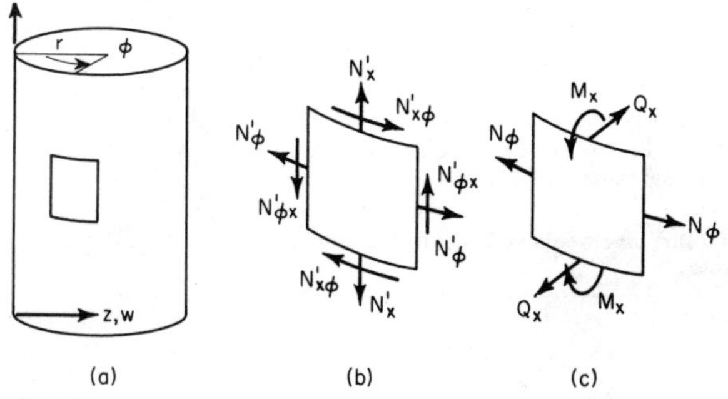

Fig. 18

TABLE 7a Values of H_ϕ in Eq. (22)

	L/D			
ϕ, deg	1	7	25	Ref. 22
0	1.0	1.0	1.0	1.0
15	0.8	0.8	0.8	0.8
30	0.1	0.1	0.1	0.2
45	−0.7	−0.8	−0.9	−0.5
60	−1.2	−1.7	−1.9	−1.2
75	−1.6	−2.2	−2.5	−1.3
90	−1.7	−2.2	−2.6	−0.9
105	−1.2	−1.7	−1.9	−0.4
120	−0.7	−0.8	−0.9	−0.4
135	−0.5	−0.6	−0.7	−0.4
150	−0.4	−0.5	−0.6	−0.4
165	−0.4	−0.5	−0.6	−0.4
180	−0.4	−0.5	−0.6	−0.4

TABLE 7b Fourier Coefficients for H_ϕ in Table 7a*

	A_n	
n	$L/D = 1$	Ref. 22
0	−0.6000	−0.3923
1	0.2979	0.2602
2	0.9184	0.6024
3	0.3966	0.5046
4	−0.0588	0.1064
5	0.0131	−0.0948
6	0.0609	−0.0186
7	−0.0179	0.0468

*From Ref. 24.

The circumferential distribution of wind pressure may be represented by the Fourier series

$$H_\phi = \Sigma A_n \cos n\phi \tag{24}$$

The first eight harmonics are generally sufficient.[24] Values of A_n corresponding to the test values of H_ϕ in Table 7*a* are given in Table 7*b*, along with values for the cylinder with $L/D = 1$.

For each term in the series Eqs. (17) gives

$$N'_\phi = p_z rA_n \cos n\phi \tag{25a}$$

$$N'_{\phi x} = p_z(H - x)A_n n \sin n\phi \tag{25b}$$

$$N'_x = \frac{p_z}{2r}(H - x)^2 A_n n^2 \cos n\phi \tag{25c}$$

Bending Theory—Axisymmetrical Loading. Edge effects are confined to a narrow zone. Therefore, except for the unusual case of a cylinder whose length is of the same order of magnitude as the affected zones, edge effects at one end can be treated independently of those at the other. Forces and displacements are given in Table 8.

TABLE 8 Forces (Fig. 18c) and Displacements in Circular Cylinders Loaded by Edge Forces Uniform around a Parallel Circle

	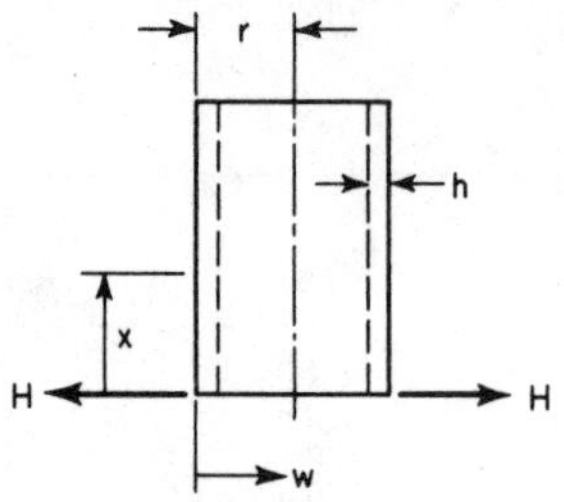 	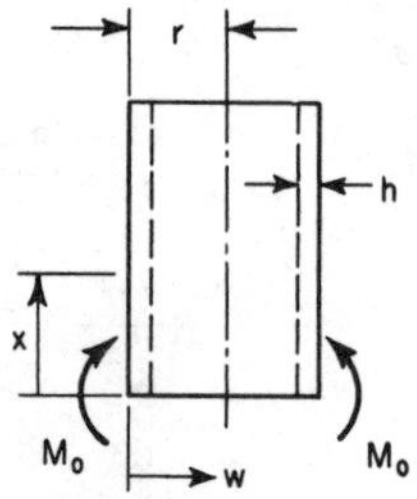
N_φ	$2C_2\beta rH$	$2C_1\beta^2 rM_0$
M_x	$\frac{C_3}{\beta}H$	C_4M_0
Q_x	C_1H	$-2C_3\beta M_0$
$w_{x=0}$	$-2\beta\frac{r^2}{Eh}H$	$-2\beta^2\frac{r^2}{Eh}M_0$
$\left(\frac{dw}{dx}\right)_{x=0}$	$2\beta^2\frac{r^2}{Eh}H$	$4\beta^3\frac{r^2}{Eh}M_0$

NOTE: $\beta^4 = 3(1 - \nu^2)/r^2h^2$. See Table 4 for *C*.

Analysis For a tank with a flat roof which is continuous with the wall, edge effects can be calculated using Tables 8 and 9.

Primary System. The roof slab is assumed to be freely supported on the wall.

Errors. Displacements are considered positive in the direction of the redundant forces (Fig. 19). Conditions at the base of the wall (i.e., whether it is fixed, hinged, free to slide, etc.) are usually immaterial in this analysis. This is because of the localized nature of edge forces on the wall. For the slab under uniform load, radial displacements are zero. Rotation of the edge is given in Table 9. Thus,

$$\delta^S_{10} = 0 \qquad \delta^S_{20} = -\frac{dw}{dR} = \frac{3qr^3}{28D} \qquad \text{for } \nu = 1/6$$

For the tank filled with liquid, from Eq. (19),

$$\delta^W_{10} = 0 \qquad \delta^W_{20} = -\frac{dw}{dx} = -\frac{\gamma r^2}{Eh}$$

TABLE 9 Symmetrical Bending of Circular Plates

	M_r uniform on circumference	Uniform q over surface	Uniform q over surface
	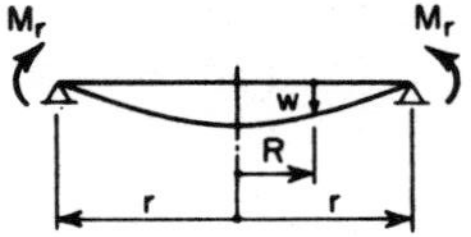	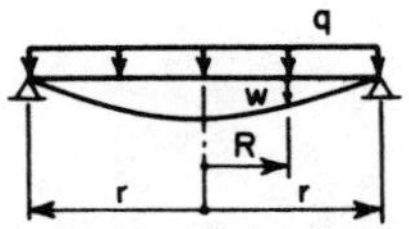	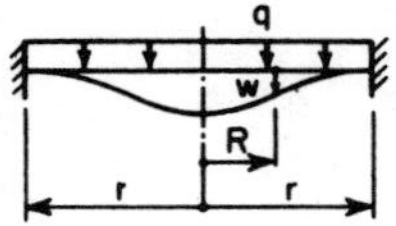
w	$\frac{M_r(r^2 - R^2)}{2D(1+\nu)}$	$\frac{q}{64D}(r^2 - R^2)\left(\frac{5+\nu}{1+\nu}r^2 - R^2\right)$	$\frac{q}{64D}(r^2 - R^2)^2$
$\frac{dw}{dR}$	$-\frac{M_r R}{D(1+\nu)}$	$-\frac{qR}{16D}\left(\frac{3+\nu}{1+\nu}r^2 - R^2\right)$	$-\frac{qR}{16D}(r^2 - R^2)$
$\frac{d^2w}{dR^2}$	$-\frac{M_r}{D(1+\nu)}$	$-\frac{q}{16D}\left(\frac{3+\nu}{1+\nu}r^2 - 3R^2\right)$	$-\frac{q}{16D}(r^2 - 3R^2)$
M_R	M_r	$\frac{q}{16}(3+\nu)(r^2 - R^2)$	$\frac{q}{16}[r^2(1+\nu) - R^2(3+\nu)]$
M_T	M_r	$\frac{q}{16}[r^2(3+\nu) - R^2(1+3\nu)]$	$\frac{q}{16}[r^2(1+\nu) - R^2(1+3\nu)]$
M_r	M_r	0	$-\frac{qr^2}{8}$
$M_{℄}$	M_r	$\frac{qr^2}{16}(3+\nu)$	$\frac{qr^2}{16}(1+\nu)$

$D = Eh^3/12(1 - \nu^2)$, h = thickness, M_R = radial moment, M_T = tangential moment.

Corrections. For the wall, using Table 8 with $H = -X_1 = -1$ and $M_0 = X_2 = 1$,

$$\delta_{11}^W = \frac{2\beta r^2}{Eh} \qquad \delta_{12}^W = -\frac{2\beta^2 r^2}{Eh} \qquad \delta_{22}^W = \frac{4\beta^3 r^2}{Eh}$$

The radial displacement of the circumference of a circular slab subjected to unit radial forces at the perimeter is given by

$$\delta_{11}^S = \frac{r}{Eh}(1 - \nu)$$

and the edge rotation is zero, so that

$$\delta_{12}^S = 0$$

The edge rotation resulting from X_2 is found from Table 9. With $M_r = X_2 = 1$,

$$\delta_{22}^S = -\frac{dw}{dR} = \frac{r}{D(1 + \nu)}$$

Compatibility is established by satisfying the simultaneous equations in X_1 and X_2, as in Art. 10, where $\delta_{11} = \delta_{11}^W + \delta_{11}^S$, etc.

With X_1 and X_2 known, the internal forces to be provided for in the tank are evaluated using $H = -X_1$ and $M_0 = X_2$ in the formulas of Table 8, to which the membrane stress resultants [Eq. (18)] must be added. The internal forces in the roof are obtained similarly from Table 9.

The analyses for a tank roofed with a cylindrical dome and for the interaction between wall and floor (if they are continuous) are identical. However, the loads acting on the floor, and the corresponding errors and corrections, may be more difficult to determine.[20]

12. Hyperboloids The type of shell wall shown in Fig. 20 has been used frequently for natural-draft cooling towers, for which the two principal loads are dead weight and wind. The membrane stress resultants for dead weight are computed from Eqs. (5) and (7) by numerical means as follows:

$$N'_{\phi j} = \frac{\sum_{i=1}^{j} r_{oi} q_i \Delta s_i}{r_{oj} \sin \phi j} \tag{26a}$$

$$N'_{\theta j} = -\frac{r_{oj}}{\sin \phi j}\left(\frac{N'_{\phi j}}{r_{1j}} + q_j \cos \phi j\right) \tag{26b}$$

where $N'_{\phi j}$ is the sum of the weights of j rings above, each of average unit weight q_i, average horizontal radius r_{oi}, and meridional length Δs_i.

Wind Load. Membrane stress resultants are not easily computed since the load is not axisymmetrical. The meridional stress resultant is critical for design. Comparisons of meridional stress resultants computed for wind distributed as described in Art. 11 with results from a bending-theory solution show very little difference except in the values of N'_θ near the base.

Because several cooling towers have collapsed during wind storms, sufficient reinforcement must be provided in both the meridional and the circumferential directions. Recommendations are given in Ref. 24.

BARREL SHELLS

Segments of cylinders, often called barrel shells, transfer load by a combination of longitudinal beam action and transverse arch action. In short barrels, the loads are carried essentially by arch action to the longitudinal edges, where they are transferred to the transverse supports by the edge sections of the shell acting as deep beams. In long barrels, the shell behaves primarily as a beam of thin, curved cross section, although there is still some arch action near the crown.

Short barrels are used for aircraft hangars and auditoriums. They have been built with transverse spans of about 150 to 330 ft, and with longitudinal spans of 20 to 50 ft between stiffening ribs. Long barrels are more commonly used for warehouses and factories, where longitudinal spans of about 50 to 150 ft are required, with transverse spans of 20 to 40 ft.

Other simply curved shells, such as truncated segments of cones, are sometimes used. They behave similarly to cylindrical segments, provided they are not too radically tapered.

Membrane stress resultants for the cylindrical barrel are given in Table 10; the boundary conditions are $N'_x = 0$ at $x = 0$ and $x = L$ (Fig. 16).

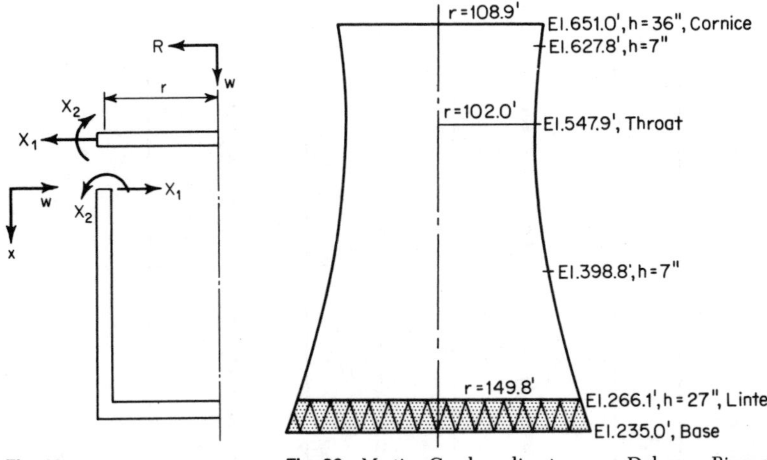

Fig. 19

Fig. 20 Martins Creek cooling tower on Delaware River at Easton.

13. Long Barrels Barrels for which the ratio r/L is less than 0.6 may be considered to be long.[25] In this case, the in-plane stresses are approximated well by

$$f_x = \frac{N_x}{h} = \frac{M_x y}{I} \tag{27}$$

$$v = \frac{N_{x\phi}}{h} = \frac{VQ}{Ib} \tag{28}$$

where M_x = bending moment about centroidal axis
I = moment of inertia of shell cross section
V = total shear at cross section
b = total cross-sectional thickness of concrete measured horizontally

Transverse moments M_ϕ (arch action) may be approximated by considering the slice from the barrel shown in Fig. 21, where the vertical load q is held in equilibrium by the vertical component of the in-plane shearing forces. The arch moments and thrusts may be determined by any of the methods for arch analysis.

Implicit in the beam-arch analysis are the assumptions that all points on a transverse cross section deflect equally in the vertical direction and not at all horizontally, and that the radial shears Q_x, the longitudinal bending moments M_x, the twisting moments $M_{x\phi}$, and the strains from in-plane shearing forces can be neglected.

Figure 22 compares the stresses from Eq. (27) with those given by the shallow-shell theory. The correspondence is good for $r/L = 0.2$. But for barrels of intermediate and short length the maximum tensile stress is considerably larger than that given by Eq. (27).

Analysis by Eq. (27) is simple, but the arch analysis tends to be lengthy.[26] Table 11, based on the beam-arch equations, gives numerical values which may be used for interior shells of multibarrel systems and for single barrels without edge beams or with relatively flexible edge beams. These values may be corrected for horizontal edge deflection by Tables 12 and 13, which are given only for uniform load p because the error for dead load q is negligible. The corrections are made by adding to the values from Table 11 the product of the horizontal displacement from Table 12 by the corresponding coefficient in Table 13. Table 14 gives values of I for calculation of deflections. Extensive tables based on the shallow-shell theory are given in Ref. 27.

TABLE 10 Membrane Forces in Cylindrical Shells (Fig. 16)

Loading	N_ϕ'	N_x'	$N_{\phi x}'$
q per ft^2 of surface	$-qr\cos(\phi_k - \phi)$	$-q\dfrac{x}{r}(L - x)\cos(\phi_k - \phi)$	$-q(L - 2x)\sin(\phi_k - \phi)$
p per ft^2 projection	$-pr\cos^2(\phi_k - \phi)$	$\frac{3}{2}p\dfrac{x}{r}(L - x)[1 - 2\sin^2(\phi_k - \phi)]$	$-\frac{3}{2}p(L - 2x)\sin(\phi_k - \phi)\cos(\phi_k - \phi)$
p_w per ft^2 projection	$-p_w r\sin(\phi_k - \phi)$	$-p_w\dfrac{x}{2r}(L - x)\sin(\phi_k - \phi)$	$p_w\left(\dfrac{L}{2} - x\right)\cos(\phi_k - \phi)$

Typical dimensions of barrel shells are given in Table 15.

An elliptical cross section substantially improves the behavior by reducing moments and longitudinal stresses. The effect of the nearly vertical edge is somewhat the same as the addition of an edge member, except that no edge moments are created. The elliptical cross section was often chosen when barrel shells were first used but was soon given up because of construction difficulties.

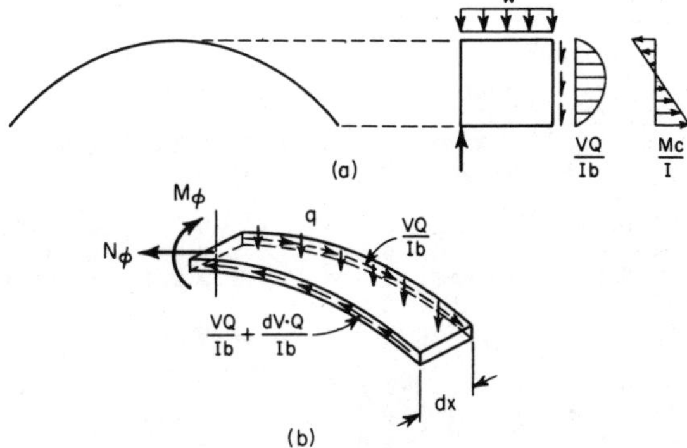

Fig. 21

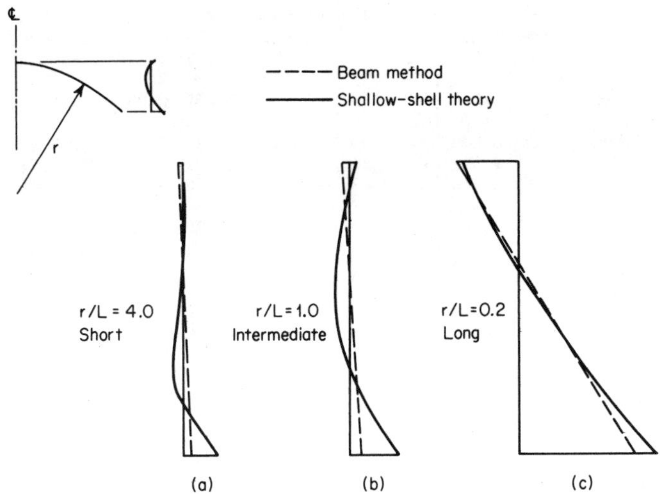

Fig. 22

Prestressing can be conveniently used in the edge members of long-span barrels, primarily to counteract longitudinal tension forces and transverse bending moments and to control deflection and cracking. Ultimate-load behavior must not be overlooked, however, because design moments increase directly with overloads whereas prestressing moments do not.

Transverse frames projecting below the roof complicate construction. Frames projecting above the roof complicate roofing and insulation. Thus, long-barrel roofs have sometimes been designed as ribless shells, in which a considerable portion of the shell is thickened and thus acts as a flat arch.[29] These arches should be investigated for partial loadings as well as for buckling and are thus used only on relatively small spans.

TABLE 11 Symmetrically Loaded Interior Circular Cylindrical Shells*

		Uniform transverse load p‡				Dead-weight load q‡			
ϕ_k, deg	$\frac{\phi\dagger}{\phi_k}$	$\frac{N_x}{pL^2/r}$	$\frac{N_\phi}{pr}$	$-\frac{N_{x\phi}}{pL}$	$\frac{M_\phi}{pr^2}$	$\frac{N_x}{qL^2/r}$	$\frac{N_\phi}{qr}$	$-\frac{N_{x\phi}}{qL}$	$\frac{M_\phi}{qr^2}$
22.5	1	−6.010	−1.411	0.000	−0.00292	−6.167	−1.433	0.000	−0.00309
	0.75	−4.875	−1.189	2.211	−0.00112	−5.003	−1.205	2.269	−0.00118
	0.50	−1.482	−0.614	3.533	0.00232	−1.521	−0.615	3.626	0.00245
	0.25	4.137	0.049	3.084	0.00235	4.245	0.065	3.165	0.00249
	0	11.927	0.361	0.000	−0.00662	12.239	0.384	0.000	−0.00702
25.0	1	−4.855	−1.402	0.000	−0.00353	−5.012	−1.430	0.000	−0.00378
	0.75	−3.937	−1.182	1.985	−0.00135	−4.064	−1.202	2.049	−0.00145
	0.50	−1.193	−0.612	3.170	0.00280	−1.232	−0.613	3.273	0.00300
	0.25	3.342	0.044	2.765	0.00282	3.451	0.064	2.855	0.00304
	0	9.617	0.347	0.000	−0.00797	9.929	0.374	0.000	−0.00857
27.5	1	−4.000	−1.393	0.000	−0.00417	−4.158	−1.426	0.000	−0.00453
	0.75	−3.242	−1.175	1.799	−0.00159	−3.370	−1.199	1.869	−0.00173
	0.50	−0.980	−0.609	2.871	0.00331	−1.018	−0.610	2.985	0.00360
	0.25	2.755	0.038	2.503	0.00332	2.863	0.063	2.602	0.00363
	0	7.908	0.331	0.000	−0.00938	8.220	0.363	0.000	−0.01025
30.0	1	−3.350	−1.383	0.000	−0.00482	−3.508	−1.422	0.000	−0.00533
	0.75	−2.714	−1.166	1.643	−0.00183	−2.842	−1.195	1.720	−0.00203
	0.50	−0.817	−0.606	2.622	0.00384	−0.856	−0.607	2.746	0.00424
	0.25	2.308	0.032	2.284	0.00383	2.417	0.061	2.392	0.00426
	0	6.608	0.314	0.000	−0.01082	6.920	0.352	0.000	−0.01204
32.5	1	−2.844	−1.372	0.000	−0.00548	−3.002	−1.418	0.000	−0.00618
	0.75	−2.303	−1.158	1.511	−0.00207	−2.431	−1.191	1.595	−0.00235
	0.50	−0.691	−0.603	2.410	0.00438	−0.729	−0.603	2.544	0.00492
	0.25	1.960	0.026	2.098	0.00434	2.069	0.060	2.215	0.00492
	0	5.596	0.297	0.000	−0.01227	5.908	0.339	0.000	−0.01393
35.0	1	−2.442	−1.361	0.000	−0.00615	−2.601	−1.414	0.000	−0.00707
	0.75	−1.977	−1.148	1.397	−0.00232	−2.105	−1.186	1.488	−0.00268
	0.50	−0.591	−0.599	2.227	0.00491	−0.629	−0.600	2.372	0.00565
	0.25	1.684	0.019	1.938	0.00484	1.793	0.058	2.064	0.00561
	0	4.794	0.278	0.000	−0.01370	5.105	0.326	0.000	−0.01591
37.5	1	−2.118	−1.349	0.000	−0.00679	−2.278	−1.409	0.000	−0.00800
	0.75	−1.174	−1.138	1.298	−0.00255	−1.842	−1.181	1.396	−0.00302
	0.50	−0.510	−0.596	2.069	0.00544	−0.548	−0.596	2.224	0.00640
	0.25	1.461	0.012	1.798	0.00532	1.571	0.057	1.933	0.00632
	0	4.146	0.260	0.000	−0.01509	4.458	0.312	0.000	−0.01796
40.0	1	−1.853	−1.335	0.000	−0.00742	−2.013	−1.404	0.000	−0.00897
	0.75	−1.498	−1.127	1.211	−0.00277	−1.627	−1.176	1.315	−0.00337
	0.50	−0.443	−0.592	1.929	0.00595	−0.482	−0.592	2.095	0.00719
	0.25	1.279	0.005	1.675	0.00578	1.389	0.055	1.819	0.00705
	0	3.616	0.241	0.000	−0.01641	3.928	0.297	0.000	−0.02006
45.0	1	−1.449	−1.307	0.000	−0.00853	−1.610	−1.393	0.000	−0.01096
	0.75	−1.170	−1.104	1.065	−0.00316	−1.299	−1.165	1.183	−0.00408
	0.50	−0.343	−0.585	1.694	0.00688	−0.381	−0.583	1.882	0.00883
	0.25	1.001	−0.011	1.468	0.00657	1.112	0.052	1.630	0.00854
	0	2.809	0.202	0.000	0.01872	3.120	0.266	0.000	−0.02437

TABLE 11 Symmetrically Loaded Interior Circular Cylindrical Shells* *(Continued)*

		Uniform transverse load p‡				Dead-weight load q‡			
ϕ_k, deg	$\frac{\phi\dagger}{\phi_k}$	$\frac{N_x}{pL^2/r}$	$\frac{N_\phi}{pr}$	$-\frac{N_{x\phi}}{pL}$	$\frac{M_\phi}{pr^2}$	$\frac{N_x}{qL^2/r}$	$\frac{N_\phi}{qr}$	$-\frac{N_{x\phi}}{qL}$	$\frac{M_\phi}{qr^2}$
50.0	1	−1.160	−1.276	0.000	−0.00939	−1.322	−1.380	0.000	−0.01301
	0.75	−0.935	−1.079	0.947	−0.00344	−1.065	−1.152	1.079	−0.00480
	0.50	−0.271	−0.578	1.504	0.00762	−0.308	−0.574	1.713	0.01054
	0.25	0.802	−0.029	1.300	0.00714	0.914	0.049	1.481	0.01002
	0	2.232	0.164	0.000	−0.02052	2.543	0.234	0.000	−0.02871
55.0	1	−0.946	−1.242	0.000	−0.00989	−1.109	−1.367	0.000	−0.01506
	0.75	−0.761	−1.053	0.849	−0.00358	−0.892	−1.139	0.995	−0.00549
	0.50	−0.217	−0.572	1.347	0.00807	−0.255	−0.563	1.578	0.01227
	0.25	0.655	−0.048	1.161	0.00742	0.767	0.045	1.360	0.01144
	0	1.805	0.128	0.000	−0.02130	2.115	0.201	0.000	−0.03293
60.0	1	−0.783	−1.205	0.000	−0.00992	−0.947	−1.352	0.000	−0.01705
	0.75	−0.629	−1.025	0.766	−0.00355	−0.761	−1.124	0.927	−0.00613
	0.50	−0.177	−0.566	1.213	0.00815	−0.214	−0.552	1.467	0.01398
	0.25	0.543	−0.068	1.043	0.00734	0.656	0.042	1.261	0.01275
	0	1.481	0.095	0.000	−0.02118	1.790	0.167	0.000	−0.03688

*From Ref. 28.
†ϕ measured from edge (Fig. 16).
‡See Fig. 16 for N_x, N_ϕ, $N_{x\phi}$; Fig. 21 for M_ϕ.

TABLE 12 Horizontal Edge Displacement Δ of Circular Cylindrical Shells under Uniform Load p*

$$\Delta = -\frac{pr^2}{Eh} \times \text{(constant from table)}$$

ϕ_k, deg	r/L						
	0.100	0.125	0.150	0.175	0.200	0.225	0.250
22.5	91.83	38.37	18.88	10.37	6.16	3.87	2.52
25.0	123.80	51.79	25.54	14.09	8.42	5.33	3.52
27.5	161.58	67.67	33.43	18.49	11.09	7.07	4.71
30.0	205.26	86.02	42.55	23.59	14.20	9.08	6.09
32.5	254.82	106.85	52.91	29.38	17.72	11.38	7.66
35.0	310.10	130.69	64.47	35.84	21.66	13.94	9.42
37.5	370.84	155.63	77.17	42.94	25.99	16.77	11.37
40.0	436.65	183.31	90.94	50.65	30.70	19.83	13.47
45.0	581.48	244.21	121.26	67.61	41.05	26.59	18.13
50.0	739.50	310.68	154.34	86.14	52.37	33.98	23.22
55.0	904.36	380.03	188.87	105.47	64.18	41.70	28.54
60.0	1,068.79	449.20	223.31	124.77	75.97	49.41	33.85

*From Ref. 28.

TABLE 13. Effect of Unit Horizontal Displacement of Circular Cylindrical Shells under Uniform Load p*

ϕ_k, deg	$\phi\dagger/\phi_k$	N_ϕ	M_ϕ	ϕ_k, deg	$\phi\dagger/\phi_k$	N_ϕ	M_ϕ
22.5	1	2,463.0	62.82	37.5	1	199.2	13.92
	0.75	2,451.1	50.96		0.75	196.5	11.26
	0.50	2,415.6	15.49		0.50	188.6	3.35
	0.25	2,356.9	−43.34		0.25	175.7	−9.60
	0	2,275.5	−124.66		0	158.0	−27.24
25.0	1	1,461.9	45.95	40.0	1	145.5	11.53
	0.75	1,453.2	37.26		0.75	143.3	9.32
	0.50	1,427.2	11.30		0.50	136.7	2.76
	0.25	1,384.3	−31.63		0.25	126.0	−7.96
	0	1,324.9	−91.02		0	111.4	−22.50
27.5	1	912.9	34.65	45.0	1	82.2	8.20
	0.75	906.4	28.09		0.75	80.7	6.62
	0.50	886.8	8.49		0.50	76.0	1.94
	0.25	854.4	−23.86		0.25	68.4	−5.66
	0	869.8	−68.50		0	58.2	−15.89
30.0	1	594.6	26.80	50.0	1	49.6	6.06
	0.75	589.5	21.71		0.75	48.4	4.88
	0.50	574.3	6.54		0.50	44.9	1.41
	0.25	549.3	−18.46		0.25	39.3	−4.19
	0	514.9	−52.86		0	31.9	−11.65
32.5	1	401.2	21.17	55.0	1	31.5	4.62
	0.75	397.2	17.14		0.75	30.6	3.72
	0.50	385.2	5.14		0.50	27.9	1.66
	0.25	365.4	−14.59		0.25	23.7	−3.20
	0	338.4	−41.66		0	18.1	−8.81
35.0	1	279.0	17.03	60.0	1	20.9	3.62
	0.75	275.8	13.79		0.75	20.2	2.91
	0.50	266.1	4.12		0.50	18.1	0.82
	0.25	250.3	−11.74		0.25	14.8	−2.51
	0	228.6	−33.43		0	10.5	−6.84

*From Ref. 28.
†ϕ measured from edge (Fig. 16).

TABLE 14 Moment of Inertia of Circular Cylindrical Shells*

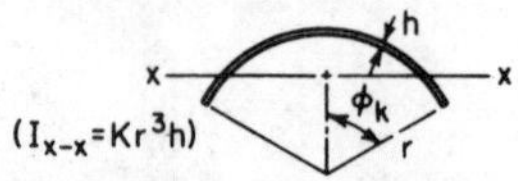

ϕ_k, deg	K	ϕ_k, deg	K
22.5	0.00041	37.5	0.00502
25.0	0.00068	40.0	0.00687
27.5	0.00110	45.0	0.01216
30.0	0.00168	50.0	0.02017
32.5	0.00249	55.0	0.03174
35.0	0.00358	60.0	0.04782

*From Ref. 28.

Continuous Shells. The behavior of beamlike thin shells continuous over three or more transverse supports is similar to that of continuous beams with regard to in-plane stresses, but transverse bending moments are affected less by longitudinal continuity.

14. Short Barrels Barrels for which $r/L > 0.6$ may be considered to be short. Such shells carry load essentially by arch action and are usually shaped to the arch pressure line. Because of the large value of r relative to L, the shell thickness is often controlled by buckling rather than by strength (Art. 5). Intermediate stiffeners have been used to prevent buckling. Here the cost of extra formwork must be weighed against the saving in cost of materials.

TABLE 15 Typical Dimensions for Barrel Shells*

Section†	Span, ft	Bay width, ft	R, ft	r, ft	h, in.	Rein-forcing‡
	80	30	8	25	3	3.5
	100	30	10	30	3	4.0
	120	35	12	30	3	4.5
	140	40	14	35	3	5.0
	160	45	16	35	3.5	6.5

* From Portland Cement Association.
† For long-span multiple barrels, the usual depth-span ratio varies from 1:10 to 1:15.
‡ Pounds per square foot of projected area.

Since most of the roof load is carried to the supporting transverse frames by beam action only near the longitudinal edges, the supporting arch is subjected to outward thrusts near the springing lines and hence may have only a small compression, or even some tension, near the crown. But the adjacent shell is under compression; so that there is incompatibility of strain or, more properly, the ideal behavior is modified and some load is transferred into the arch by bending forces. The effect is to increase the compression in the supporting arch and reduce the compression in the shell adjacent to the arch or, in other words, to force part of the shell to act as a flange for the arch in T-beam behavior.

15. General Procedure for Shallow Shells The theory of shallow shells is based on the following assumptions: (1) the slope of the shell is small compared with some plane of reference (usually the horizontal plane for roofs); (2) the curvature of the surface is small; (3) the shell boundaries are such that the loads are carried primarily by the in-plane stress resultants N_x, N_ϕ, and $N_{x\phi}$; and (4) changes in curvature of the surface are small. The problem can be reduced to the solution of an eighth-order partial differential equation in one unknown.

In the following analysis the shell is assumed to be supported at $x = 0$ and $x = L$ by transverse frames which are rigid in their vertical planes, so that displacements v and w are zero at each end, and completely flexible in respect to out-of-plane displacements, so that N_x and M_x are zero at each end.

Complementary Function. With the displacement w (Fig. 16b) as the unknown, the complementary function has the form

$$w = \sum_{n=1,3\ldots} A_n e^{Mn\phi} \sin kx$$

where $k = n\pi/L$. Table 16 gives the complementary functions for the displacements and resultants for a single, circular, cylindrical shell with symmetrical geometry and loading. Displacements u, v, and w are positive in the positive directions of x, y, and z, respectively (Fig. 16a), while θ corresponds to Δ_ϕ in Fig. 16b. Note that ϕ is measured from the crown, rather than from the edge as in Fig. 16.

Except for Q'_ϕ and Q'_x, the stress resultants and couples in this table can be identified in Fig. 4a and b by substituting ϕ for y (see also Fig. 16a). Q'_ϕ and Q'_x are combinations analogous to those arising in the theory of plates:

$$Q'_\phi = Q_\phi + \frac{\partial M_{\phi x}}{\partial x}$$

$$Q'_x = Q_x + \frac{\partial M_{\phi x}}{R\partial \phi}$$

Thus, although there are five stress resultants at the longitudinal edge ($M_\phi, N_\phi, Q_\phi, N_{\phi x}$, and $M_{\phi x}$) with only four boundary conditions to be satisfied, Q'_ϕ reduces the number to four.

The shell constants Q and γ in Table 16 are defined as follows:

$$Q^8 = 3(kr)^4\left(\frac{r}{h}\right)^2 \qquad \gamma = \left(\frac{kr}{Q}\right)^2$$

Values of the coefficients of ϕ are

$$\alpha_1 = Q\left(\frac{\sqrt{(1+\gamma)^2+1}+(1+\gamma)}{2}\right)^{1/2} = Qm_1 \tag{29a}$$

$$\beta_1 = Q\left(\frac{\sqrt{(1+\gamma)^2+1}-(1+\gamma)}{2}\right)^{1/2} = Qn_1 \tag{29b}$$

$$\alpha'_1 = Q\left(\frac{\sqrt{(1-\gamma)^2+1}-(1-\gamma)}{2}\right)^{1/2} = Qm_2 \tag{29c}$$

$$\beta'_1 = Q\left(\frac{\sqrt{(1-\gamma)^2+1}+(1-\gamma)}{2}\right)^{1/2} = Qn_2 \tag{29d}$$

Particular Integral. To determine the particular integral, the load is expressed in a Fourier series

$$p' = \frac{4}{\pi} q \sum_{n=1,3\ldots} \frac{1}{n} \sin kx \qquad k = \frac{n\pi}{L}$$

where q = uniformly distributed load per square foot of shell surface. The first term of this

TABLE 16 Shell Coefficients (ϕ Measured from the Crown)*

$$F = 2\bar{R}\begin{bmatrix} (aB_1 - bB_2)\cos\beta_1\phi\cosh\alpha_1\phi - (aB_2 + bB_1)\sin\beta_1\phi\sinh\alpha_1\phi \\ (cB_3 - dB_4)\cos\beta_1'\phi\cosh\alpha_1'\phi - (cB_4 + dB_3)\sin\beta_1'\phi\sinh\alpha_1'\phi \end{bmatrix}$$

F	$\bar{R}$	B_1	B_2	B_3	B_4
M_ϕ	$-\frac{2D}{r^2}\sin kx$	$Q^2(1+\gamma)$	Q^2	$Q^2(\gamma-1)$	Q^2
M_x	$2Dk^2 \sin kx$	1	0	1	0
Q_x	$-\frac{2Dk^3}{\gamma}\cos kx$	1	1	−1	1
N_ϕ	$\frac{4Drk^4}{\gamma^2}\sin kx$	0	1	0	−1
N_x	$-\frac{4Drk^4}{\gamma^3}\sin kx$	−1	$1+\gamma$	1	$1-\gamma$
Q_x'	$-\frac{2Dk^3}{\gamma}\cos kx$	$\gamma+2$	2	$\gamma-2$	2
u	$\frac{4Drk^3}{hE\gamma^3}\cos kx$	−1	$1+\gamma$	1	$1-\gamma$
w	$2 \sin kx$	1	0	1	0

TABLE 16 Shell Coefficients (ϕ Measured from the Crown)* *(Continued)*

$$F = 2\bar{R}\begin{bmatrix} (aB_1 - bB_2)\cos\beta_1\phi\sinh\alpha_1\phi - (aB_2 + bB_1)\sin\beta_1\phi\cosh\alpha_1\phi \\ (cB_3 - dB_4)\cos\beta_1'\phi\sinh\alpha_1'\phi - (cB_4 + dB_3)\sin\beta_1'\phi\cosh\alpha_1'\phi \end{bmatrix}$$

F	$\bar{R}$	B_1	B_2	B_3	B_4
Q_ϕ	$-\dfrac{2Dk^3}{(\sqrt{\gamma})^3}\sin kx$	$m_1 - n_1$	$m_1 + n_1$	$-(m_2 + n_2)$	$m_2 - n_2$
Q_ϕ'	$-\dfrac{2Dk^3}{(\sqrt{\gamma})^3}\sin kx$	$m_1(1-\gamma) - n_1$	$m_1 + n_1(1-\gamma)$	$-m_2(1+\gamma) - n_2$	$m_2 - n_2(1+\gamma)$
$N_{x\phi}$	$\dfrac{4Drk^4}{(\sqrt{\gamma})^5}\cos kx$	$-n_1$	m_1	n_2	$-m_2$
v	$\dfrac{4Drk^3}{Eh(\sqrt{\gamma})^7}\sin kx$	$m_1 + n_1(1-\gamma)$	$n_1 - m_1(1-\gamma)$	$-m_2 + n_2(1+\gamma)$	$-n_2 - m_2(1+\gamma)$
θ†	$\sin kx$	$\dfrac{2\alpha_1}{r} + \dfrac{(\bar{R}B_1)_v}{r}$	$\dfrac{2\beta_1}{r} + \dfrac{(\bar{R}B_2)_v}{r}$	$\dfrac{2\alpha_1'}{r} + \dfrac{(\bar{R}B_3)_v}{r}$	$\dfrac{2\beta_1'}{r} + \dfrac{(\bar{R}B_4)_v}{r}$
$M_{x\phi}$	$\dfrac{2Dk}{r}\cos kx$	α_1	β_1	α_1'	β_1'

*After Ref. 30 with $\nu = 0$, so that $D = Eh^3/12$.

†Observe that one part of coefficients B_1, etc., for θ is obtained from v, that is, $(\bar{R}B_1)_v/r = [4Dk^3/Eh(\sqrt{\gamma^7}]\,[m_1 + n_1(1-\gamma)]$.

series is sufficiently dominant that good results are obtained (except possibly for short shells) by retaining only it:

$$p' = \frac{4}{\pi} q \sin \frac{\pi x}{L}$$

The particular integral is approximated quite closely by the membrane stress resultants. Thus, in Fig. 16a,

$$p_x = 0 \qquad p_\phi = -p' \sin(\phi_k - \phi) \qquad p_z = p' \cos(\phi_k - \phi)$$

and Eqs. (17) give

$$N'_\phi = -\frac{4q}{\pi} r \cos(\phi_k - \phi) \sin kx \tag{30a}$$

$$N'_{\phi x} = -\frac{4q}{\pi}\frac{2}{k} \sin(\phi_k - \phi) \cos kx \tag{30b}$$

$$N'_x = -\frac{4q}{\pi}\frac{2}{rk^2} \cos(\phi_k - \phi) \sin kx \tag{30c}$$

where $k = \pi/L$. Both $f_1(\phi)$ and $f_2(\phi)$ vanish, the former because $N'_{x\phi} = 0$ at $L/2$, the latter because $N'_x = 0$ at $s = 0$ and $x = L$.

The corresponding displacements are found by Eqs. (17):

$$u = \frac{1}{Eh}\frac{4q}{\pi}\frac{2}{rk^3}\left(1 - \frac{\nu r^2 k^2}{2}\right) \cos(\phi_k - \phi) \cos kx \tag{31a}$$

$$v = -\frac{1}{Eh}\frac{4q}{\pi}\frac{2}{r^2k^4}\left[1 + \left(2 + \frac{3\nu}{2}\right) r^2k^2\right] \sin(\phi_k - \phi) \sin kx \tag{31b}$$

$$w = \frac{2}{Eh}\frac{4q}{\pi}\frac{1}{r^2k^4}\left[1 + \left(2 + \frac{\nu}{2}\right) r^2k^2 + \frac{r^4k^4}{2}\right] \cos(\phi_k - \phi) \sin kx \tag{31c}$$

$$\theta = \frac{1}{Eh}\frac{4q}{\pi} r \left(1 - \frac{2\nu}{r^2k^2}\right) \sin(\phi_k - \phi) \sin kx \tag{31d}$$

Both $f_3(\phi)$ and $f_4(\phi)$ vanish, the former because it represents a lengthwise rigid-body translation of the shell, the latter because v is zero at each end.

Similarly, for uniform load p on the horizontal projection of the surface,

$$p' = \frac{4}{\pi} p \sin \frac{\pi x}{L}$$

$$p_x = 0 \qquad p_\phi = -p' \sin(\phi - \phi_k) \cos(\phi - \phi_k) \qquad p_z = p' \cos^2(\phi_k - \phi)$$

$$N'_\phi = -\frac{4p}{\pi} r \cos^2(\phi_k - \phi) \sin kx \tag{32a}$$

$$N'_{\phi x} = -\frac{4p}{\pi}\frac{3}{k} \sin(\phi_k - \phi) \cos(\phi_k - \phi) \cos kx \tag{32b}$$

$$N'_x = -\frac{4p}{\pi}\frac{3}{rk^2} [1 - 2\sin^2(\phi_k - \phi)] \sin kx \tag{32c}$$

and the displacements

$$u = \frac{1}{Eh}\frac{4p}{\pi}\frac{3}{rk^3}\left[1 - 2\sin^2(\phi_k - \phi) - \frac{\nu r^2k^2}{3}\cos^2(\phi_k - \phi)\right] \cos kx \tag{33a}$$

$$v = -\frac{1}{Eh}\frac{4p}{\pi}\frac{12}{r^2k^4}\left[1 + \left(\frac{1}{2} + \frac{\nu}{3}\right) r^2k^2\right] \sin(\phi_k - \phi) \cos(\phi_k - \phi) \sin kx \tag{33b}$$

$$w = \frac{1}{Eh}\frac{4p}{\pi}\frac{12}{r^4k^4}\left\{\left[1 + \left(\frac{1}{2} + \frac{\nu}{12}\right) r^2k^2\right] [1 - 2\sin^2(\phi_k - \phi)] + \frac{r^4k^4}{12}\cos^2(\phi_k - \phi)\right\} \sin kx \tag{33c}$$

$$\theta = \frac{1}{Eh}\frac{4p}{\pi}\frac{36}{r^3k^4}\left(1 + \frac{r^2k^2}{2} + \frac{r^4k^4}{18}\right) \sin(\phi_k - \phi) \cos(\phi_k - \phi) \sin kx \tag{33d}$$

For concrete ν is about ⅙ and appears to have little influence on deformations; so that Eqs. (31) and (33) are usually simplified by taking $\nu = 0$.

Solution. The complete solution is found by adding the complementary function and the particular integral as given by the membrane stress resultants. The four arbitrary constants a, b, c, and d in the function F of Table 16 are found by solving four simultaneous equations given by the boundary conditions. The boundary conditions are obtained by an appropriate choice of four among the stress resultants M_ϕ, N_ϕ, Q'_ϕ, and $N_{\phi x}$ and the displacements u, v, w, and θ. These conditions may range from those for the unsupported edge

$$M_\phi = N_\phi = Q'_\phi = N_{\phi x} = 0$$

to those for the fixed edge

$$u = v = w = \theta = 0$$

A computer program (MULEL) for the solution of these equations is available.[43] More general (finite-element) programs that can be used for the barrel shell are NASTRAN and EASE. Extensive tabulated results have been published.[27]

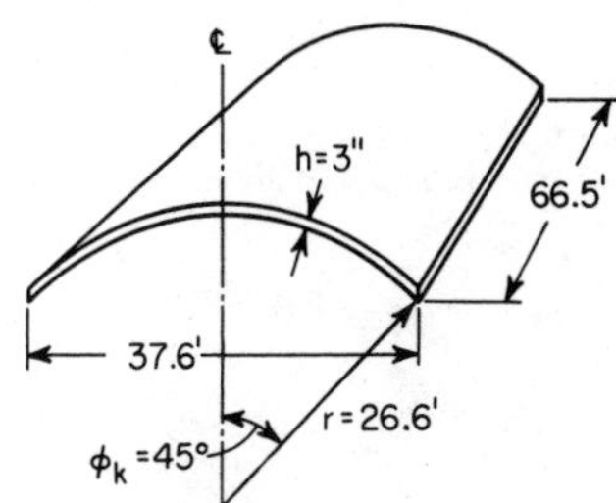

Fig. 23

16. Example The loads for the shell of Fig. 23 are:

Shell dead load	40
Roofing and mechanical equipment	10
Snow	30
	q = 80 psf

The first two are distributed essentially uniformly over the surface of the shell. Distribution of snow load is not usually specified in codes. Live load is often assumed to be distributed uniformly over the horizontal projection of the surface, but it is assumed here to be distributed in the same manner as the dead load.

The longitudinal edges are assumed to be free. Three stress resultants N_ϕ, N_x, and $N_{\phi x}$ and one stress couple M_ϕ are important in the proportioning of the shell. In the case of the shell with free edge, Q'_ϕ also is needed to establish one of the boundary conditions.

Only the first term of the Fourier series for w and p' is used, i.e., $n = 1$, so that $k = \pi/L$. Equations (30) give the membrane stress resultants

$$\begin{aligned} N'_\phi &= -2709 \cos(\phi_k - \phi) \sin kx \\ N'_{\phi x} &= -4313 \sin(\phi_k - \phi) \cos kx \\ N'_x &= -3430 \cos(\phi_k - \phi) \sin kx \end{aligned}$$

For the complementary functions,

$$\begin{aligned} kr &= \frac{\pi r}{L} = \frac{26.6\pi}{66.5} = 1.2566 \\ Q^8 &= 3 \times 1.2566^4 \times \left(\frac{26.6}{0.25}\right)^2 \\ Q &= 4.1303 \\ \gamma &= \left(\frac{1.2566}{4.1303}\right)^2 = 0.09257 \end{aligned}$$

Substituting the values of Q and γ into Eqs. (29), we get

$$\alpha_1 = 4.6854 \qquad \beta_1 = 1.8205 \qquad \alpha'_1 = 1.9437 \qquad \beta'_1 = 4.3884$$

The required coefficients B_1, B_2, B_3, and B_4 are, from Table 16,

	B_1	B_2	B_3	B_4
M_ϕ	18.6384	17.0593	−15.4801	17.0593
N_ϕ	0	1	0	−1
N_x	−1	1.0926	1	0.9074
$N_{\phi x}$	−0.4408	1.1344	1.0625	−0.4706
Q_ϕ'	0.5886	1.5344	−1.5766	−0.6903

Because ϕ is measured from the crown in Table 16, while in Eqs. (30) to (33) it is measured from the edge, it must be taken negative in evaluating the functions in Table 16. It will be noted that this results in no changes in F for M_ϕ to w, inclusive, but changes the sign of F for Q_ϕ to $M_{x\phi}$, inclusive.

Values of cos $\beta_1\phi$ cosh $\alpha_1\phi$, etc., must be determined for the values of ϕ at which the internal forces are to be investigated. Only those at the edge, where $\phi = \phi_k = -45°$, are given here:

$$\cos \beta_1\phi_k \cosh \alpha_1\phi_k = 2.7870 \qquad \cos \beta_1\phi_k \sinh \alpha_1\phi_k = -2.7835$$
$$\sin \beta_1\phi_k \sinh \alpha_1\phi_k = 19.6115 \qquad \sin \beta_1\phi_k \cosh \alpha_1\phi_k = -19.6365$$
$$\cos \beta_1'\phi_k \cosh \alpha_1'\phi_k = -2.2986 \qquad \cos \beta_1'\phi_k \sinh \alpha_1'\phi_k = 2.0913$$
$$\sin \beta_1'\phi_k \sinh \alpha_1'\phi_k = -0.6585 \qquad \sin \beta_1'\phi_k \cosh \alpha_1'\phi_k = 0.7238$$

With these values and the values of B, the required edge stress resultants are determined from Table 16, to which must be added the membrane stress resultants (evaluated at the edge). With the boundary conditions

$$M_{\phi k} = N_{\phi k} = Q'_{\phi k} = N_{\phi kx} = 0$$

we get

$$M_{\phi k} = -2D(-0.7988a - 1.1676b + 0.1323c + 0.08202d) \sin kx = 0$$
$$N_{\phi k} = -2D(1.2130a + 0.1724b + 0.04073c + 0.1422d) \sin kx - 1915 \sin kx = 0$$
$$N_{\phi kx} = -2D(-4.7777a + 1.1176b - 0.5209c - 0.04374d) \cos kx - 3050 \cos kx = 0$$
$$Q'_{\phi k} = 2D(-0.2133a - 0.1185b + 0.02095c - 0.01935d) \sin kx = 0$$

from which

$$a = \frac{1177}{D} \qquad b = \frac{-2989}{D} \qquad c = \frac{-13{,}504}{D} \qquad d = \frac{-9292}{D}$$

Computation of Forces. With a, b, c, and d known, values of M_ϕ, N_ϕ, N_x, and $N_{\phi x}$ can now be determined throughout the shell. Thus, from Table 16,

$$M_\phi = -\frac{4}{26.6^2}(72{,}927 \cos \beta_1\phi \cosh \alpha_1\phi + 35{,}631 \sin \beta_1\phi \sinh \alpha_1\phi + 367{,}560 \cos \beta_1'\phi \cosh \alpha_1'\phi + 86{,}528 \sin \beta_1'\phi \sinh \alpha_1'\phi) \sin kx$$

N_ϕ, N_x, and $N_{\phi x}$ are found similarly. The results are given in Table 17, where M_ϕ, N_ϕ, and N_x are given for $x = L/2$ and $N_{\phi x}$ for $x = 0$. Values for any other value of x are obtained by multiplying by sin kx or cos kx, as the case may be.

TABLE 17 Stress Resultants and Couples for Barrel of Fig. 23

ϕ, deg from edge	M_ϕ, ft-kips/ft, $x = L/2$	N_ϕ, kips/ft, $x = L/2$	N_x, kips/ft, $x = L/2$	$N_{\phi x}$, kips/ft, $x = 0$
45	−2.49	−3.49	+0.57	0
40	−2.44	−3.50	−1.47	−0.01
30	−2.02	−3.45	−14.63	−1.63
20	−1.15	−2.80	−24.33	−6.21
10	−0.16	−1.18	+1.09	−9.77
0	0	0	+105.25	0

17. Shell with Edge Beams Vertical edge beams (Fig. 24) are usually used for long shells, where the principal structural action is longitudinal bending. Horizontal beams are commonly used with short shells, where the principal action is transverse arching.

If the vertical edge beam is slender, it is reasonable to assume that it offers negligible resistance to rotation and horizontal displacement, so that $M_{\phi k} = 0$ and $H_b = 0$ become two boundary conditions. The third boundary condition is found by equating the vertical displacement of the shell edge to the corresponding vertical deflection of the edge beam. The fourth condition is obtained from compatibility of edge displacements u_k of the shell and the corresponding displacements of the edge beam or, what amounts to the same thing, equality of edge stress in the shell and the corresponding stress in the beam.

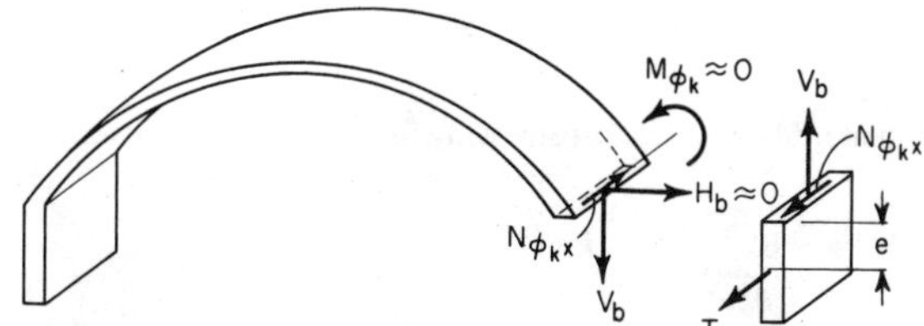

Fig. 24

The vertical deflection of the shell edge (positive upward) is given by

$$\Delta_{vs} = v_k \sin \phi_k - w_k \cos \phi_k$$

where v_k and w_k are found by adding the membrane displacements [Eq. (31) or (33), depending upon the distribution of load] to those of Table 16.

The reaction components V_b and H_b are given by

$$V_b = N_{\phi k} \sin \phi_k - Q'_{\phi k} \cos \phi_k \tag{34}$$

$$H_b = N_{\phi k} \cos \phi_k + Q'_{\phi k} \sin \phi_k \tag{35}$$

Note that $N_{\phi k}$ and $Q'_{\phi k}$, and therefore V_b and H_b, vary sinusoidally from $x = 0$ to $x = L$.

The axial tension T (Fig. 24) at any section of the beam is

$$T = -\int_0^x N_{\phi kx} dx \tag{36}$$

Because $N_{\phi kx}$ varies as $\cos kx$, T varies as $\sin kx$.

The bending moments and deflections of the beam due to V_b and T are found by successive integrations of $EId^4y/dx^4 = w$. The combined effect is

$$M = \frac{V_b}{k^2} + Te \tag{37}$$

$$\Delta_{vb} = \frac{1}{k^2EI}\left(\frac{V_b}{k^2} + Te\right) \tag{38}$$

where e is the eccentricity of T. The bending moment M is considered positive with the top fiber in tension and Δ is positive upward.

The stress at the top fiber of the beam is

$$f = \frac{T}{A} + \frac{V_b/k^2 + Te}{Z} \tag{39}$$

Using the above relations, the four boundary conditions become

$$M_{\phi k} = 0 \tag{40a}$$

$$H_b = 0 \tag{40b}$$

$$v_k \sin \phi_k - w_k \cos \phi_k = \frac{1}{k^3EI}\left(\frac{V_b}{k} - eN_{\phi kx}\right) \tag{40c}$$

$$\frac{N_{xk}}{h} = \frac{1}{kZ}\left(\frac{V_b}{k} - eN_{\phi kx}\right) - \frac{N_{\phi kx}}{kA} \tag{40d}$$

It should be noted that the factor $\sin kx$ is common to every term in Eq. (40). Therefore, when v, w, and the various resultants are taken from Table 16 and Eq. (30), (31), and/or (32), (33), the trigonometric term should be omitted.

Following the solution of Eq. (40) for the constants a, b, c, and d, computation of forces and moments throughout the shell is carried out as in Art. 16.

The computer program MULEL can be used to solve the equations for the shell with edge beams.[43] Also, an analysis that requires the solution of only two simultaneous equations (involving V and H) can be made with the use of tables.[20,27]

Prestressed Edge Beam. The bending moment and deflection caused by a prestressing force F in a tendon draped to a parabolic curve passing through the centroids at each end of the edge beam can be computed by considering the equivalent uniform load

$$w = \frac{8Fe_c}{L^2}$$

where e_c = eccentricity of tendon at midspan. The prestressing force must be represented by a Fourier series

$$F_x = \frac{4}{\pi}F \sum_{n=1,3\ldots} \frac{1}{n} \sin kx$$

and the moment and deflection determined by successive integration of $EId^4y/dx^4 = w$.

With the effects of F added to Eq. (40), the four boundary conditions for the prestressed edge beam are

$$M_{\phi k} = 0 \tag{41a}$$

$$H_b = 0 \tag{41b}$$

$$v_k \sin \phi_k - w_k \cos \phi_k = \frac{1}{k^3EI}\left(\frac{V_b}{k} - eN_{\phi kx} + \frac{4}{\pi}\frac{8Fe_c}{kL^2}\right) \tag{41c}$$

$$\frac{N_{xk}}{h} = \frac{1}{kZ}\left(\frac{V_b}{k} - eN_{\phi kx} + \frac{4}{\pi}\frac{8Fe_c}{kL^2}\right) - \frac{N_{\phi kx}}{kA} - \frac{4}{\pi}\frac{F}{A} \tag{41d}$$

The factor $\sin kx$ is omitted in the terms involving F because it is common to all other terms, as was noted in connection with Eq. (40).

18. Transverse Frames Figure 25 shows an arch rib loaded by the in-plane forces $N_{x\phi}$ and the radial shears Q_x. The latter are so small that they may be neglected, so that the rib may be analyzed as an arch for the forces $N_{x\phi}$ alone. Tables[31] or standard computer programs simplify the analysis.

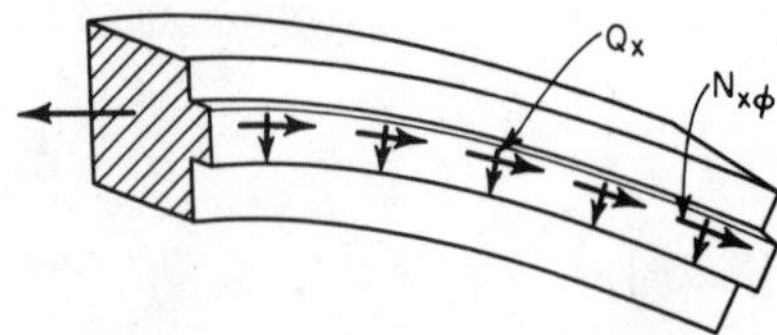

Fig. 25

19. Barrel-Shell Reinforcement Figure 26 shows a layout of reinforcement for the barrel shell of Fig. 27. Reinforcement is provided to take all the principal tensile stresses, and may be placed either in the direction of the stress trajectories or in two directions, usually orthogonal. Principal stresses are computed from the values of the stress resultants N_ϕ and N_x. Stress trajectories for the barrel of Fig. 27 are shown in Fig. 28.

At the corners, where the shear is a maximum, reinforcement is generally placed at 45°. Transverse (hoop) reinforcement (No. 3 bars in Fig. 26) is based on M_ϕ, usually neglecting N_ϕ. If the torsional and lateral stiffnesses of the edge beam are neglected in the analyses, $M_\phi = 0$ at the edge. However, it is advisable to provide reinforcement for some positive moment caused by edge-beam stiffness. This can be done by placing the principal-tension reinforcement in this region near the underside of the shell (the No. 4 bars at 11 in. in Fig. 26).

Longitudinal reinforcement at the juncture of the shell and the transverse rib is usually based on the assumption that the rib permits no radial movement and no rotation of the shell. This produces a moment $M_x = -0.29hN_\phi$. At $x = 0$, $N_\phi = 0$ because of the

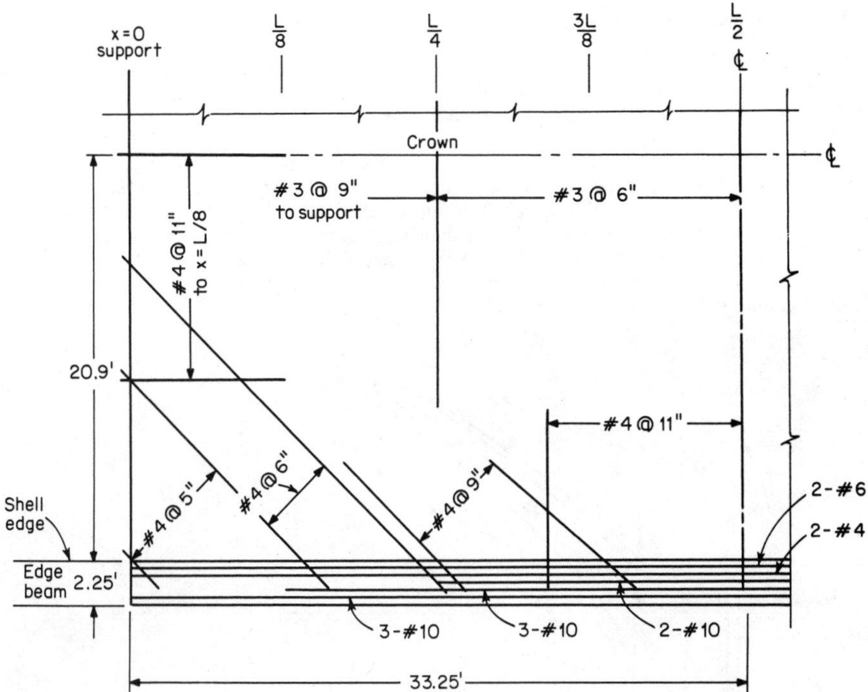

Fig. 26 Developed plan of shell reinforcement (welded wire fabric not shown).

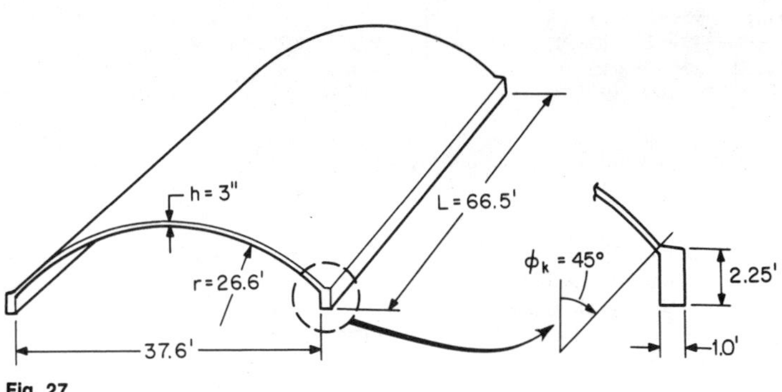

Fig. 27

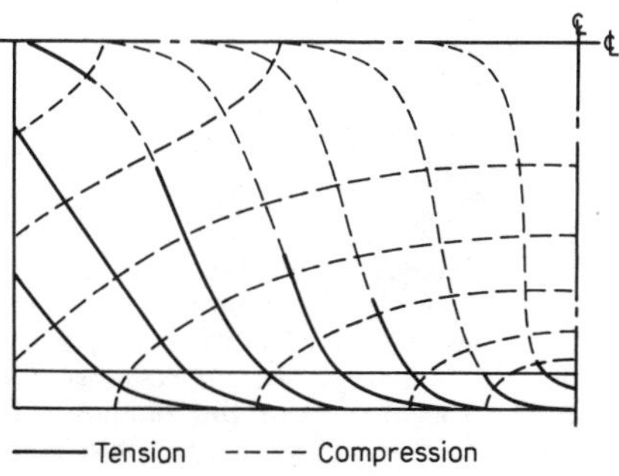

Fig. 28 Stress trajectories for shell and edge beams.

assumption of sinusoidal loading, but since the actual loading is uniform, N_ϕ must be constant along the span. Therefore, the value of N_ϕ at $x = L/2$ should be used to determine M_x at $x = 0$. An allowance for additional moment resulting from arch deflection of the rib can be made.[20] The No. 4 bars at 11 in. in Fig. 26 are proportioned on this basis.

Minimum reinforcement of 0.35 percent in each of two directions in tensile zones and 0.18 percent in other zones, spaced no farther apart than five times the shell thickness, is recommended. This can be supplied in the form of fabric.

Edge-beam reinforcement is usually sized for the tensile force in the beam. This can be determined by using Simpson's rule to evaluate Eq. (36), or by computing the stresses at the top and bottom of the beam. Tests show that it is advisable to place most of the reinforcement near the bottom (Fig. 29).

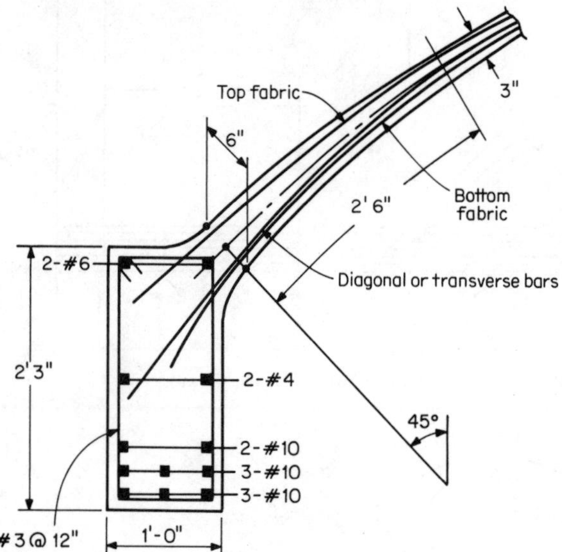

Fig. 29 Section at edge beam.

FOLDED PLATES

Figure 30 shows typical folded-plate cross sections. Typical dimensions for folded-plate roof structures are given in Table 18. These structures have no curvature but can be considered as approximations of shell geometry; i.e., the shape of Fig. 30*a* approximates a barrel shell with vertical edge beams and the shape in Fig. 30*c* a multiple-barrel shell. An

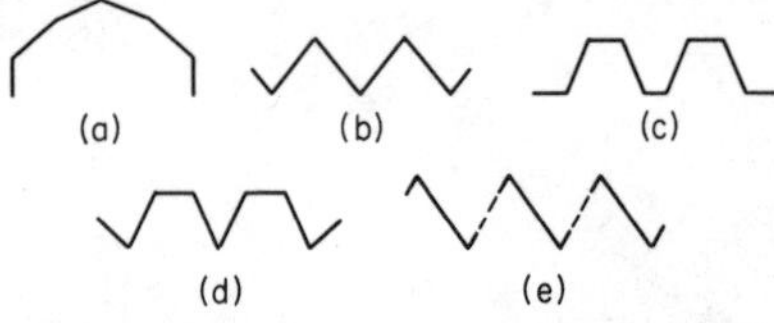

Fig. 30

important difference between folded plates and barrel shells is the difference in the transverse bending moments M_ϕ (Fig. 31), which results from the plates acting as one-way slabs to carry roof loads transversely to the ridges by bending. The entire cross section then carries the loads longitudinally as a deep beam, with high in-plane tension stresses at the bottom and in-plane compression stresses near the top. Because of the longitudinal bending, the ridges deflect. Therefore, the one-way slabs rest on yielding supports and

Table 18 Typical Dimensions for Folded Plates*

	Span, ft	Bay† width, ft	*D*, ft-in.		*h*, in.‡	Rein-forcing§
			Max	Min		
Two-segment plate, 25°< φ < 45° 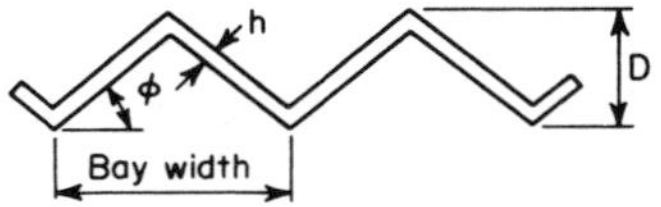	40	15	4-0	2-9	4	1.2–1.6
	60	20	6-0	4-0	4–6	1.9–2.7
	75	25	7-6	5-0	4–6	2.6–3.7
	100	30	10-0	6-9	5–6	4.0–5.2
Four-segment plate 30°< φ < 45° 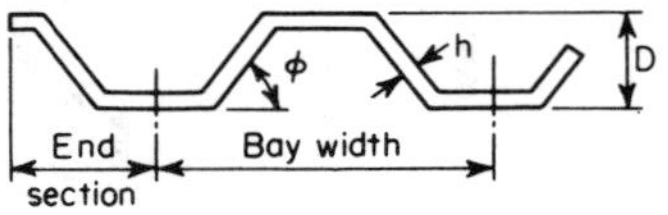	40	20	5-0	2-6	3	1.5–2.0
	60	25	6-0	4-0	3–3.5	2.0–3.0
	75	30	7-7	5-0	3–4	2.5–4.0
	100	40	10-0	6-6	4–5	4.0–6.0

* From Portland Cement Association.
† Varies with design.
‡ Average thickness.
§ In psf of projected area.
Usual ratio of depth to span varies from 1:10 to 1:15. Cantilevers can help counterbalance the span.

their bending moments depart radically from those that would be obtained by assuming the ridges unyielding as in Fig. 31.

20. Analysis of Folded Plates* The classical method of analysis of folded plates is based upon a simplification similar to the beam-arch method for long barrels.[32] A number of computer programs, such as MULEL, MULTPL, MUPDI (direct stiffness harmonic analyses), and FINPLA (a finite-element analysis) are available.[43]

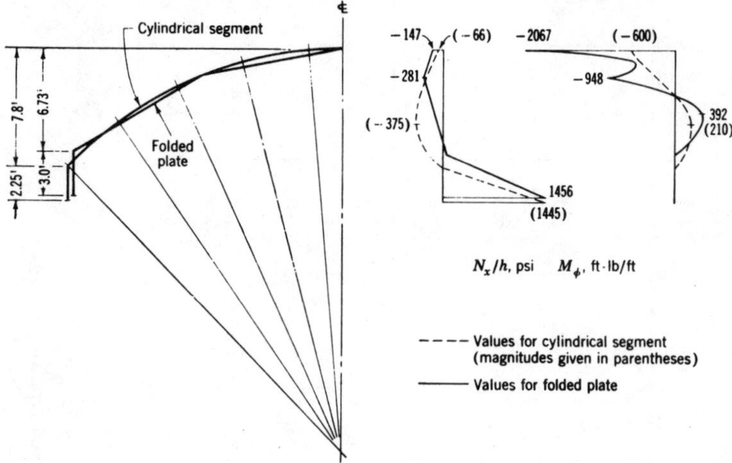

Fig. 31

The system is considered in two parts, (1) a continuous one-way slab spanning transversely between joints, and (2) a series of simple beams spanning longitudinally between end diaphragms, ribs, or trusses, etc.

1. *The primary system* consists of the plates, considered hinged at the joints, loaded with the reactions R from the continuous one-way slab analysis (Fig. 32). The resulting in-plane plate loads are given by

$$P_n = -R_{n-1}\frac{\cos\phi_{n-1}}{\sin\alpha_{n-1}} + R_n\frac{\cos\phi_{n+1}}{\sin\alpha_n} \tag{42}$$

R_n is positive downward, P_n is positive in the direction from joint n to joint $n-1$, and ϕ is positive counterclockwise from the horizontal at joint $n-1$. Furthermore,

$$\alpha_n = \phi_n - \phi_{n+1} \tag{43}$$

The uniformly distributed loads P_n produce stresses in the beam of

$$f = \pm\frac{M_n}{Z_n} = \pm\frac{3P_nL^2}{4h_nd_n^2} \tag{44}$$

if the edges are free to slide longitudinally at the joints. Tension is positive. These stresses ordinarily will not be equal in the plates common to a given joint. Since they must be equal if the plates are monolithic, adjustments are required. The differences can be removed by a method of stress distribution in which, for plates of constant thickness,

$$\text{Stress stiffness factor} = \frac{4}{A_n} = \frac{4}{h_nd_n}$$

$$\text{Carryover factor} = -\tfrac{1}{2}$$

The free-edge stresses correspond to the fixed-end moments of moment distribution.

*A review of various methods of analysis and a comprehensive bibliography are given in Ref. 32.

2. *The errors* arise from relative rotation of the plates common to a joint, because of the in-plane deflections of the plates. With the edge stresses due to the loading P_n, the plate beam deflections y (Fig. 32c) are given by

$$y_n = (f_{n-1,n} - f_{n,n-1}) \frac{5L^2}{48d_nE} \tag{45}$$

where y_n is positive in the direction of positive P_n, and $f_{n-1,n}$ denotes the stress at joint $n - 1$ in the plate bounded by joints $n - 1$ and n (plate n). The rotation of each plate is given by

$$\delta_n^B = -\frac{1}{d_n}\left[\frac{y_{n-1}}{\sin \alpha_{n-1}} - y_n(\cot \alpha_{n-1} + \cot \alpha_n) + \frac{y_{n+1}}{\sin \alpha_n}\right] \tag{46}$$

where a positive rotation is taken as clockwise. The errors at any joint will then be

$$\delta_{n0} = \delta_n^B - \delta_{n+1}^B \tag{47}$$

The rotations given by Eqs. (46) and (47) occur at midspan and reduce to zero at the supports.

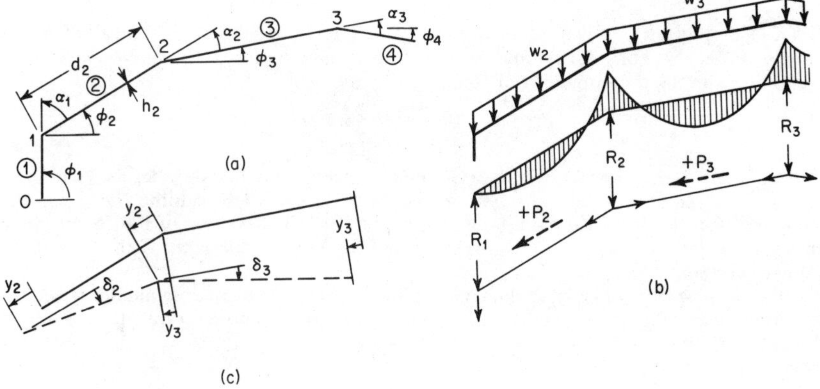

Fig. 32

3. *The corrections* consist of transverse joint moments, which must have the same longitudinal variation as the errors. Such a variation is closely approximated by a sine wave. A correction moment $X_n = 1$ at joint n produces plate rotations which consist of the sum of the rotation δ^S due to the action of the plate as a slab and the relative rotations δ^B of the plates due to their in-plane beam deflections. The rotations δ^S are given by

$$\delta_{n,n}^S = \frac{4d_n}{Eh_n^3} + \frac{4d_{n+1}}{Eh_{n+1}^3} \tag{48a}$$

$$\delta_{n-1,n}^S = \frac{2d_n}{Eh_n^3} \tag{48b}$$

$$\delta_{n+1,n}^S = \frac{2d_{n+1}}{Eh_{n+1}^3} \tag{48c}$$

The rotations δ^B are found by computing the joint reactions

$$R_n = +\frac{1}{d_n \cos \phi_n} + \frac{1}{d_{n+1} \cos \phi_{n+1}} \tag{49a}$$

$$R_{n-1} = -\frac{1}{d_n \cos \phi_n} \tag{49b}$$

$$R_{n+1} = -\frac{1}{d_{n+1} \cos \phi_{n+1}} \tag{49c}$$

from which the corresponding plate loads are found from Eq. (42). The stresses f are given by

$$f = \pm \frac{6P_nL^2}{\pi^2 h_n d_n^2} \tag{50}$$

As is the case in the primary system, these stresses ordinarily will not be equal in the plates common to a given joint. Corrections by the stress-distribution procedure mentioned previously are required.

The deflections y are found from the corrected stresses f by

$$y_n = (f_{n-1,n} - f_{n,n-1})\frac{L^2}{\pi^2 d_n E} \tag{51}$$

Equations (50) and (51) differ from the corresponding equations (44) and (45) only because of the assumed sinusoidal variation of the correction moment.

With y known, the rotations δ^B are found from Eq. (46). The total correction at joint n is then

$$\delta_{nn} = \delta_{nn}^S + \delta_{nn}^B \tag{52}$$

in which

$$\delta_{nn}^B = \delta_n^B - \delta_{n+1}^B \tag{53}$$

4. *Compatibility of displacements* is obtained in the usual way for statically indeterminate structures. Thus, for symmetrical loading of the structure of Fig. 32a, and assuming joint 1 simple, the equations of compatibility are

$$\delta_{22}X_2 + \delta_{23}X_3 + \delta_{20} = 0$$
$$\delta_{32}X_2 + \delta_{33}X_3 + \delta_{30} = 0$$

where X_2 and X_3 are the statically indeterminate correction moments at joints 2 and 3.

5. *Final values* of the transverse moments are obtained by adding the correction moments X to those of the continuous slab analysis. Final values of the longitudinal stresses are found by adding the stresses f resulting from the moments X to those of the primary system.

6. The significant results of the analysis are the transverse bending moments and the longitudinal principal stresses. The latter are obtained by combining the stresses f with the shearing stresses v given by

$$v = \frac{4N}{h_n L}\left(1 - \frac{2x}{L}\right) \tag{54}$$

where, at the joints,

$$N_n = N_{n-1} - \frac{h_n d_n}{2}(f_{n-1,n} + f_{n,n-1}) \tag{55}$$

and, midway between joints,

$$N_{d_n/2} = \frac{N_{n-1} + N_n}{2} - \frac{h_n d_n}{8}(f_{n-1,n} - f_{n,n-1}) \tag{56}$$

Equation (54) is based on uniformly distributed loading. An equation corresponding to the assumed sine distribution of the correction loads can be derived, but since the correction loads are usually small relative to the loads on the primary system, it is sufficient for practical purposes to use Eq. (54) for both.

21. Example* The folded-plate system of Fig. 33 will be analyzed for a uniformly distributed load consisting of

Live load	20
Roofing, insulation, etc.	10
Plate at 4 in.	50
	80 psf

Table 19 gives the results of the one-way slab analysis, from which $R_1 = 535$, $R_2 = 908$, $R_3 = 764$ plf.

*From Ref. 20.

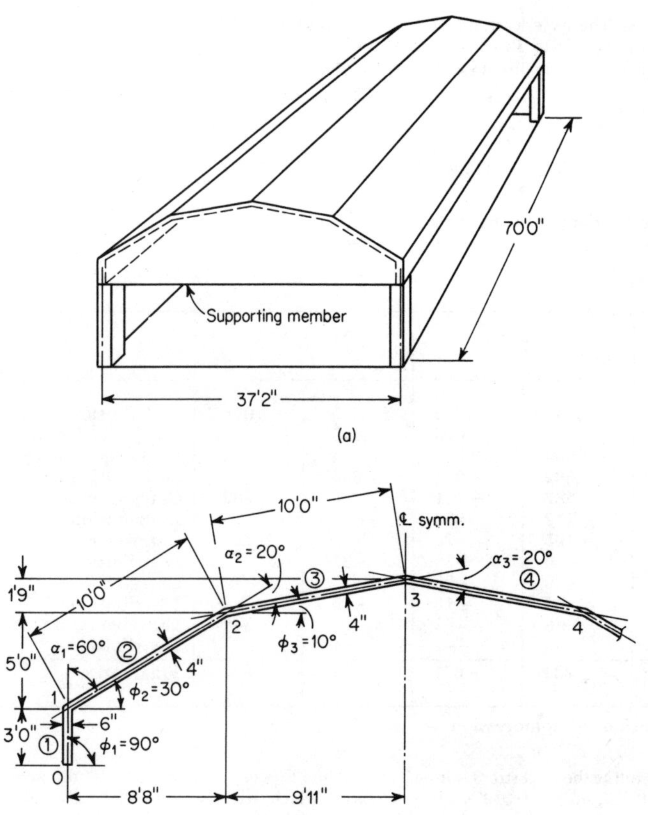

Fig. 33

TABLE 19 Elementary Analysis of One-Way Slab at Midspan

0	1		2		3 Joint
1	2		3		Plate
	1	0.428	0.572	0.500	Distribution factor
		867	−656	656	Fixed-end moment, ft-lb/ft
0	0	−90	−121		Distribute
				−60	Carryover
0	0	777	−777	596	Final moment, ft-lb/ft
0	−90	90	18	−18	$M/(d_n \cos \phi_n)$, lb/ft
225*	400	400	400	400	$wd_n/2$, lb/ft
225	310	490	418	382	Total shear, lb/ft
535		908		764	Joint reaction, lb/ft

* Weight of edge beam = wd_1.

1. PRIMARY SYSTEM. The plate loads are obtained by substituting the joint reactions R into Eq. (42), which gives $P_1 = 535$, $P_2 = 2615$, $P_3 = -97$ plf. Because of symmetry, only half the system need be considered. The resulting free-edge stresses are, from Eq. (44),

$$\begin{aligned} f_{01} &= -f_{10} = +3034 \text{ psi} \\ f_{12} &= -f_{21} = +2003 \\ f_{23} &= -f_{32} = -74 \end{aligned}$$

Distribution to balance these stresses is shown in Table 20.

TABLE 20 Stress Distribution for Elementary Analysis

0		1		2		3 Joint
1		2		3		Plate
0	0.69	0.31	0.5	0.5	0	Distribution factor
	−0.5	−0.5	−0.5	−0.5		Carryover factor
+3,034	−3,034	+2,003	−2,003	−74	+74	Free-edge stress*
	+3,475	−1,562	+964	−965		Distribution
−1,738		−482	+781		+482	Carryover
	−333	+149	−390	+391		Distribution
+166		+195	−75		−195	Carryover
	+135	−60	+38	−37		Distribution
−67		−19	+30		+19	Carryover
	−14	+5	−15	+15		Distribution
+7		+8	−2		−8	Carryover
	+5	−3	+1	−1		Distribution
+1,402	+234	+234	−671	−671	+372	Final stress*

* Stress in psi, tension +, compression −.

2. ERRORS. Substituting the final stresses from Table 20 into Eqs. (45), with $E = 2 \times 10^6$ psi, gives $y_{10} = +1.193$, $y_{20} = +0.277$, $y_{30} = -0.319$, $y_{40} = +0.319$ in., which, when substituted into Eq. (46), gives

$$\begin{aligned} \delta_2^B &= -\frac{1}{120}\left[\frac{1.193}{\sin 60°} - 0.277(\cot 60° + \cot 20°) + \frac{-0.319}{\sin 20°}\right] \\ &= 0.398 \times 10^{-2} \\ \delta_3^B &= -\frac{1}{120}\left[\frac{0.277}{\sin 20°} - (-0.319)(\cot 20° + \cot 20°) + \frac{0.319}{\sin 20°}\right] \\ &= -2.914 \times 10^{-2} \end{aligned}$$

Then, from Eq. (47),

$$\begin{aligned} \delta_{20} &= +0.398 \times 10^{-2} - (-2.914 \times 10^{-2}) = +3.312 \times 10^{-2} \\ \delta_{30} &= -2.914 \times 10^{-2} \end{aligned}$$

Note that, because of symmetry, joint 3 does not rotate, so that the error is only the rotation due to plate 3.

3. The *corrections* are first obtained for a unit moment at joint 2. From Eq. (48),

$$\begin{aligned} \delta_{22}^S &= 2\frac{4 \times 10}{2 \times 10^6 \times 144 \times (1/3)^3} = +0.75 \times 10^{-2} \\ \delta_{32}^S &= \frac{2 \times 10}{2 \times 10^6 \times 144 \times (1/3)^3} = +0.1875 \times 10^{-2} \end{aligned}$$

The joint reactions from Eq. (49) are $R_1 = -115$, $R_2 = +217$, $R_3 = -203$ plf, and from Eq. (42), the plate loads are $P_1 = -115$, $P_2 = +625$, $P_3 = -1138$ plf. The free-edge stresses, from Eq. (50), are

$$\begin{aligned} f_{01} &= -f_{10} = -530 \text{ psi} \\ f_{12} &= -f_{21} = +388 \\ f_{23} &= -f_{32} = -704 \end{aligned}$$

Table 21 shows the stress distribution. The plate deflections are obtained by substituting the final stresses from this table into Eq. (51) to give $y_{11} = -0.985$, $y_{21} = +0.308$, $y_{31} = -0.351$ in. The plate rotations are found from Eq. (46):

$$\delta_2^B = -\frac{1}{120}\left[\frac{-0.985}{\sin 60°} - 0.308(\cot 60° + \cot 20°) + \frac{-0.351}{\sin 20°}\right]$$
$$= 2.65 \times 10^{-2}$$
$$\delta_3^B = -\frac{1}{120}\left[\frac{0.308}{\sin 20°} - (-0.351)(\cot 20° + \cot 20°) + \frac{0.351}{\sin 20°}\right]$$
$$= -3.212 \times 10^{-2}$$

The joint rotations are, from Eq. (53),

$$\delta_{22}^B = +2.65 \times 10^{-2} - (-3.212 \times 10^{-2}) = 5.862 \times 10^{-2}$$
$$\delta_{32}^B = -3.212 \times 10^{-2}$$

The total joint rotations are, from Eq. (52),

$$\delta_{22} = +0.75 \times 10^{-2} + 5.862 \times 10^{-2} = +6.612 \times 10^{-2}$$
$$\delta_{32} = +0.188 \times 10^{-2} - 3.212 \times 10^{-2} = -3.024 \times 10^{-2}$$

The same order of computations for a unit moment at joint 3 yields

$$\delta_{33}^S = +0.375 \times 10^{-2} \qquad \delta_{23}^S = +0.1875 \times 10^{-2}$$
$$R_1 = 0 \qquad R_2 = -102 \qquad R_3 = +203 \text{ plf}$$
$$P_1 = 0 \qquad P_2 = -292 \qquad P_3 = +842 \text{ plf}$$
$$f_{01} = -f_{10} = 0 \qquad f_{12} = -f_{21} = -182 \qquad f_{23} = -f_{32} = +522 \text{ psi}$$

The distribution to balance these stresses is shown in Table 22. From the final stresses in this table $y_{12} = +0.285$, $y_{22} = -0.155$, $y_{32} = +0.225$ in. The plate rotations are

$$\delta_2^B = -1.25 \times 10^{-2} \qquad \delta_3^B = +1.957 \times 10^{-2}$$

and the joint rotations are

$$\delta_{33}^B = 1.957 \times 10^{-2} \qquad \delta_{23}^B = -3.207 \times 10^{-2}$$

so that the total joint rotations are

$$\delta_{33} = 2.332 \times 10^{-2} \qquad \delta_{23} = -3.019 \times 10^{-2}$$

4. The equations of compatibility are

$$6.612X_2 - 3.02X_3 + 3.312 = 0$$
$$-3.02X_2 + 2.332X_3 - 2.914 = 0$$

from which $X_2 = 0.172$, $X_3 = 1.472$ ft-kips/ft.

TABLE 21 Stress Distribution for Correction Moment $X_2 = 1$

0	1	1	2	2	3 Joint	
1		2		3		Plate
0	0.69	0.31	0.5	0.5	0	Distribution factor
	−0.5	−0.5	−0.5	−0.5		Carryover factor
−530.4	+530.4	+387.8	−387.7	−704.0	+704.0	Free-edge stress*
	−98.5	+44.2	−158.1	+158.2		Distribution
+49.2		+79.0	−22.1		−79.1	Carryover
	+54.5	−24.5	+11.1	−11.0		Distribution
−27.2		−5.5	+12.2		+5.6	Carryover
	−3.8	+1.7	−6.1	+6.1		Distribution
+1.9		+3.0	−0.9		−3.0	Carryover
	+2.1	−0.9	+0.4	−0.5		Distribution
−1.0		−0.2	+0.4		+0.2	Carryover
	−0.2		−0.2	+0.2		Distribution
−507.5	+484.5	+484.5	−551.0	−551.0	+627.7	Final stress*

* Stress in psi, tension +, compression −.

These corrections are entered in Table 23 to determine the final values of the transverse bending moments. Final values of the longitudinal stresses are computed in Table 24. The uncorrected stresses are from Table 20, while the corrections are stresses from Tables 21 and 22 proportional to the respective correction moments. The final stresses in this table enable the longitudinal shears to be determined from Eqs. (55) and (56). These are entered in Table 25, together with the corresponding stresses from Eq. (54).

Principal stresses throughout the system are given in Table 26. The stress trajectories are plotted in Fig. 34.

TABLE 22 Stress Distribution for Correction Moment X_3 = 1

0	1		2		3	Joint
1		2		3		Plate
0	0.69	0.31	0.5	0.5	0	Distribution factor
	−0.5	−0.5	−0.5	−0.5		Carryover factor
		−181.5	+181.5	+521.5	−521.5	Free-edge stress*
	−125.2	+56.3	+170.0	+170.0		Distribution
+62.6		−85.0	−28.2		+85.0	Carryover
	−58.6	+26.4	+14.1	−14.1		Distribution
+29.3		−7.0	−13.2		+7.0	Carryover
	−4.8	+2.2	+6.6	−6.6		Distribution
+2.4		−3.3	−1.1		+3.3	Carryover
	−2.3	+1.0	+0.5	−0.6		Distribution
+1.2		−0.3	−0.5		+0.3	Carryover
	−0.2	+0.1	+0.3	−0.2		Distribution
+95.5	−191.1	−191.1	+330.0	+330.0	−425.9	Final stress*

* Stresses in psi, tension +, compression −.

TABLE 23 Transverse Bending Moments at Midspan, Ft-lb/Ft

	Joint 1	Plate 2	Joint 2	Plate 3	Joint 3
Elementary analysis.........	0	+478	−777	+299	−596
Correction..................	0	−86	−172	−822	−1,472
Final value.................	0	+392	−949	−523	−2,068

TABLE 24 Longitudinal Stresses at Midspan*

Joints	0	1	2	3
Elementary analysis........	0	0	0	0
Uncorrected stresses........	+1,402	+234	−671	+372
Corrections:				
X_2 = 0.172 ft-kip........	−87	+83	−95	+108
X_3 = 1.472 ft-kips.......	+141	−281	+485	−627
Final stress..............	+1,456	+36	−281	−147

* Stresses in psi, tension +, compression −.

Reinforcement The pattern of reinforcement is shown in Fig. 35. The requirements are based on the principal stresses of Table 26. Along the support, starting at joint 1 in plate 1 and proceeding to the longitudinal centerline, the required reinforcement at 45° is

Plate 1, Joint 1:

$$A_s = \frac{129 \times 6 \times 12}{20{,}000} = 0.463 \text{ in.}^2/\text{ft}$$

Plate 2:

Joint 1: $A_s = \dfrac{193 \times 4 \times 12}{20{,}000} = 0.463$ in.²/ft, No. 4 at 5 in.

Middle: $A_s = \dfrac{179 \times 48}{20{,}000} = 0.43$ in.²/ft, No. 4 at 5 in.

Joint 2: $A_s = \dfrac{123 \times 48}{20{,}000} = 0.295$ in.²/ft, No. 4 at 8 in.

Plate 3:

Middle: $A_s = \dfrac{51 \times 48}{20{,}000} = 0.122$ in.²/ft, No. 3 at 10 in.

Joint 3: $A_s = 0$

Next, the steel required along joint 1 is computed. The angle of the line of principal stress is always 45° ± 15° to very near midspan, so that we may place the steel at 45°.

At support: $A_s = 0.463$ in.²/ft, No. 4 at 5 in.

At $L/8$: $A_s = \dfrac{153 \times 48}{20{,}000} = 0.368$ in.²/ft, No. 4 at 7 in.

At $L/4$: $A_s = \dfrac{111 \times 48}{20{,}000} = 0.267$ in.²/ft, No. 4 at 9 in.

At $3L/8$: $A_s = \dfrac{68 \times 48}{20{,}000} = 0.164$ in.²/ft, No. 4 at 15 in.

TABLE 25 Longitudinal Shearing Stresses at Midspan

Location	0	01*	1		12	2	23	3
Final shear force, kips......	0	−119	−161		−150	−102	−43	0
Final shear stress, psi.......	0	−95	−128†	−192‡	−179	−121	−51	0

* Denotes point midway between joints 0 and 1.
† Stress in plate 1 (h = 6 in.).
‡ Stress in plate 2 (h = 4 in.).

TABLE 26 Principal Stresses, psi, Tension +, Compression −

Plate	Location	$x = 0$	$L/8$	$L/4$	$3L/8$	$L/2$
3	Joint 3	0	0	0	0	0
		0	−64	−110	−138	−147
	Middepth	+51	+13	+4	+1	0
		−51	−107	−164	−199	−214
	Joint 2	+123	+45	+17	+4	0
		−123	−167	−227	−268	−281
2	Joint 2	+123	+45	+17	+4	0
		−123	−167	−227	−268	−281
	Middepth	+179	+110	+56	+16	0
		−179	−164	−146	−130	−122
	Joint 1	+193	+153	+111	+68	36
		−193	−137	−83	−34	0
1	Joint 1	+129	+105	+79	+53	+36
		−129	−89	−51	−19	0
	Middepth	+95	+341	+564	+700	+745
		−95	−15	−4	0	0
	Joint 0	0	+635	+1,090	+1,362	+1,456
		0	0	0	0	0

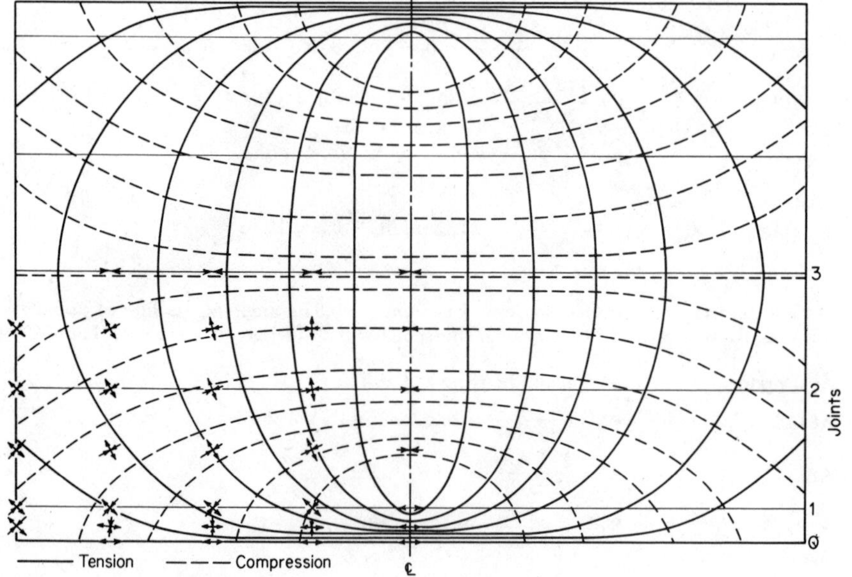

Fig. 34 Stress trajectories in developed folded plate of Fig. 33.

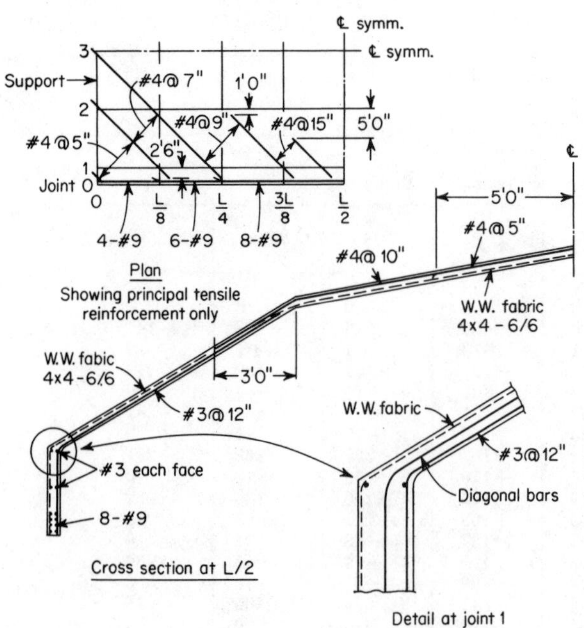

Fig. 35

From $7L/16$ to $L/2$, the minimum reinforcement required by ACI 318 is sufficient. For simplicity, these bar spacings are continued through to the support, even though somewhat less reinforcement would suffice.

In plate 2 at $L/2$ there is a longitudinal tension of

$$T = \frac{f_{12}^2}{f_{12} - f_{21}} \frac{h_2 d_2}{2} = \frac{36^2}{36 + 281} \frac{4 \times 120}{2} = 985 \text{ lb}$$

for which a minimum welded-wire-fabric reinforcement, as required for slabs in ACI 318, Sec. 7.12.2, of $0.0018 \times 12h = 0.087$ in.²/ft each way, is sufficient. The fabric 4 × 4–6/6 is provided.

In plate 1 at $L/2$ the total tensile force is

$$T = \frac{f_{01} + f_{10}}{2} h_1 d_1 = \frac{1456 + 36}{2} \times 6 \times 36 = 161 \text{ kips}$$

and $A_s = {}^{161}\!/_{20} = 8.05$ in.², for which eight No. 9 bars are provided in the bottom 16 in. A minimum tension reinforcement of $0.0035 \times 12h = 0.126$ in.²/ft is provided in the remaining portion by two No. 3 bars in each face.

The tension forces at other sections in plate 1 are

$$T_x = 4 \frac{x}{L} \left(1 - \frac{x}{L}\right) T_{L/2}$$

from which $T_x = 151$ kips at $3L/8$, 121 kips at $L/4$, and 71 kips at $L/8$. These are all nearly horizontal. The forces will be taken by eight No. 9, six No. 9, and four No. 9 bars, respectively. The four No. 3 and four No. 9 bars are carried into the support.

Reinforcement for Transverse Moments. At the middle of plate 2, $M = +392$ ft-lb/ft (Table 23) and $h = 4$ in. With a cover of ½ in., $d = 3$ in. Then, with $f_s = 20{,}000$ psi and $f'_c = 4000$ psi,

$$A_s = \frac{392 \times 12}{20{,}000 \times \frac{7}{8} \times 3} = 0.09 \text{ in.}^2\text{/ft}$$

for which No. 3 at 12 in. is provided at the bottom of the plate. At joints 2 and 3, respectively,

$$A_s = \frac{949 \times 12}{20{,}000 \times \frac{7}{8} \times 3} = 0.217 \text{ in.}^2\text{/ft}$$

$$A_s = \frac{2{,}068 \times 12}{20{,}000 \times \frac{7}{8} \times 3} = 0.473 \text{ in.}^2\text{/ft}$$

for which No. 4 at 10 in. and No. 4 at 5 in. are provided. At the middle of plate 3 the moment is 55 percent of that at joint 2, so that half the No. 4 bars coming from joint 3 are cut off at that point.

The longitudinal distribution of these moments is not well defined in the analysis. For cylindrical-segment thin shells, it appears that the moment drops off toward the supports somewhat more slowly than a sine curve. In this case we shall provide the full amount of steel in the center half of the span and reduce it to roughly 75 percent in the outer quarter span lengths. The reinforcement is shown in Fig. 35.

Influence of Span. Figure 36 shows how the longitudinal stresses from in-plane plate bending and transverse moments from one-way slab bending are influenced by the span L. The dashed line in Fig. 36*c* shows values of M for unyielding ridge lines, which is the limiting case as $L \to 0$. When $L = 35$ ft, the moments have approximately the same values as for $L = 0$, but for $L = 105$ ft the maximum moments are from 2.5 to 6.5 times the values for $L = 0$. On the other hand, the in-plane stresses (Fig. 36*b*) exhibit the same general behavior whether the span is 35 or 105 ft, with relatively high tension at the bottom and much lower compression near the top.

Significance of Joint Rotations. The effects of joint rotations ordinarily cannot be neglected in monolithic concrete folded plates, as is clearly shown in the example. They should be neglected only where previous analysis by more rigorous theory has demonstrated that it is safe to do so.[32] The relative magnitude of the corrections depends on the longitudinal and transverse rigidities of the component plates as well as on the shape of the structure.

Figure 36 compares the uncorrected slab moments M and beam stresses f for the structure of Fig. 33 with the values corrected for plate rotation. Values for the 70-ft span are from Tables 23 and 24. It is noted that transverse behavior approximates the ideal behavior fairly well for the 35-ft span but differs drastically as the span increases. This is because the increased flexibility of the edges permits a large relative deflection between

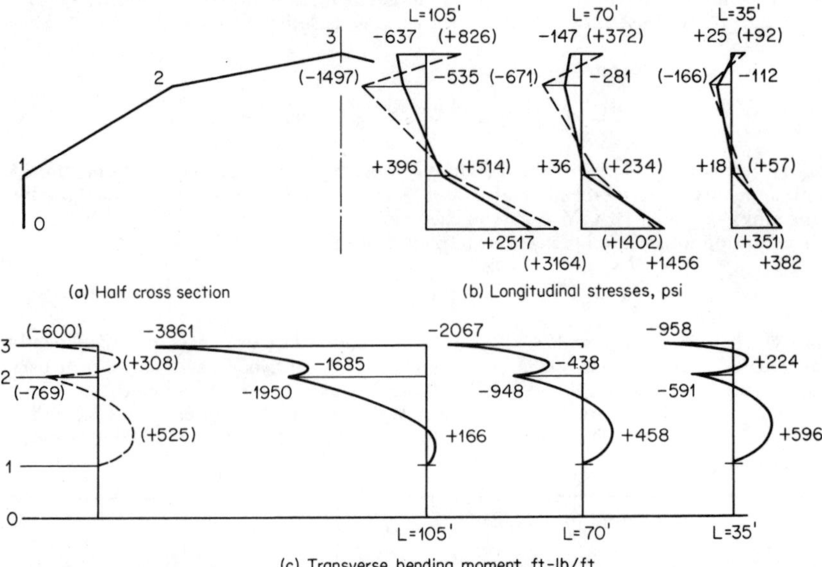

Fig. 36 Longitudinal stresses and transverse bending moments for folded plate of Fig. 33. Dashed lines (values in parentheses) denote plate rotation ignored; final values shown in solid lines.

the exterior and interior joints, which gives rise to large negative moments at the crown. Longitudinal beam action is concentrated in the region close to the edge for the 35-ft span but tends to spread over the entire cross section as the span increases.

In Fig. 30*b*, each plate tends to act in an ideal manner except for the edge plates, where there tends to be larger relative deflection between exterior and interior joints, with a resulting increase in negative moment at the first interior ridge and some deviation from longitudinal beam stresses in the exterior plates. However, ideal behavior is approached if an edge beam is used.

Where joints are not monolithic, as is usually the case in folded-plate structures in metal and wood, the analysis can be simplified to that of a one-way slab under surface loads and a series of inclined beams acted upon by the loads P_n.

Transverse Frames. The T-beam behavior of the transverse frame is similar to that of the transverse arch for cylindrical shells (Art. 18). The behavior of the longitudinally continuous folded-plate structure is also similar to that of the continuous cylindrical shell.

Other Configurations. Folded plates of other cross sections are common, e.g., Fig. 30*b* and *c*. For such cases, where there are many folds, the analysis presented in this article would lead to a large number of equations of compatibility. However, the interior plates will normally behave nearly as beams, which enables the analysis to be simplified. The analysis can be completed after one moment distribution, as in Table 19, and, with the resulting values of R and P, one computation from Eq. (44) for the beam stresses. Only the exterior plates need be analyzed by stress distribution and compatibility of plate rotations. For this simplification in analysis, it is necessary to establish a joint interior to which the beam method is applicable. For trapezoidal cross sections with four plates per repeated unit, it is suggested in Ref. 44 that the beam method be used for units interior to the sixth joint.

22. Continuous Folded Plates When folded plates are continuous over transverse supports, design practice consists of determining stresses by using the ratios of continuous beam moments to simple beam moments. For example, for a folded plate with the cross section shown in Fig. 36*a*, continuous for two spans of 70 ft each, the midspan tension of 1456 psi at the lower edge (Fig. 36*b*) would be reduced to 728 psi (ratio of $wL^2/16$ to $wL^2/8$). The lower-edge compression at the interior support would be 1456 psi (the same as the midspan tension for a simple span of 70 ft) because the negative moment at the interior support of a fully loaded two-span continuous beam is $wL^2/8$. Results of more accurate analyses confirm this practice as a reasonable basis for design.[33]

23. Prestressed Folded Plates If the longitudinal tensile stress in a folded-plate system is large, prestressing can be used to advantage (Fig. 37). The analysis proceeds as in Art. 21. The equivalent uniform load for a prestressing force F in a parabolic profile at an eccentricity e is $p = 8Fe/L^2$. The resulting free-edge plate bending stresses are given by Eq. (44). They must be brought into balance by stress distribution as in Table 20. The axial compressive stresses $f = F/A = F/hd$ must be balanced separately by a second distribution. This is because they are constant throughout the span, while the bending stresses vary parabolically.

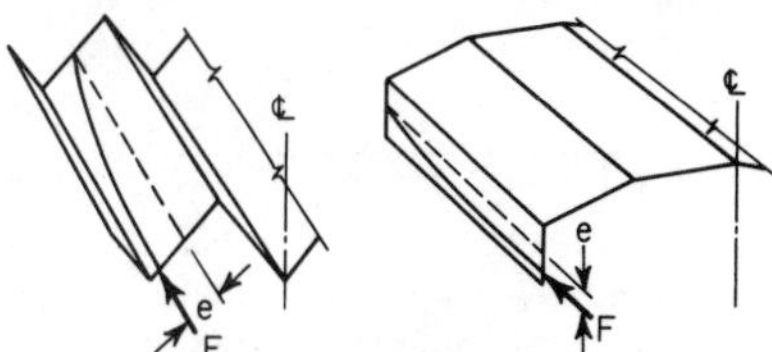

Fig. 37

With the stresses in balance, deflections y are computed, using Eq. (45) for the bending stresses and the following equation for the uniform compression:

$$y_n = (f_{n-1,n} - f_{n,n-1}) \frac{L^2}{8d_nE}$$

The plate rotations for each set of stresses are determined next from Eq. (46) and the errors at the joints from Eq. (47).

Results of prestressing the edge beam of the folded plate of Art. 21 with a force of 146 kips are given in Table 27. They show the prestressing essentially eliminates the in-plane tension and reduces substantially the maximum negative transverse moments, but increases the positive moment in plate 2.

TABLE 27

	DL and LL	Prestress	Combined
f_3	−147	−83	−230
f_2	−281	+85	−196
f_1	34	−31	3
f_0	1456	−1456	0
M_3	−2067	+954	−1113
M_{32}	−523	+888	+365
M_2	−984	+822	126
M_{21}	+392	+411	+803
M_1	0	0	0

TRANSLATION SHELLS OF DOUBLE CURVATURE

24. Membrane Theory Figure 38 shows the stress resultants on a differential element. The relations between them and their projections on a horizontal plane are given by

$$N'_x = \overline{N}_x \frac{\cos\theta}{\cos\phi} \qquad N'_y = \overline{N}_y \frac{\cos\phi}{\cos\theta} \qquad N'_{xy} = \overline{N}_{xy} \tag{57}$$

where $\tan\phi = \partial z/\partial x$ and $\tan\theta = \partial z/\partial y$.

The horizontal projections are given by

$$\overline{N}_x = \frac{\partial^2 F}{\partial y^2} - \int \overline{p}_x dx \qquad \overline{N}_y = \frac{\partial^2 F}{\partial x^2} - \int \overline{p}_y dy \qquad \overline{N}_{xy} = -\frac{\partial^2 F}{\partial x \partial y} \tag{58a}$$

where F is a stress function which satisfies the equation

$$\frac{\partial^2 F}{\partial x^2}\frac{\partial^2 z}{\partial y^2} - 2\frac{\partial^2 F}{\partial x \partial y}\frac{\partial^2 z}{\partial x \partial y} + \frac{\partial^2 F}{\partial y^2}\frac{\partial^2 z}{\partial x^2} = q \tag{58b}$$

in which

$$q = -\overline{p}_z + \overline{p}_x \frac{\partial z}{\partial x} + \overline{p}_y \frac{\partial z}{\partial y} + \frac{\partial^2 z}{\partial x^2}\int \overline{p}_x dx + \frac{\partial^2 z}{\partial y^2}\int \overline{p}_y dy \tag{58c}$$

In Eq. (58), $\overline{p}_x$, $\overline{p}_y$, and $\overline{p}_z$ are loads per unit of area of the horizontal projection of the surface. The relation between these components and the load p_x, p_y, and p_z per unit of area of the shell surface is

$$\frac{\overline{p}_x}{p_x} = \frac{\overline{p}_y}{p_y} = \frac{\overline{p}_z}{p_z} = \frac{\sqrt{1 - \sin^2\phi \sin^2\theta}}{\cos\phi\cos\theta} = \sqrt{1 + \tan^2\phi + \tan^2\theta} \tag{59}$$

Where only uniform vertical loads need be considered, $\overline{p}_x = \overline{p}_y = 0$, so that $q = -\overline{p}_z$ and Eq. (58) are greatly simplified. Furthermore, the vertical load can be assumed to be uniform over the horizontal projection for fairly flat shells, so that Eq. (59) is eliminated, i.e., $\overline{p}_z = p_z$.

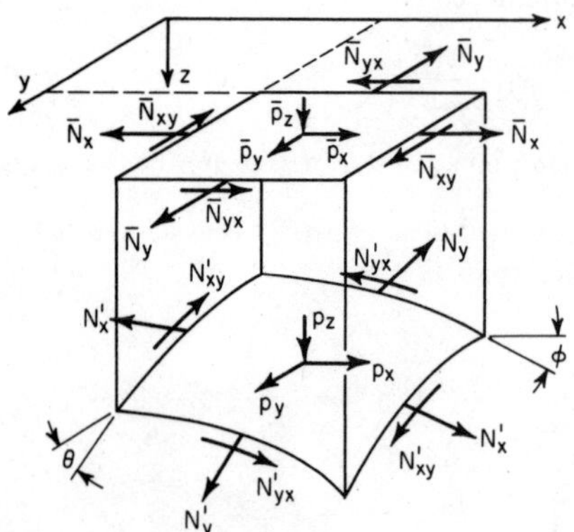

Fig. 38

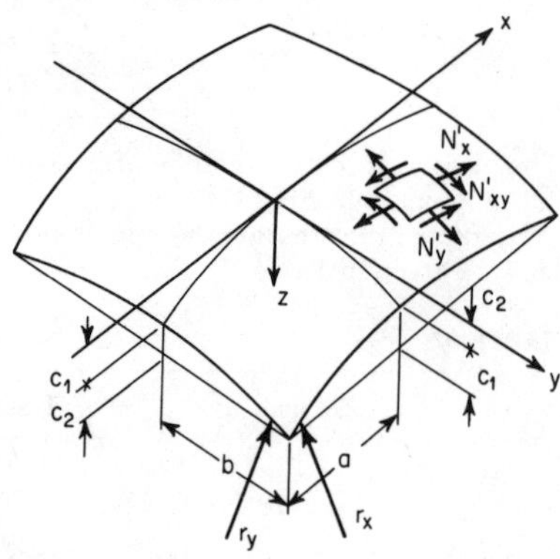

Fig. 39 Elliptical paraboloid shell.

25. Elliptic Paraboloids The elliptic paraboloid (Fig. 39) is defined by

$$z = c_1 \frac{x^2}{a^2} + c_2 \frac{y^2}{b^2} \tag{60}$$

This surface intersects a horizontal plane in an ellipse. Sections cut by vertical planes parallel to the xz and yz planes are parabolas.

Substitution of Eq. (60) into Eq. (58b) yields

$$\frac{2c_2}{b^2}\frac{\partial^2 F}{\partial x^2} + \frac{2c_1}{a^2}\frac{\partial^2 F}{\partial y^2} = q$$

Once a stress function F which satisfies this equation is found, the membrane stress resultants can be determined by Eqs. (57) and (58a). Various solutions are possible, depending upon the assumptions relative to the boundaries.[20] Most important is the case where all four edges are supported so that they are subjected only to in-plane shears N'_{xy} (Fig. 39). In this case, the shell needs only four point supports, with edge members to carry the shear. Derivation of the stress resultants is given in Ref. 34:

$$N'_x = -\frac{\bar{p}_z a^2 k}{c_1} \times \text{coefficient} \tag{61a}$$

$$N'_y = -\frac{\bar{p}_z b^2}{k c_2} \times \text{coefficient} \tag{61b}$$

$$N'_{xy} = -\frac{\bar{p}_z ab}{\sqrt{c_1 c_2}} \times \text{coefficient} \tag{61c}$$

where $k^2 = \dfrac{1 + (2c_1/a)^2(x/a)^2}{1 + (2c_2/b)^2(y/b)^2}$

$\bar{p}_z$ = uniform load per unit area of horizontal projection

For shallow shells, k approaches unity and usually may be taken as 1 in practical applications since N'_x and N'_y are not large.

Values of the coefficients in these equations are given in Table 28. Table 29 gives values of the coefficients to determine the edge shears at closer intervals. It will be noted that these shears tend to become infinitely large at the corners. This means that transverse shearing forces and bending moments develop so that edge shears can remain finite. It is suggested in Ref. 34 that N'_{xy} can be considered to be maximum at

$$x = a - 0.4\sqrt{r_x h} \qquad y = b - 0.4\sqrt{r_y h} \tag{62}$$

where r_x and r_y are the corner radii (Fig. 39). This assumption is based on the fact that edge disturbances damp out rapidly.

The Geckeler approximation (Art. 9) may be used for an approximate evaluation of the bending near the edges. If the edge arch is much stiffer than the shell, the shell edge can be assumed fixed. Thus, for the edges $x = \pm a$,

$$w = -2\beta \frac{r^2}{Eh} Q_{xa} - 2\beta^2 \frac{r^2}{Eh} M_{xa} + \frac{\bar{p}_z r^2}{Eh} = 0$$

$$\frac{dw}{dx} = 2\beta^2 \frac{r^2}{Eh} Q_{xa} + 4\beta^3 \frac{r^2}{Eh} M_{xa} = 0$$

In these equations, the terms involving the edge shear Q_{xa} and the edge moment M_{xa} are obtained from Table 8. The last term in the first equation is the radial (hoop) displacement due to $\bar{p}_z$. The solution is

$$Q_{xa} = \frac{\bar{p}_z}{\beta} \qquad M_{xa} = -\frac{\bar{p}_z}{2\beta^2} \tag{63a}$$

With the edge shear and moment known, moments at points interior to the boundary can be determined from Table 8. Thus

$$M_x = \frac{C_3}{\beta} Q_{xa} + C_4 M_{xa} = \frac{\bar{p}_z}{\beta^2}\left(C_3 - \frac{C_4}{2}\right) \tag{63b}$$

If the torsional stiffness of the supporting arch is small, the shell edge can be assumed to be simply supported, in which case $w = M_{xa} = 0$, so that

$$Q_{xa} = \frac{\bar{p}_z}{2\beta} \tag{64a}$$

TABLE 28 Coefficients for Computing Stress Resultants in Elliptical Paraboloids [Eqs. (61)]*

Value of y/b

x/a	Stress resultants	(a) $c_1/c_2 = 1.0$: 0	0.25	0.50	0.75	1.0	(d) $c_1/c_2 = 0.8$: 0	0.25	0.50	0.75	1.0
0.00	N_y	0.250	0.233	0.182	0.101	0	0.289	0.270	0.213	0.119	0
	N_x	0.250	0.267	0.318	0.399	0.500	0.211	0.230	0.287	0.381	0.500
	N_{xy}	0	0	0	0	0	0	0	0	0	0
0.25	N_y	0.267	0.250	0.199	0.111	0	0.304	0.285	0.228	0.130	0
	N_x	0.233	0.250	0.301	0.389	0.500	0.196	0.215	0.272	0.370	0.500
	N_{xy}	0	0.029	0.068	0.096	0.108	0	0.034	0.069	0.100	0.114
0.50	N_y	0.318	0.301	0.250	0.150	0	0.347	0.331	0.277	0.169	0
	N_x	0.182	0.199	0.250	0.350	0.500	0.153	0.169	0.223	0.331	0.500
	N_{xy}	0	0.068	0.140	0.210	0.244	0	0.065	0.139	0.215	0.255
0.75	N_y	0.399	0.389	0.350	0.250	0	0.416	0.406	0.369	0.270	0
	N_x	0.101	0.111	0.150	0.250	0.500	0.084	0.094	0.131	0.230	0.500
	N_{xy}	0	0.096	0.210	0.356	0.465	0	0.091	0.201	0.353	0.480
1.0	N_y	0.500	0.500	0.500	0.500	0	0.500	0.500	0.500	0.500	0
	N_x	0	0	0	0	0	0	0	0	0	0
	N_{xy}	0	0.108	0.243	0.465	∞	0	0.101	0.229	0.443	∞

x/a	Stress resultants	(b) $c_1/c_2 = 0.6$: 0	0.25	0.50	0.75	1.0	(e) $c_1/c_2 = 0.4$: 0	0.25	0.50	0.75	1.0
0.00	N_y	0.336	0.316	0.252	0.143	0	0.395	0.374	0.307	0.180	0
	N_x	0.164	0.184	0.248	0.357	0.500	0.105	0.126	0.193	0.320	0.500
	N_{xy}	0	0	0	0	0	0	0	0	0	0
0.25	N_y	0.348	0.329	0.267	0.155	0	0.403	0.383	0.319	0.192	0
	N_x	0.152	0.171	0.233	0.345	0.500	0.097	0.117	0.181	0.308	0.500
	N_{xy}	0	0.031	0.067	0.103	0.120	0	0.026	0.060	0.101	0.125
0.50	N_y	0.383	0.367	0.312	0.197	0	0.425	0.410	0.357	0.235	0
	N_x	0.117	0.133	0.188	0.304	0.500	0.075	0.090	0.143	0.265	0.500
	N_{xy}	0	0.060	0.132	0.216	0.265	0	0.049	0.115	0.208	0.274
0.75	N_y	0.436	0.426	0.392	0.296	0	0.459	0.451	0.419	0.331	0
	N_x	0.064	0.074	0.108	0.204	0.500	0.041	0.049	0.081	0.169	0.500
	N_{xy}	0	0.081	0.185	0.342	0.494	0	0.065	0.156	0.316	0.506
1.00	N_y	0.500	0.500	0.500	0.500	0	0.500	0.500	0.500	0.500	0
	N_x	0	0	0	0	0	0	0	0	0	0
	N_{xy}	0	0.089	0.208	0.413	∞	0	0.070	0.173	0.363	∞

x/a	Stress resultants	(c) $c_1/c_2 = 0.2$: 0	0.25	0.50	0.75	1.0
0.00	N_y	0.462	0.446	0.388	0.248	0
	N_x	0.038	0.054	0.112	0.252	0.500
	N_{xy}	0	0	0	0	0
0.25	N_y	0.465	0.451	0.396	0.261	0
	N_x	0.035	0.049	0.104	0.239	0.500
	N_{xy}	0	0.014	0.040	0.088	0.128
0.50	N_y	0.473	0.462	0.414	0.303	0
	N_x	0.027	0.038	0.086	0.197	0.500
	N_{xy}	0	0.027	0.074	0.174	0.280
0.75	N_y	0.485	0.480	0.456	0.383	0
	N_x	0.015	0.020	0.044	0.117	0.500
	N_{xy}	0	0.034	0.098	0.246	0.510
1.00	N_y	0.500	0.500	0.500	0.500	0
	N_x	0	0	0	0	0
	N_{xy}	0	0.038	0.108	0.262	∞

*From Ref. 34.

TABLE 29 Shear along Edges of Elliptical Paraboloids [Eq. (61c)]*

y/b	c_1/c_2 1.0	0.8	0.6	0.4	0.2
	At $x = \pm a$				
0.0	0.0000	0.0000	0.0000	0.0000	0.0000
0.1	0.0419	0.0389	0.0342	0.0307	0.0137
0.2	0.0854	0.0793	0.0701	0.0550	0.0286
0.3	0.1319	0.1231	0.1096	0.0872	0.0481
0.4	0.1836	0.1721	0.1546	0.1254	0.0731
0.5	0.2432	0.2294	0.2081	0.1728	0.1075
0.6	0.3204	0.3066	0.2859	0.2493	0.1818
0.7	0.4071	0.3897	0.3627	0.3173	0.2296
0.8	0.5363	0.5178	0.4887	0.4400	0.3443
0.85	0.6279	0.6090	0.5791	0.5292	0.4306
0.9	0.7570	0.7378	0.7074	0.6667	0.5659
0.95	0.9777	0.9582	0.9276	0.8763	0.7741
1.0	∞	∞	∞	∞	∞
x/a	At $y = \pm b$				
0.0	0.0000	0.0000	0.0000	0.0000	0.0000
0.1	0.0419	0.0444	0.0468	0.0488	0.0500
0.2	0.0854	0.0903	0.0950	0.0990	0.1014
0.3	0.1319	0.1391	0.1460	0.1519	0.1553
0.4	0.1836	0.1930	0.2019	0.2095	0.2140
0.5	0.2432	0.2545	0.2652	0.2743	0.2798
0.6	0.3204	0.3317	0.3425	0.3516	0.3571
0.7	0.4071	0.4213	0.4348	0.4463	0.4532
0.8	0.5363	0.5515	0.5659	0.5782	0.5855
0.85	0.6279	0.6434	0.6582	0.6707	0.6782
0.9	0.7570	0.7728	0.7878	0.8005	0.8081
0.95	0.9777	0.9935	1.0087	1.0215	1.0290
1.0	∞	∞	∞	∞	∞

*From Ref. 34.

With the edge shear known, moments at interior points are determined from Table 8. Thus,

$$M_x = C_3 \frac{\overline{p}_z}{2\beta^2} \tag{64b}$$

Example Given $a = 40$ ft, $b = 50$ ft, $c_1 = 8$ ft, $c_2 = 10$ ft, $h = 3$ in., $p_z = 60$ psf. From Eqs. (61), assuming $k = 1$,

$$N'_x = -\frac{60 \times 40^2}{8} = -12{,}000 \times \text{coefficient}$$

$$N'_y = -\frac{60 \times 50^2}{10} = -15{,}000 \times \text{coefficient}$$

$$N'_{xy} = -\frac{60 \times 50 \times 40}{\sqrt{80}} = -13{,}400 \times \text{coefficient}$$

Stress resultants along the edge $x = a = 40$ ft will be calculated. From Table 28,

$$N'_x = 0 \qquad N'_y = -0.5 \times 15{,}000 = -7500 \text{ lb/ft}$$

The edge shears N'_{xy} are determined from the coefficients in Table 29, following which principal tensions are calculated from

$$N' = \frac{N'_y}{2} \pm \sqrt{\left(\frac{N'_y}{2}\right)^2 + N'^2_{xy}}$$

The results are shown in Table 30.
From Eq. (60)

$$\frac{\partial z}{\partial y} = \frac{2c_2 y}{b^2} \qquad \frac{\partial^2 z}{\partial y^2} = \frac{2c_2}{b^2}$$

from which the radius of curvature r_y is

$$r_y = \frac{[1 + (\partial z/\partial y)^2]^{3/2}}{\partial^2 z/\partial y^2} = 125\left[1 + \left(\frac{y}{125}\right)^2\right]^{3/2}$$

At the corner, $r_y = 125 \times 1.16^{3/2} = 156$ ft, and from Eq. (62) the edge shear can be considered maximum at

$$y = 50 - 0.4\sqrt{156 \times 0.25} = 47.5 \text{ ft} = 0.95b$$

Thus, according to Table 30, the largest principal tension along the edge is 9.64 kips/ft.

Reinforcement for the tension at the corners is usually placed diagonally. The controlling tension is usually that at the edge, but principal stresses at several interior points should also be computed to determine the extent of the area to be reinforced. Minimum reinforcement of at least $0.0018bh$ for welded-wire fabric (ACI 318, Sec. 7.12.2) should be supplied throughout.

The transverse shear and moment along the edge $x = a = 40$ ft can be estimated by Eq. (63) or (64), depending upon the assumption as to relative stiffnesses of the shell and its supporting arch. The corner radius r_x is found to be 125 ft, from which $\beta = 0.24$. Equation (63*a*) gives $Q_{xa} = 250$ lb/ft and $M_{xa} = -520$ ft-lb/ft. However, if the edge is assumed to be simply supported, $Q_{xa} = 125$ lb/ft [Eq. (64*a*)] and the corresponding maximum moment is, from Eq. (64*b*)

$$M_x = 0.32 \frac{60}{2 \times 0.24^2} = 167 \text{ ft-lb/ft}$$

The maximum value 0.32 of C_3 is from Table 8, and corresponds to $\beta s = 0.8$. Therefore, this maximum moment is located about $0.8/0.24 = 3.3$ ft from the edge.

The supporting arches should be designed for the edge shears N'_{xy}, as in the case of the barrel shell (Art. 18).

26. Hyperbolic Paraboloids These shells have a simplicity of geometry and a potential for shape that is economical and attractive. The simplicities, however, may mislead the designer into geometries or details that violate good practice in reinforced concrete. Specifically, if the rise is too slight, compressions are high and moments due to creep can cause instabilities. Also, simplified procedures for determining stresses often provided no criterion for reinforcement except for the minimum dictated by good practice.

Many hypars designed on the basis of membrane theory have performed well. However, numerical analyses of bending have shown that membrane-theory stresses can be misleading at times, especially with regard to edge beams. For example, Schnobrich has shown that in a 3-in. hypar roof shaped as in Fig. 42*d*, spanning 80 ft on a side and with a crown rise of 8 ft, the axial force in the horizontal ridge beams at midspan is between 5 and 35 percent of that computed by the membrane theory, depending upon the beam size.[36]

TABLE 30 Shear and Principal Tension, Kips per Ft, on Edge $x = a$

y/b	0	0.1	0.2	0.3	0.4	0.5	0.6	0.7	0.8	0.85	0.90	0.95
N_{xy}'	0	−0.52	−1.06	−1.65	−2.31	−3.08	−4.11	−5.23	−6.95	−8.17	−9.90	−12.85
N'	0	+0.04	+0.15	+0.35	+0.65	+1.10	+1.82	+2.68	+4.14	+5.24	+6.84	+9.64

The design of hypars must be based on a study of successful structures as well as those that have experienced difficulties.[37,38,39]

The equation of the hyperbolic paraboloid is (Fig. 40*a*)

$$z = c_2 \frac{y^2}{b^2} - c_1 \frac{x^2}{a^2} \tag{65}$$

For x (or y) constant, z describes a parabola in a plane parallel to yz (or xz), so that the surface can be generated by translating the parabola EOF along the parabola GOH.

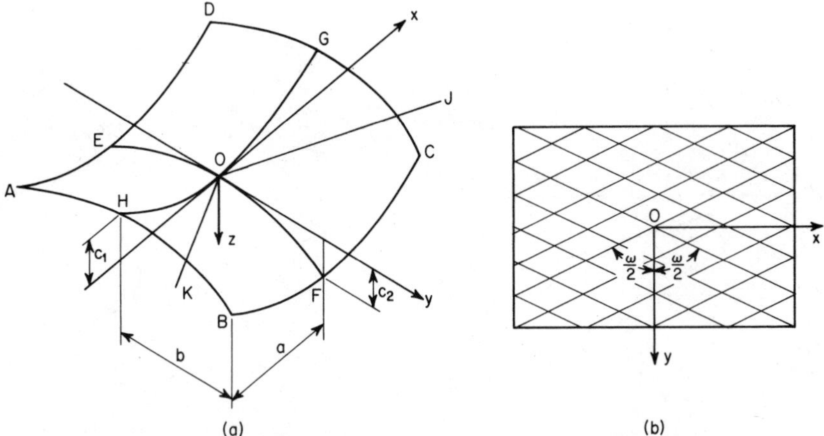

Fig. 40 Hyperbolic paraboloid shell.

The surface contains two sets of straight generators; OJ in Fig. 40*a* represents one set, OK the other. The projections of these generators on the xy plane are shown in Fig. 40*b*, where ω is given by

$$\tan^2 \frac{\omega}{2} = \frac{c_2}{c_1} \frac{a^2}{b^2} \tag{66}$$

Alternatively, the surface can be referred to axes x,y parallel to the projections of the straight generators (Fig. 41*a*). The equation is

$$z = \frac{c_0}{a_0 b_0} xy \tag{67}$$

In this position, the surface can be considered to be generated by allowing the line OJ to rotate around OK as it moves from OJ to KL, always remaining parallel to the xz plane.

Orthogonal Straight Boundaries. Where the boundaries of the shell are straight generators, and the latter are orthogonal, Eq. (67) can be substituted into Eq. (58*b*) of Art. 24, which gives

$$\frac{\partial^2 F}{\partial x \partial y} = \frac{q a_0 b_0}{2 c_0}$$

Once a stress function F which satisfies this equation is found, the membrane stress resultants can be determined by Eqs. (57) and (58*a*). Various solutions are possible, depending upon the assumptions for boundary conditions.[20] When the load is $\overline{p}_z$ uniform over the horizontal projection of the surface, a solution of practical interest is (Fig. 41*b*)

$$\overline{N}_{xy} = -\frac{\overline{p}_z a_0 b_0}{2c_0} \qquad \overline{N}_x = \overline{N}_y = 0 \tag{68a}$$

The edges are subjected only to shearing-stress resultants, which must be resisted by edge members. The projected principal-stress resultants are inclined 45° with the edges:

$$\overline{N}_1 = -\overline{N}_2 = \frac{\overline{p}_z a_0 b_0}{2c_0} \tag{68b}$$

Various combinations of the surface of Fig. 41*b* can be made. Continuing it into the other three quadrants gives the saddle surface in Fig. 42*a*. Figure 42*b* shows four

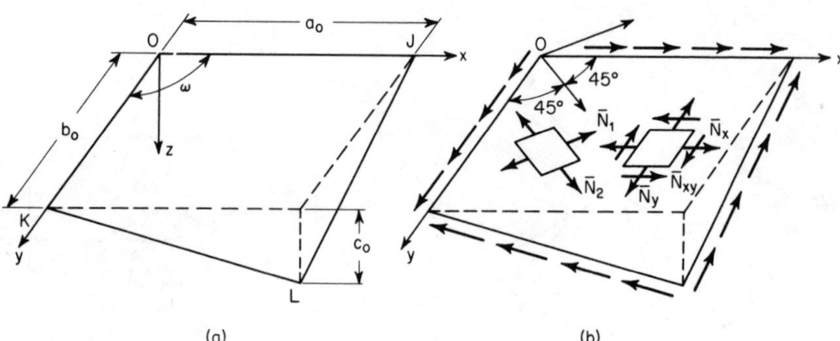

Fig. 41

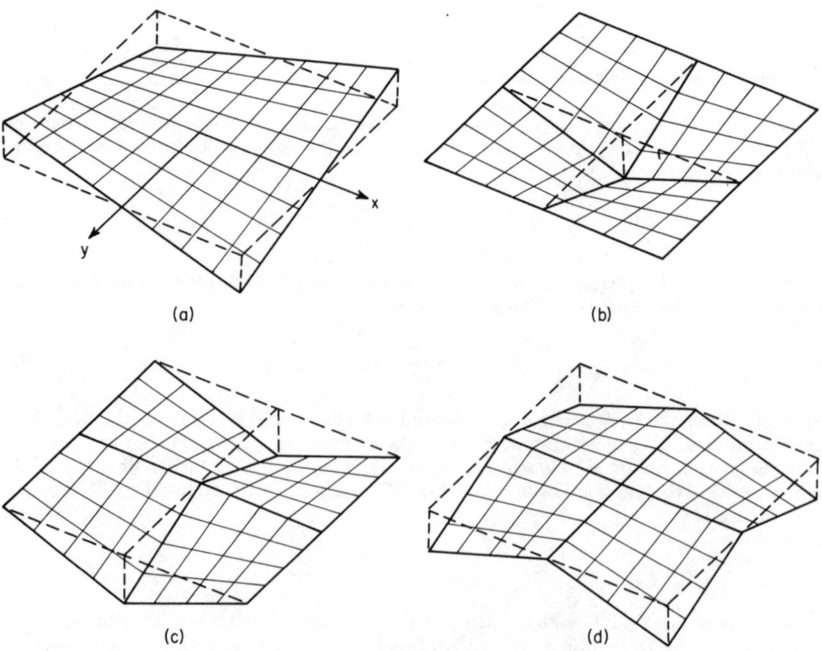

Fig. 42 Hyperbolic paraboloid roof surfaces.

quadrants combined to produce a surface which is usually called an inverted umbrella. Other combinations are shown in *c* and *d*.

Skewed Surfaces. If the surface is bounded by straight generators and $\omega \neq 90°$ (Fig. 41*a*), a load $\overline{p}_z$ uniform over the horizontal projection can be carried by

$$\overline{N}_{xy} = \overline{N}_{yz} = -\frac{\overline{p}_z a_0 b_0}{2c_0} \sin \omega \tag{69a}$$

$$\overline{N}_x = \overline{N}_y = 0 \tag{69b}$$

where $\overline{N}_{xy} = \overline{N}_{yz}$ are shearing stress resultants on an element bounded by the (nonorthogonal) projections on the plane xy of the straight generators. $\overline{N}_x$ and $\overline{N}_y$, being parallel to x and y, respectively, are nonorthogonal.

The corresponding principal stress resultants in the surface lie approximately in vertical, orthogonal planes, one of which bisects the quadrant xy,

$$N_1' = -N_2' = \frac{\overline{p}_z a_0 b_0}{2c_0} \sin \omega \tag{69c}$$

Unsymmetrical Load. If some quadrants of the surfaces in Fig. 42 are unloaded, stress resultants for the loaded quadrants can be determined by Eq. (68). However, this results in unbalanced forces in certain edge members. If only two quadrants of the surface in Fig. 42d are loaded (Fig. 43), the edge-member forces F must be reacted. Unless there is external restraint at these points, F must be equilibrated in each case by the other half of the edge member. This subjects the unloaded panels to shearing forces along their edges. This problem is discussed in Ref. 35.

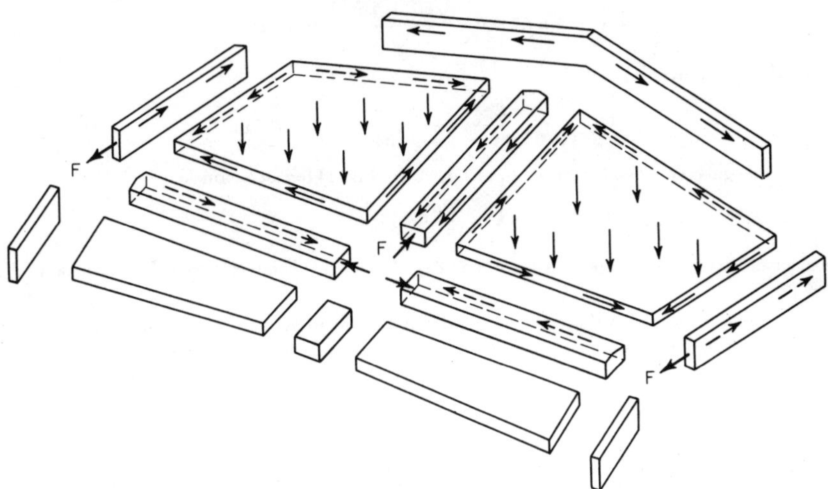

Fig. 43

Example* For the inverted umbrella shown in Fig. 44, $a_0 = b_0 = 20$ ft and $c_0 = 5.5$ ft. The weight of the 3-in. shell is 37.5 psf, to which is added 5 psf to account for the weight of the edge beams. The live load is 30 psf. Then, from Eqs. (68)

$$-\overline{N}_{xy} = \overline{N}_1 = -\overline{N}_2 = \frac{72.5 \times 20 \times 20}{2 \times 5.5} = 2640 \text{ lb/ft}$$

The required reinforcement in the direction of $\overline{N}_1$ is

$$A_s = \frac{2640}{20{,}000} = 0.132 \text{ in.}^2\text{/ft}$$

Minimum reinforcement in the orthogonal direction for temperature and shrinkage is $0.002 \times 36 = 0.072$ in.²/ft (ACI 313-7.12.2).

*From Ref. 35.

For easier placement, steel is often placed along the straight-line generators. Assuming $A_{sx} = A_{sy}$, Eq. (72) gives, since $\theta = 45°$,

$$A_{sx} = A_{sy} = \frac{N_p}{f_s} = \frac{2640}{20{,}000} = 0.132 \text{ in.}^2\text{/ft}$$

Thus, where it follows the straight-line generators, the steel required is $2 \times 0.132 = 0.264$ in.^{2}/ft. On the other hand, only $0.132 + 0.072 = 0.204$ in.^{2}/ft is required if it is placed along the parabolas.

The compressive stress is only

$$f_c = \frac{2640}{3 \times 12} = 74 \text{ psi}$$

The tension T at midlength of the edge members is

$$T = 2640 \times 20 = 52{,}800 \text{ lb}$$

for which

$$A_s = \frac{52{,}800}{20{,}000} = 2.64 \text{ in.}^2$$

The shearing forces on both sides of the sloped edge members contribute to its axial force. The compression C is

$$C = 2 \times 52{,}800 \times \frac{\sqrt{20^2 + 5.5^2}}{20} = 109{,}560 \text{ lb}$$

For a tied column with $p_g = 0.01$,

$$A_g = \frac{C}{0.8(0.225f'_c + f_s p_g)} = \frac{109{,}560}{540 + 160} = 157 \text{ in.}^2$$

For the triangular cross section shown in section AA of Fig. 44 the area furnished is

$$A_g = \frac{20d^2}{5.5} = 157 \text{ in.}^2$$

from which $d = 7$ in. This depth is made 9 in. in Ref. 35 to provide bending strength for unsymmetrical loading.

Nonuniform Load. Although dead load is not uniform over the horizontal projection of sloping surfaces, it is a good approximation for shells of uniform thickness, provided the surface is not too steep. Equations (68) must be modified for steep surfaces.

The true load on the horizontal projection is given by Eq. (59) (see Fig. 38):

$$\overline{p}_z = p_z\sqrt{1 + \tan^2\phi + \tan^2\theta}$$

For the hyperbolic paraboloid with orthogonal straight boundaries on axes x, y [Eq. (67)],

$$\overline{N}_{xy} = -\frac{p_z a_0 b_0}{2c_0}\sqrt{1 + k^2x^2 + k^2y^2} \tag{70a}$$

$$\overline{N}_x = \frac{p_z y}{2}\log\frac{kx + \sqrt{1 + k^2x^2 + k^2y^2}}{\sqrt{1 + k^2y^2}} + f_1(y) \tag{70b}$$

$$\overline{N}_y = \frac{p_z x}{2}\log\frac{ky + \sqrt{1 + k^2x^2 + k^2y^2}}{\sqrt{1 + k^2x^2}} + f_2(x) \tag{70c}$$

where $k = c_0/a_0b_0$ and $f_1(y)$ and $f_2(x)$ are constants of integration. With $\overline{N}_{xy}$, $\overline{N}_x$, and $\overline{N}_y$ known, the stress resultants in the surface are determined from Eq. (57).

Equations (70) show that edge members are no longer subjected only to shear. However, by assigning appropriate values to f_1 and f_2, two adjoining edges can be freed of normal stress. The other two edges will be subjected to normal forces, which must be carried by the edge members or by bending in the shell.

Equations (70) cannot be used for the skewed surface, for which the solution is more difficult.[40]

Parabolic Boundaries. Possible systems of stress resultants for a surface bounded as in Fig. 40*a*, and with *x,y* directed as in that figure, are

$$\overline{N}_x = \frac{\overline{p}_z a^2}{2c_1} \qquad \overline{N}_y = \overline{N}_{xy} = 0 \tag{71a}$$

$$\overline{N}_y = -\frac{\overline{p}_z b^2}{2c_2} \qquad \overline{N}_x = \overline{N}_{xy} = 0 \tag{71b}$$

$$\overline{N}_x = \frac{\overline{p}_z a^2}{4c_1} \qquad \overline{N}_y = -\frac{\overline{p}_z b^2}{4c_2} \qquad \overline{N}_{xy} = 0 \tag{71c}$$

The first system requires anchorage of the surface at $x = \pm a$, the second at $y = \pm b$, and the third at both $x = \pm a$ and $y = \pm b$. The resultants $\overline{N}_x$ and $\overline{N}_y$ normal to the vertical planes *AHB, AED,* etc., constitute heavy load which it will usually be impracticable to support.

Combinations of the surface can be made. An example is the groined vault of Fig. 45. Reference 35 contains tables of coefficients for this structure.

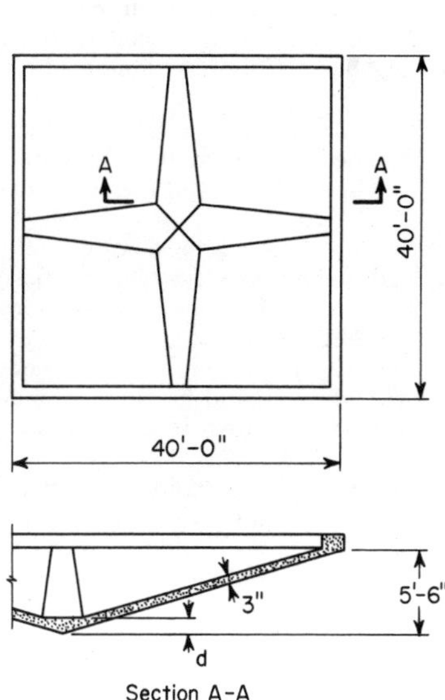

Fig. 44 Example of Art. 26.

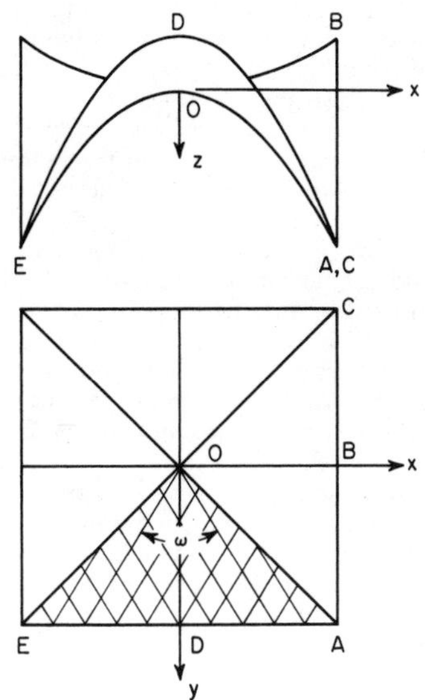

Fig. 45 Groined vault.

DIMENSIONING

The following comments on proportioning of concrete and reinforcement in thin shells are based on Ref. 25.

Thickness is rarely based on allowable stress, but usually on construction requirements or stability.

Reinforcement should be provided to resist completely the principal tensile stresses, assumed to act at the middle surface. It may be placed either in the general direction of the lines of principal stress or in two or three directions. In regions of high tension it is advisable to place it in the general direction of the principal stress. Whenever possible, such reinforcement may run along lines practical for construction, such as straight lines.

Reinforcement which does not deviate in direction more than 15° from the direction of the principal stress may be considered to be parallel to it. A slightly greater deviation can be tolerated in areas where the stress in the reinforcement is two-thirds, or less, of the allowable.

For reinforcement placed in two directions at right angles, as along x and y axes, and where A_{sy} forms the angle θ with the direction of the principal stress resultant N_p, reinforcement is required such that

$$N_p = f_{sy}(A_{sy} \cos^2 \theta + A_{sx} \sin^2 \theta \tan \theta) \quad (72a)$$

$$\tan 2\theta = \frac{2N_{xy}}{N_x - N_y} \quad (72b)$$

For positive values of $\tan 2\theta$, θ is measured counterclockwise from the face on which N_p acts.

Minimum reinforcement should be provided as required in ACI 318 even where not required by analysis.

In areas where the computed tensile stress in the concrete exceeds 300 psi, at least one layer of reinforcement must be parallel to the principal tensile stress, unless it can be proved that a deviation is permissible because of the geometrical characteristics of the shell and because, for reasons of geometry, only insignificant and local cracking could develop.

The allowable stress for reinforcement may be used at any point in the shell independently of the magnitude of the stress in the concrete at that point.

Additional reinforcement to resist bending moments must be proportioned and provided in the conventional manner. Generally, where moments are significant in thin shells, the effect of direct compression forces may be neglected. Either working-load or ultimate-load analysis may be used.

Where the computed principal tensile stress (psi) exceeds $2\sqrt{f'_c}$ (f'_c in psi), the spacing of reinforcement should not exceed three times the thickness of the shell. Otherwise, reinforcement should be spaced not more than five times the thickness of the shell, or more than 18 in.

Splices in principal tensile reinforcement should be kept to a practical minimum. Splices should be staggered, with not more than one-third of the bars spliced at any one cross section. Bars should be lapped only within the same layer. The minimum lap for draped reinforcing bars should be 30 diameters, with a minimum of 18 in. unless more is required by ACI 318, except that the minimum may be 12 in. for reinforcement not required by analysis. The minimum lap for welded-wire fabric should be 8 in. or one mesh, whichever is greater, except that ACI 318 governs where wire fabric at a splice must carry the full allowable stress.

Concrete cover should be at least ½ in. for bars, ⅜ in. where precast and for welded-wire fabric, and 1 in. for prestressing tendons, provided the concrete surfaces are protected from weather and are not in contact with the ground. In no case should the cover be less than the diameter of the bar or tendon. If greater cover is required for fire protection, it need apply only to principal tensile and moment reinforcement whose yielding would cause failure.

CONSTRUCTION

A thin shell should be as thin as is practicable in order to induce in-plane stress as opposed to bending. However, the high cost of labor in the United States usually makes it more economical to use a thicker shell than to enforce the careful casting techniques essential to construction of very thin shells. Three layers of steel are usually used, and to place concrete properly a 3-in. thickness is about minimum.

Slope of the shell should be less than 45° to avoid top forms, which increase the difficulty of eliminating honeycombing. Where the slope is 30°, concrete with a slump of 1 to 3 in. can be cast to reasonable tolerances without a top form. Where the slope exceeds 45° it may be possible to cast without top forms if the slump is low, but the cost of placing is increased. In some cases concrete has been shot onto vertical surfaces successfully.

Location of ribs can be important. Application of insulation and roofing is relatively simple when the surface is free of ribs. However, ribs projecting below the soffit complicate movement of forms. Form movement may constitute a large item of cost in continuous, cylindrical-segment roofs, in which case ribs are built above the roof. These projections must be carefully flashed. Ribless shells, i.e., shells with wide, flat ribs, have been successfully built for short spans where buckling was not a major factor in design.[29]

Maximum aggregate size should not exceed one-half the shell thickness, or the clear distance between bars, or 1½ times the cover. Where top forms are required, maximum size of aggregate should not exceed one-fifth the minimum clear distance between forms, or the cover over the reinforcement.

Forms should be carefully built. The designer should consider the structural effect of small deviations from the plans and should set tolerances. The stability of the shell depends upon its radii of curvature, and relatively small radial deviations in surface dimensions can cause large variations in radius. For example, a variation of ½ in. radially in an arc length of 10 ft in the dome of Art. 10 causes a 45 percent change in radius if the surface remains spherical.

Concrete should be cast in a symmetrical pattern to avoid bracing the scaffolding for the effect of unbalanced loads. It is recommended that concreting commence at the low point, or points, and proceed upward. Concrete should be deposited as nearly as possible in its final position. Vibration of thin sections is difficult. It has been done by vibrating the reinforcement or the forms, but these systems must be rigid to withstand such treatment. Vibrating screeds have been successfully used.

Thin shells are susceptible to shrinkage cracking if curing of the concrete is faulty. In hot weather, the use of retarders, preliminary fog-spray curing, and wet burlap or water curing is advisable. In cold weather, accelerators and special precautions against freezing are usually required. In moderate weather (40 to 70°), ordinary methods such as membrane-curing compounds are usually satisfactory, although wet curing may produce better results.

The method of form removal (decentering) is usually specified by the designer, in order particularly to avoid any unwanted temporary supports or concentrated reactions on the shell. It is best to begin decentering at points of maximum deflections and progress toward points of minimum deflection, with decentering of edge members proceeding simultaneously with that of the adjoining shell. It is important to control deflections at the time of decentering, and it is common to specify a modulus of elasticity that must be obtained before permission to decenter is granted. Small, lightly reinforced beams tested in flexure have been used successfully to determine E.[41]

Thin shells of dramatic shape and lightness, which can be successfully built in warm, dry climates, cannot always be duplicated in harsher regions. Furthermore, it is not easy to predict whether a particular construction scheme will be economical.

REFERENCES

1. Kalinka, J. E.: Monolithic Concrete Construction for Hangars, *Mil. Eng.*, January–February 1940.
2. Koiter, W. T.: A Consistent First Approximation in the General Theory of Thin Elastic Shells, *Proc. IUTAM Symp. Theory of Thin Elastic Shells* (Delft), North-Holland Publishing Company, Amsterdam, 1959.
3. Bouma, A. L., A. C. Van Riel, H. Van Koten, and W. J. Beranek: Investigations of Models of Eleven Cylindrical Shells Made of Reinforced and Prestressed Concrete, *Proc. Symp. Shell Research* (Delft), North-Holland Publishing Company, Amsterdam, 1961.
4. Timoshenko, S. P., and S. Woinowski-Krieger: "Theory of Plates and Shells," 2d ed., McGraw-Hill Book Company, New York, 1959.
5. Pflüger, Alf: "Elementary Statics of Shells," 2d ed. (English translation by Ervin Galantay), McGraw-Hill Book Company, New York, 1961.
6. Medwadowski, S., W. C. Schnobrich, and A. C. Scordelis (eds.): "Concrete Thin Shells," ACI Publication SP-2B, Detroit, 1971.
7. Flugge, W.: "Stresses in Shells," Springer-Verlag OHG, Berlin, 1960.
8. Timoshenko, S. P., and J. M. Gere: "Theory of Elastic Stability," 2d ed., McGraw-Hill Book Company, New York, 1961.
9. Collected Papers on Instability of Shell Structures, 1962 NASA TN-D1510, 1962.
10. Schmidt, H.: Ergebnisse von Beulversuchen mit doppelt gekrummten Schalen-modellen aus Aluminium, *Proc. Symp. Shell Research* (Delft), North-Holland Publishing Company, Amsterdam, 1961.
11. Wang, Y-S., and D. P. Billington: Buckling of Cylindrical Shells by Wind Pressure, *J. Eng. Mech. Div. ASCE*, October 1974.
12. Kundurpi, P. S., et al.: Stability of Cantilever Shells under Wind Loads, *J. Eng. Mech. Div. ASCE* October 1975.
13. Cole, P. C., et al.: Buckling of Cooling-Tower Shells: State of the Art, *J. Struct. Div. ASCE*, June 1975.

14. Der, T. J., and R. Fidler: A Model Study of the Buckling Behavior of Hyperbolic Shells, *Proc. Inst. Civil Eng.*, vol. 41, London, September 1968.
15. Ewing, D. J. F.: The Buckling and Vibration of Cooling Tower Shells, Part II: Calculations, *Lab. Rept.* RD/L/R 1764, Central Electricity Research Laboratories, Leatherhead, England, November 1971.
16. Krätzig, W., Statische und dynamische Stabilitat der Kühlturmschale, *Konstr. Ingenieurbau Ber.*, vol. 1, Vortage der Tagung Naturzug-Kühlturme, Haus der Technik, Essen, Germany, Apr. 19, 1968.
17. Cole, P. P., J. F. Abel, and D. P. Billington: Buckling of Cooling-Tower Shells: Bifurcation Results, *J. Struct. Div. ASCE,* June 1975.
18. Billington, D. P., and J. F. Abel: Design of Cooling Towers for Wind, *Proc. Specialty Conf. ASCE,* Madison, Wis., Aug. 22–25, 1976.
19. Hanna, M. M.: Thin Spherical Shells under Rim Loading, Fifth Congress IABSE, Lisbon, 1956.
20. Billington, D. P.: "Thin Shell Concrete Structures," McGraw-Hill Book Company, New York, 1965.
21. Wind Forces on Structures, *Trans. ASCE,* vol. 126, part II, 1961.
22. Niemann, H. J.: "Zur stationären Windbelastung rotationssymmetrischer Bauwerke im Bereich Transkritischer Reynoldszahlen," Technischwissenschaftliche Mitteilung, no. 71-2, des Instituts für Konstruktiven Ingenieurbau der Ruhr-Universität, Bochum, 1971.
23. Sollenberger, N. J., and Robert H. Scanlan: Pressure Differences across the Shell of a Hyperbolic Natural Draft Cooling Tower, *Proc. Int. Conf. Full Scale Testing of Wind Effects,* London, Ontario, June 1974.
24. Reinforced Concrete Cooling Tower Shells—Practice and Commentary, Report by ACI-ASCE Committee 334, *J. ACI,* January 1977.
25. Concrete Shell Structures, Practice and Commentary, *J. ACI,* September 1964.
26. Chinn, J.: Cylindrical Shell Analysis Simplified by Beam Method, *J. ACI,* May 1959.
27. Design of Cylindrical Concrete Shell Roofs, "ASCE Manual of Engineering Practice 31," New York, 1952.
28. Parme, A. L., and H. W. Conner: Discussion of Ref. 26, *J. ACI,* December 1959.
29. Tedesko, A.: Multiple Ribless Shells, *J. Struct. Div. ASCE,* October 1961.
30. Gibson, J. E.: "The Design of Cylindrical Shell Roofs," 2d ed., D. Van Nostrand Company, Inc., Princeton, N.J., 1961.
31. Design Constants for Circular Arch Bents, *Adv. Eng. Bull.* 7, Portland Cement Association, Chicago, 1963.
32. Phase I Report of the Task Committee on Folded Plate Construction, *J. Struct. Div. ASCE,* December 1963.
33. Pultar, M., et al.: Folded Plates Continuous over Flexible Supports, *J. Struct. Div. ASCE,* October 1967.
34. Parme, A. L.: Shells of Double Curvature, *Trans. ASCE,* vol. 123, 1958.
35. Elementary Analysis of Hyperbolic Paraboloid Shells, Portland Cement Association, Chicago, 1960.
36. Schnobrich, W. C.: Analysis of Hipped Roof Hyperbolic Paraboloid Structures, *J. Struc. Div. ASCE,* July 1972.
37. Students Clear Gym Moments Before Roof Fails, *Eng. News-Rec.,* Sept. 24, 1970.
38. 15-Year-Old H.P. Roof Fails, Injuring 18, *Eng. News-Rec.,* July 10, 1975.
39. Tedesko, A.: Shell at Denver—Hyperbolic Paraboloid Structure of Wide Span, *J. ACI,* October 1960.
40. Candela, F.: General Formulas for Membrane Stresses in Hyperbolic Paraboloid Shells, *J. ACI,* October 1960.
41. Tedesko, A.: Construction Aspects of Thin Shell Structures, *J. ACI,* February 1953.
42. Building Code Requirements for Minimum Design Loads in Buildings and Other Structures, American National Standards Institute, Inc. New York, 1972.
43. "Concrete Thin Shells," ACI Publication SP-28, Detroit, 1971.
44. Whitney, C. S., B. G. Anderson, and H. Birnbaum: Reinforced Concrete Folded Plate Construction, *J. Struct. Div. ASCE,* October 1959.

Section 6

Reinforced-Concrete Bunkers and Silos

GERMAN GURFINKEL
Professor of Civil Engineering, University of Illinois, Urbana

1. Introduction A bin is an upright container for the storage of bulk granular materials. Shallow bins are usually called bunkers; tall bins are called silos. Typical bunkers are shown in Fig. 1. Vertical cross sections of silos are shown in Fig. 2. While bins may be erected individually, they are commonly grouped for increased efficiency in operation. Typical configurations of grouped silos are shown in Fig. 3.

Janssen[1] in 1895 was the first to develop a theory to predict pressures from stored material at rest. Design using his formulas produced acceptable structures for many years. Theimer[2] notes, however, that early designers used low allowable stresses that rendered a factor of safety of about 2.5. With the use of higher-strength steel and concrete, together with higher allowable stresses, the safety margin is reduced considerably. Thus old silos can withstand pressures substantially greater than those predicted by the Janssen formulas, while some newer ones may not.

A number of investigations throughout Europe and the United States have reported pressures generated during material emptying higher than those predicted by Janssen's formulas for the material at rest.[3–11] Provisions for the calculation of these effects have appeared for many years in the German Specification[12] DIN 1055, Sheet 6. In the United States, ACI Committee 313 published a standard where provisions for the calculation of overpressures are given.[13]

Pieper[14] explains emptying phenomena by identifying three zones within the cell (Fig. 4). In zone 3 the stored material is likely to move as a mass, although with different velocities across the section. In zone 2, funnel flow develops. This contrasts with mass flow in that only the central portion of the mass moves toward the outlet, within a contracting channel formed within a stagnant mass. While the latter has poorly defined boundaries in zone 2, it is sharply defined in zone 1, where the material underlying the funnel is at a complete standstill. The relative heights of the zones may be quite different, and any of the three may vanish.

The largest lateral pressures occur in region *A*, Fig. 4, because of the abrupt change in material cross section. In region *B* development of domes may inhibit material flow. These domes are short-lived. Following the collapse of a dome the entire material above falls rapidly and creates a shock. Periodic repetition of this phenomenon creates a hammering effect. In region C the firm walls of the funnel offer the necessary support for development of a large dome which may halt material flow.

Material flow is influenced by the inclination α of the hopper walls with the horizontal. Hoppers with steep inclinations ($\alpha > 70°$) promote major mass flow; flat bottoms and hoppers with gentle inclinations generate conditions for funnel flow of the material.

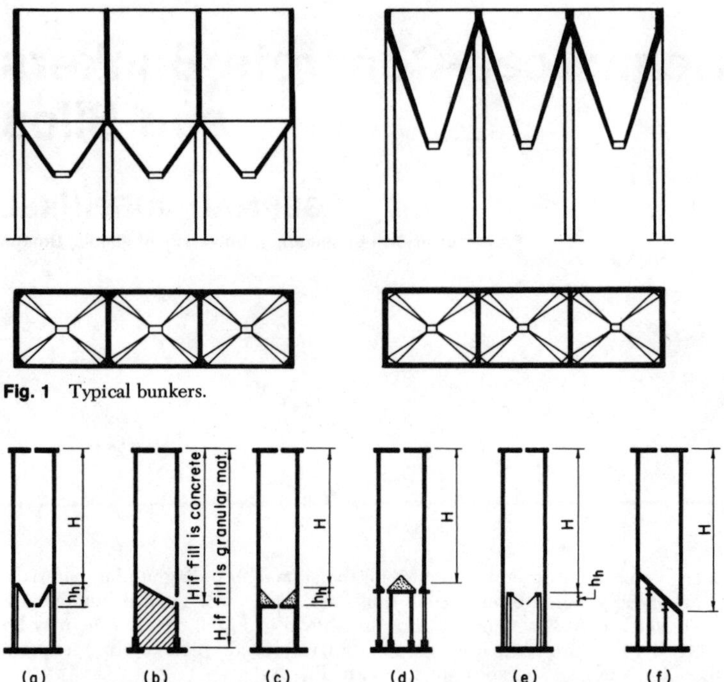

Fig. 1 Typical bunkers.

Fig. 2 Typical silos: (*a*) On raft foundation, independent hopper on pilasters; (*b*) with wall footings and independent bottom slab on fill; (*c*) with hopper-forming fill and bottom slab on thickened lower walls; (*d*) with multiple discharge openings and hopper-forming fill on bottom slab, all supported by columns; raft foundation has stiffening ribs; (*e*) on raft foundation with hopper on ring beam and columns; (*f*) on raft foundation with hopper on concrete walls and steel frames.

2. Bin Pressures Formulas by Janssen[1] and Reimbert[6] for computing bin pressures are given in Table 1. Notation is defined in the table and in Fig. 5. These formulas give pressures for the stored material at rest (static pressures). They give substantially the same results, and are commonly used in the United States.

The Janssen formulas are used in DIN 1055 (Table 1), but with two sets of values of μ' and k, one for filling pressures and one for emptying pressures, and with a coefficient c_1 by which the Janssen values must be multiplied to obtain emptying pressures. Values of c_1 are given by

$$c_1 = 1 + 0.2\left(c_2 + \frac{eL}{1.5A}\right) \tag{1}$$

where c_2 = 1 for organic materials
c_2 = 0 for inorganic materials
e = eccentricity of outlet, measured from centroid of cross section
L = perimeter of silo
A = cross-sectional area of silo

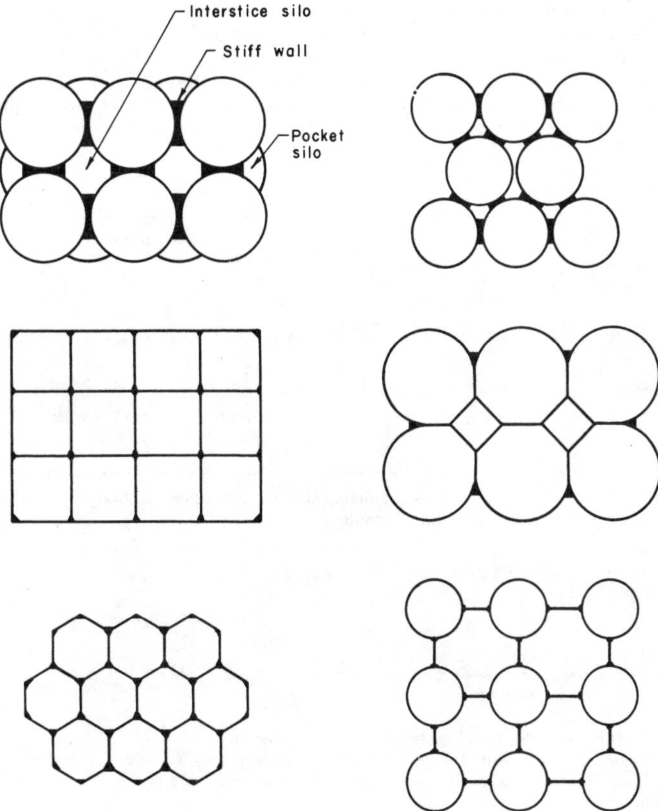

Fig. 3 Grouped silo configurations.

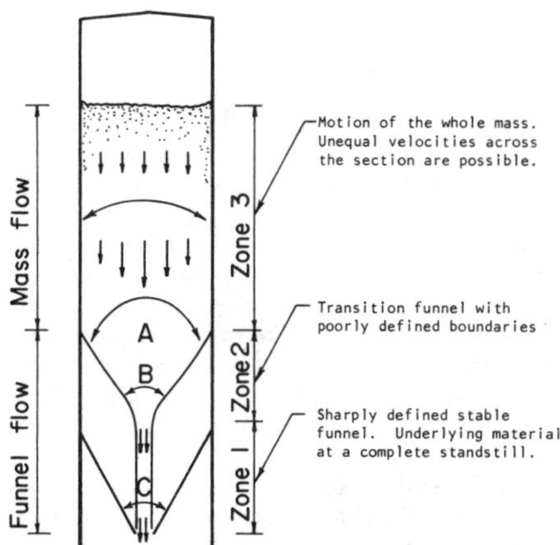

Fig. 4 Flow patterns during emptying. *(From Ref. 14.)*

Equation (1) is not to be used for sugar; instead, use $c_1 = 1$. Also, values of c_1 by Eq. (1) must be multiplied by 1.3 for corn. This is because European experience shows that corn may contain substantial amounts of dust and broken kernels, which probably results from the number of loadings and unloadings during shipment from foreign sources. This situation would not be expected with corn shipped from the field to storage, but might exist to some extent in dockside elevators.

TABLE 1 Material Pressures in Bins

Value at depth Y for	Janssen (static values only)	Reimbert (static values only)	DIN 1055, Sheet 6 (dynamic effects included) Filling	DIN 1055, Sheet 6 (dynamic effects included) Emptying
Vertical pressure q	$\gamma Y_0(1 - e^{-Y/Y_0})$	$\gamma\left[Y\left(\frac{Y}{C}+1\right)^{-1} + \frac{h_s}{3}\right]$	$\gamma Y_{0f}(1 - e^{-Y/Y_{0f}})$	$\gamma Y_{0e}(1 - e^{-Y/Y_{0e}})$
Lateral pressure p	qk	$p_{max}\left[1 - \left(\frac{Y}{C}+1\right)^{-2}\right]$	$q_f k_f$	$c_1 q_e k_e$
Vertical frictional force per unit width of wall V	$(\gamma Y - q)R$*	$(\gamma Y - q)R$	$(\gamma Y - q_f)R$	$c_1(\gamma Y - q_e)R$

*Reference 13 gives this as $(\gamma Y - 0.8q)R$. The coefficient 0.8 was introduced so that the formula would give about the same values as the Reimbert formulas.

Notation:

γ = weight per unit volume of stored material (Tables 4 and 5)
Y = depth defined in Fig. 5
$Y_0 = R/\mu' k$
$Y_{0f} = R/\mu'_f k_f$, filling
$Y_{0e} = R/\mu'_e k_e$, emptying
R = hydraulic radius of horizontal cross section of storage space = A/L (Table 6)
A = cross-sectional area of silo (Table 6)
L = perimeter of silo (Table 6)
μ' = coefficient of friction between stored material and wall (Table 5)
μ'_f = coefficient of friction between stored material and wall during filling (Table 2)
μ'_e = coefficient of friction between stored material and wall during emptying (Table 2)
c = roughness factor (Table 3)
c_1 = coefficient given by Eq. (1)
$k = p/q = (1 - \sin\rho)/(1 + \sin\rho)$, Janssen and Reimbert
$k_f = p_f/q_f$ (Table 2)
$k_e = p_e/q_e$ (Table 2)
ρ = angle of internal friction or (approximately) angle of repose ϕ (Tables 4 and 5)
C = characteristic abscissa for Reimbert's formula (Table 8)
h_s = height of sloping top surface of stored material (Fig. 5)
p_{max} = Reimbert's maximum static pressure (Table 8)

To account for the load-reducing effect of the bin bottom, DIN allows the emptying pressure to be reduced over a height $1.2D$, but not to exceed $0.75h$, above the outlet. This is accomplished by prescribing a linear variation from the filling pressure at the outlet to the emptying pressure at the top of the reduced-pressure zone.

Pieper[14] also suggests two sets of values of μ' and k for use in the Janssen formula, with a roughness factor c to determine μ' (Table 2). Five categories of roughness are defined (Table 3). However, it has been observed that rough walls of grain silos have been made smoother by the development with time of a skin of wax and fat-like material. Because of the resulting reduced friction, less weight of grain is transferred to the walls, so that more acts on the hopper. As a result, hoppers have sometimes failed unexpectedly after a number of years of operation.[15] Where this effect is to be expected, a reduction in the value of c in Table 3 is recommended (footnote a).

Bunkers. Although the formulas in Table 1 can be used for shallow bins, Rankine's active pressure is sometimes used for bunkers. In this case $q = \gamma Y$ and $p = kq$. Coulomb's formula for pressure on retaining walls has also been used for bunkers.

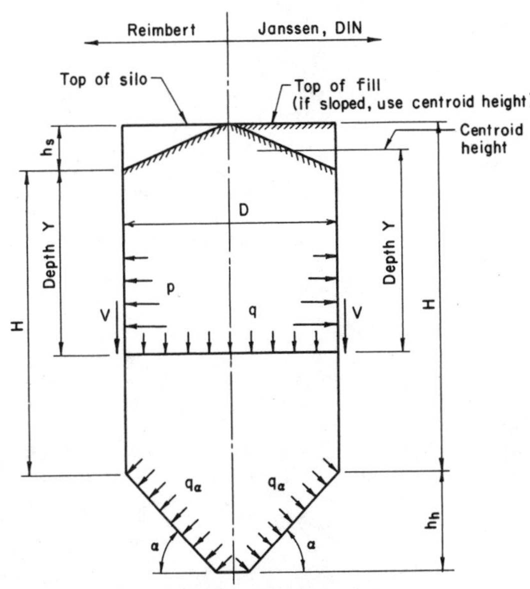

Fig. 5 Silo dimensions and pressures.

TABLE 2 Values of k and μ': DIN 1055[12] and Pieper[14]

	DIN 1055		
	Grain material, avg diam > 0.2 mm	Powder material, avg diam < 0.06 mm	Pieper
k_f	0.5	0.5	$1 - \sin \rho_f$*
k_e	1	1	$1.5\ k_f$
μ'_f	$\tan 0.75\ \rho_f$	$\tan \rho_f$	$\tan \rho_f c$
μ'_e	$\tan 0.60\ \rho_f$	$\tan \rho_f$	$\tan 0.8\ \rho_f c$

See Table 1 for notation, Table 3 for values of c, Table 4 for values of p_f.
*Use $1 - \sin 1.5\ \rho_f$ for flour.

TABLE 3 Values of Roughness Factor c (From Ref. 14)

	Roughness category[a]				
Stored material	1[b]	2[c]	3[d]	4[e]	5[f]
Dustlike[g]	0.85	0.90	0.95	0.95	
Fine-grained[h]	0.75	0.80	0.85	0.90	0.95
Coarse-grained[i]		0.75	0.80	0.85	0.95

[a]For stored material that makes silo walls smoother, the value of c for the calculation of q and p should be reduced by 0.05, except for roughness category 1. When designing against buckling of silo walls, increase c by 0.05.

[b]Glass and enamel with smooth joints, few in number, welded aluminum, coatings of synthetic materials.

[c]Plywood with few joints, riveted or bolted (round heads) aluminum, welded steel, finished concrete, wood (planed boards with vertical joints).

[d]Riveted or bolted steel, perforated steel sheets, wood (unplaned boards with horizontal joints), unfinished concrete.

[e]Corrugated steel sheets. fine-mesh wire screen.

[f]Coarse-mesh wire screen.

[g]Grain diameter < 0.1 mm. Grain flour, finely ground rock, cement.

[h]Maximum grain diameter < 1 cm. Sand, grain, soybeans, granulated artificial materials.

[i]Gravel, cement clinker, coke, ore.

Angle of Internal Friction. This depends on whether the material is at rest or in motion. The value ρ_f for material at rest can be determined experimentally, approximately as the angle of repose ϕ or exactly by using a shear box. In the latter method, ρ_f is the slope of the line obtained by plotting pairs of values of shear τ and normal stress σ at which the material fails by shear, that is, $\rho_f = \tan^{-1} \tau/\sigma$.

The angle of internal friction ρ_e during emptying of model silos has been measured at approximately 0.8 ρ_f.

The static value of ρ_f is used in computing pressures by the Reimbert formulas and, except for DIN and Pieper, Janssen's formulas.

Values of ρ_f for various materials are given in Table 4. Values of ϕ are given in Table 5.

TABLE 4 Physical Properties of Storage Materials[a]

	Unit weight, lb/ft³						
	γ_{min}	γ_{max}	γ_1[b]	γ_2[b]	ρ_f[c]	k_f[d]	k_e[e]
Grain	50	56	54	52	31	0.485	0.727
Grain flour	37	50	46	42	29	0.312	0.467
Rice	50	56	54	52	33	0.455	0.683
Corn	46	52	50	48	28	0.530	0.796
Quartz sand	91	104	100	96	34	0.441	0.661
Cement	75	112	100	87	28	0.530	0.796
Cement clinker	100	128	119	109	36	0.412	0.618
Limestone powder	66	82	77	71	27	0.546	0.819
Gravel	87	125	112	100	32	0.470	0.705
Sugar, refined	52	62	59	56	34	0.441	0.661

[a]Adapted from Ref. 14.
[b]Use is determined by h; see Eqs. (2).
[c]For the angle of internal friction during emptying, take $\rho_e = 0.8\ \rho_f$.
[d]$k_f = 1 - \sin \rho_f$, except for grain flour where $k_f = 1 - \sin 1.5\ \rho_f$.
[e]$k_e = 1.5\ k_f$.

Weight of Material. The unit weight γ of the material stored depends not only on its specific weight but also on its compressibility. The latter is influenced by storage height and time, and by vibrations to which the material may be subjected. Pieper defines lower and upper values γ_{min} and γ_{max}, which are determined experimentally. The design value is determined by the height h of the silo (Fig. 5) as follows:

$$h > 10\ \text{m}(32.8\ \text{ft}) \qquad \gamma_1 = \frac{\gamma_{min} + 2\gamma_{max}}{3} \tag{2a}$$

$$h < 5\ \text{m}(16.4\ \text{ft}) \qquad \gamma_2 = \frac{2\gamma_{min} + \gamma_{max}}{3} \tag{2b}$$

Values of γ for $5 < h < 10$ m are determined by linear interpolation.

Values of γ_{min}, γ_{max}, γ_1, and γ_2 for various materials are given in Table 4. Values of γ are also given in Table 5.

Coefficient of Friction. **Values of μ' for various materials are given in Table 5. These values are not to be used in the DIN and Pieper procedures, for which μ' depends on the internal angle of friction ρ_f (Table 2).**

Values of k. Except for the DIN and Pieper procedures, values of k are computed from $k = (1 - \sin \rho)/(1 + \sin \rho)$ (Table 1). DIN and Pieper values are given in Table 2. Values of k for various materials according to Pieper's formulas are given in Table 4.

Geometric Properties. Formulas for the cross-sectional area A, perimeter L, and hydraulic radius R for various cross sections are given in Table 6.

Janssen Formulas (Table 1). As Y approaches infinity (practically, and within 1 percent error, when $Y/Y_0 = 4.6$), the maximum pressures are

$$q_{\max} = \gamma Y_0 = \frac{\gamma R}{\mu' k} \tag{3a}$$

$$p_{\max} = k q_{\max} = \frac{\gamma R}{\mu'} \tag{3b}$$

Table 7 gives values of $1 - e^{-Y/Y_0}$ by which $q_{\max}$ and $p_{\max}$ are multiplied for the region $0 < Y/Y_0 < 4.9$. Linear interpolation is acceptable.

Reimbert Formulas (Table 1). Equations (3) also give the maximum pressures by the Reimbert formulas. Values of $p_{\max}$ and C for various cross sections are given in Table 8.

TABLE 5 Physical Properties of Granular Materials*

	γ, lb/ft^3	ϕ, deg	μ' Against concrete	μ' Against steel
Cement, clinker	88	33	0.6	0.3
Cement, portland	84–100	24–30	0.36–0.45	0.30
Clay	106–138	15–40	0.2–0.5	0.36–0.7
Coal, bituminous	50–65	32–44	0.50–0.60	0.30
Coal, anthracite	60–70	24–30	0.45–0.50	0.30
Coke	38	40	0.80	0.50
Flour	38	40	0.30	0.30
Gravel	100–125	25–35	0.40–0.45	
Grains, small†	44–62	23–37	0.29–0.47	0.26–0.42
Gypsum in lumps, limestone	100	40	0.5	0.3
Iron ore	165	40	0.50	0.36
Lime, burned (pebbles)	50–60	35–55	0.50–0.60	0.30
Lime, burned, fine	57	35	0.5	0.3
Lime, burned, coarse	75	35	0.5	0.3
Lime, powder	44	35	0.50	0.30
Manganese ore	125	40		
Sand	100–125	25–40	0.40–0.70	0.35–0.50
Sugar, granular	63	35	0.43	

*From Table 4A of Ref. 13. The properties listed here are illustrative of values which might be determined from physical testing. Ranges of values show the variability of some materials. Design parameters should preferably be determined by test on samples of the actual materials to be stored.

†Wheat, corn, soybeans, barley, peas, beans (navy, kidney), oats, rice, rye.

3. Emptying Pressures in Funnel-Flow Silos. For the purpose of computing emptying pressures, Pieper[14] classifies silos as shallow, intermediate in height, and tall.

Shallow Silos (Fig. 6a). If $h < h_{F1}$, where $h_{F1} = \frac{1}{2}(D - d) \tan \alpha_1$, with $\alpha_1 = 29(\sqrt[4]{\rho_f c})$, emptying does not increase the pressures that were generated during filling. Therefore, only filling pressures need be considered in design.

Tall Silos (Fig. 6b). If $h \geq 2h_{F2}$, where $h_{F2} = \frac{1}{2}(D - d) \tan \alpha_2$, with $\alpha_2 = 32(\sqrt[4]{\rho_f c})$, design is governed by emptying pressures. Between the level of the outlet and the height h_{F2} above it, emptying pressures reduce linearly from the value at height h_{F2} to the value of the filling pressure at the outlet.

Silos of Intermediate Height. If $h_{F1} < h < 2h_{F2}$, the height of the reduced pressure zone is determined by linear interpolation between h_{F1} and h_{F2} by

$$h_F = h_{F1} + \frac{h_{F2} - h_{F1}}{2h_{F2} - h_{F1}} (h - h_{F1}) \tag{4}$$

There is a zone immediately above the reduced pressure zone in tall and intermediately tall silos in which emptying pressures exceed p_e. Figure 7c shows a simplified vertical distribution of the pressure increase for tall silos. The height h_{F2} for a central opening (Fig. 7b) also applies for eccentric outlets (Fig. 7a). Lateral distributions of the pressure increase for a central outlet, an outlet at one side, and outlets at opposite sides are shown in Fig. 7d.

TABLE 6 Geometry of Silo Cross Sections

Cross section		Area A	Perimeter L	Hydraulic radius R
Circle 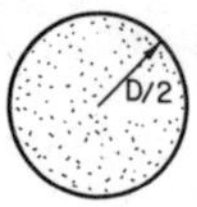		$\frac{\pi}{4}D^2$	πD	$\frac{D}{4}$
Circular sector 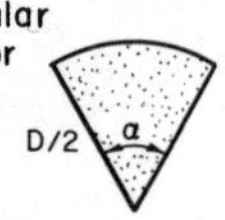	α, Radians	$\frac{\alpha D^2}{8}$	$(\frac{\alpha}{2}+1)D$	$\frac{D/4}{1+2/\alpha}$
	α, Degrees	$0.00218\,\alpha D^2$	$(0.0087\alpha+1)D$	$\frac{D/4}{1+114.6/\alpha}$
Regular polygon 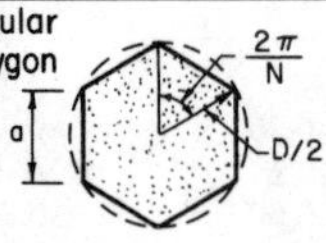		$\frac{ND^2}{8}\sin\frac{2\pi}{N}$	$ND\sin\frac{\pi}{N}$	$\frac{D}{4}\cos\frac{\pi}{N}$
		$\frac{ND^2}{4}\cot\frac{\pi}{N}$	Na	$\frac{a}{4}\cot\frac{\pi}{N}$
Rectangle 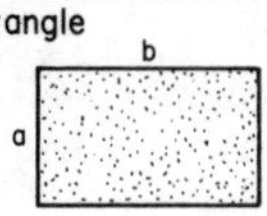		ab	$2(a+b)$	$\frac{ab}{2(a+b)}$, Long Side $\frac{a}{4}$, Short Side
Long rectangle 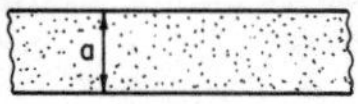		—	—	$\frac{a}{2}$
Ring 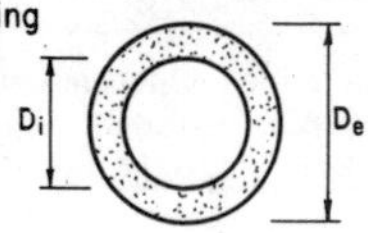		$\frac{\pi}{4}(D_e^2-D_i^2)$	$\pi(D_e+D_i)$	$\frac{(D_e-D_i)}{4}$
Interstice 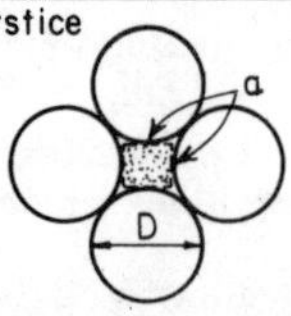		$(1-\frac{\pi}{4})D^2$	πD	$\frac{a}{4}=0.104D$

TABLE 7 Values of $1 - e^{-Y/Y_0}$

Y/Y_0	0	0.1	0.2	0.3	0.4	0.5	0.6	0.7	0.8	0.9
0	0	0.095	0.181	0.259	0.333	0.393	0.451	0.503	0.551	0.593
1	0.632	0.667	0.699	0.727	0.753	0.777	0.798	0.817	0.835	0.850
2	0.865	0.878	0.889	0.900	0.909	0.918	0.926	0.933	0.939	0.945
3	0.950	0.955	0.959	0.963	0.967	0.970	0.973	0.975	0.978	0.980
4	0.982	0.983	0.985	0.986	0.988	0.989	0.990	0.991	0.992	0.993

TABLE 8 Values of p_{max} and C in the Reimbert Formulas

Silo	p_{max}	C	Remarks
Circular	$\frac{\gamma D}{4\mu'}$	$\frac{D}{4\mu' k} - \frac{h_s}{3}$	
Polygonal, of more than four sides	$\frac{\gamma R}{\mu'}$	$\frac{L}{\pi}\frac{1}{4\mu' k} - \frac{h_s}{3}$	R from Table 6
Rectangular, on shorter wall a	$\frac{\gamma a}{4\mu'}$	$\frac{a}{\pi\mu' k} - \frac{h_s}{3}$	
Rectangular, on longer wall b	$\frac{\gamma a'}{4\mu'}$	$\frac{a'}{\pi\mu' k} - \frac{h_s}{3}$	$a' = \frac{2ab}{a + b}$

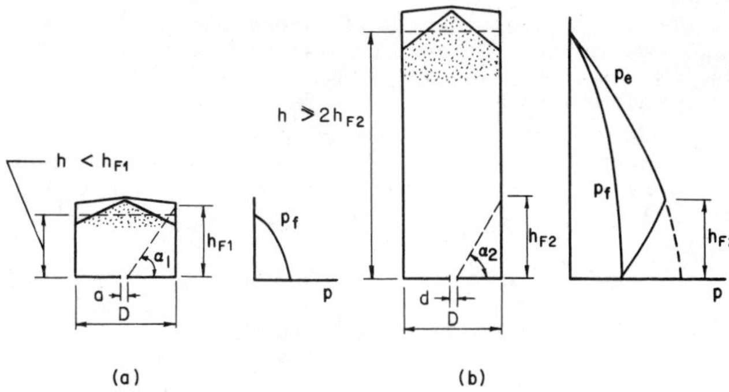

Fig. 6 (*a*) Shallow silo; (*b*) tall silo.

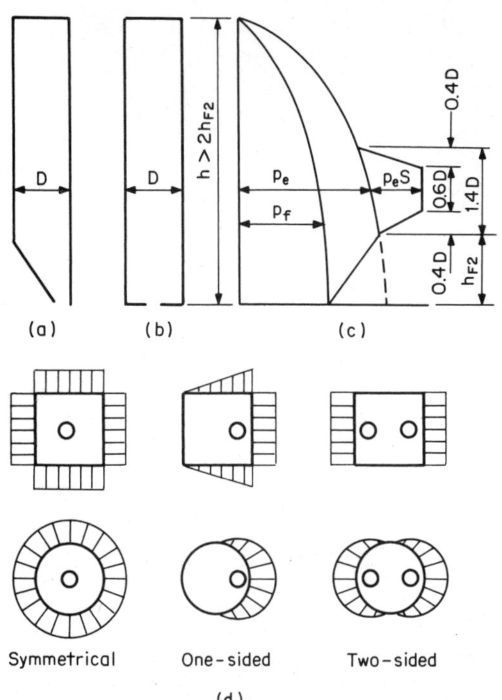

Fig. 7 Emptying pressure increase in tall silos.

Values of the factor S in Fig. 7*c* have been determined for quartz sand and wheat in square silos with a variety of outlet locations. Values ranged from 0.15 for a central outlet to 0.25 for edge and corner outlets. A value 0.4 was reached for a bin with a long slot outlet on each of the four edges. The pressure increase appears to be material-dependent, since an experiment with rice in a silo with a slot outlet on one side gave $S = 0.90$.

Values of S and the height over which the corresponding pressure increase is distributed in intermediately tall silos can be determined by linear interpolation between zero for $h = h_{F1}$ and the values for $h = h_{F2}$.

The increased pressures are assumed to be distributed over the entire width of the wall and the height shown, and pressures at some points in the region of increased pressure may be several times larger than the values shown. However, the average values can be used to dimension reinforced-concrete silos because their walls are stiff enough to distribute load concentrations.

Effect of Eccentric Outlets on Pressures below h_{F2}. Figure 8*a* shows the flow funnel in a circular silo with an eccentric outlet.[14] The emptying pressure p_e in the funnel is less than the filling pressure in the silo, and varies from p_f at the top of the funnel to zero at the level of the outlet (Fig. 8*b*). The differential pressure $\Delta p = p_f - p_e$ at the conical surface of the funnel is assumed to be transferred to the silo walls by arching, which causes a concentrated increase in pressure on the wall at the intersection of the funnel with the wall (Fig. 8*c*). For design, Δp is assumed to be most unfavorable at the center of the height $h_F/2$ from the top of the funnel, the corresponding value of p_e being given by

$$p_e = \frac{r_F}{D} p_f$$

where r_F is the radius of the funnel from point A on the silo wall.

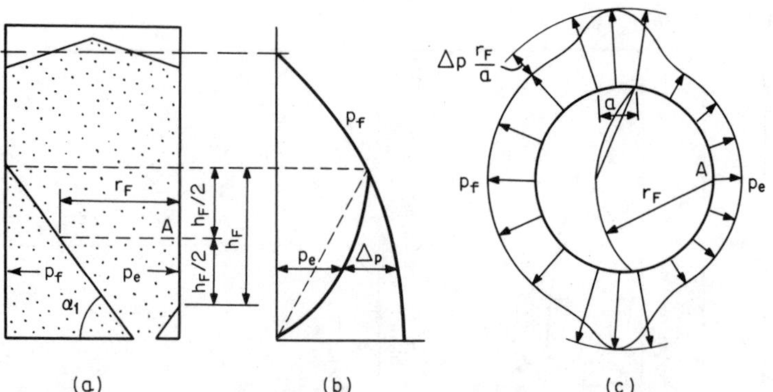

Fig. 8 Emptying pressures in silo with eccentric outlet.

The circumferential distribution of pressure is shown in Fig. 8*c*. At each intersection of the funnel boundary with the silo wall there is a zone where the pressure exceeds p_f. The width a of this zone is given by $D/10 \leqq a \leqq r_F/2$ and the average pressure over the zone by $r_F \Delta p / a$.

The angle α in Fig. 8*a* should be taken equal to α_1 except that the resulting intercept on the silo wall should not be taken larger than h_{F2}. Also, in low silos the funnel may intersect the surface of the stored material instead of the silo wall.

Pressures due to eccentric outlets in rectangular silos may be determined similarly.

4. Emptying Pressures in Funnel-Flow Silos—ACI 313. Emptying pressure for silos with central outlets are obtained by multiplying filling pressures by *overpressure factors* C_d (Table 9) or *impact factors* C_i (Table 10). The larger values are to be used for design. Values of C_d for the Janssen filling pressures are from the Soviet Silo Code,[16] with some slight modifications. Safarian[17] computed the values for the Reimbert formulas so as to give the same pressures as the Janssen formula using the Soviet Code values.

TABLE 9 Recommended Minimum Overpressure Factor C_d*

Portion of silo	$\frac{H}{D} < 2$		$2 = \frac{H}{D} < 3$		$3 = \frac{H}{D} < 4$		$4 = \frac{H}{D} < 5$		$\frac{H}{D} \geq 5$	
	J	*R*	*J*	*R*	*J*	*R*	*J*	*R*	*J*	*R*
Upper zone, H_1 high	1.35	1.10	1.45	1.20	1.50	1.25	1.60	1.30	1.65	1.35
Next zone, $(H\text{-}H_1)/4$ high	1.45	1.20	1.55	1.30	1.60	1.35	1.70	1.40	1.75	1.50
Next zone, $(H\text{-}H_1)/4$ high	1.55	1.45	1.65	1.55	1.75	1.60	1.80	1.70	1.90	1.75
Next zone, $(H\text{-}H_1)/4$ high	1.65	1.65	1.75	1.75	1.85	1.85	1.90	1.90	2.00	2.00
Lower zone, $(H\text{-}H_1)/4$ high	1.65	1.65	1.75	1.75	1.85	1.85	1.90	1.90	2.00	2.00
Hopper†										
Bottom:										
Concrete	1.35	1.50	1.35	1.50	1.35	1.50	1.35	1.50	1.35	1.50
Steel	1.50	1.75	1.50	1.75	1.50	1.75	1.50	1.75	1.50	1.75

*Adapted from Ref. 13.

†For hoppers use pressure for zone above, uniform throughout h_h, or reduce pressure in accordance with *R*. Pressures for hopper-forming fill may be reduced linearly from top of fill to top of flat slab.

J = Janssen, R = Reimbert.

H = height defined in Fig. 2

$H_1 = D \tan \rho$ for circular silos

$H_1 = a \tan \rho$ and $b \tan \rho$ for rectangular silos

C_d applies to the lateral pressure at bottom of each height zone.

Bottom pressures need not be considered larger than the pressure from the weight of silo contents.

Values in this table are too small for mass flow.

In the region of a flow-correcting insert (e.g., Buhler Nase) lateral pressures may be much larger than static pressures, and values in this table are too small.

Eccentric Outlets. It is suggested in the commentary that the increase in lateral pressure due to eccentric discharge be taken to be at least 25 percent of the static (filling) pressure at the bottom of the silo if the outlet is adjacent to the wall. For outlets with smaller eccentricities this value may be reduced linearly to zero for a central outlet. The increase may also be reduced linearly to zero at the top of the silo, and need not be multiplied by the overpressure factor C_d. Thus the design pressure at depth Y is

$$p = C_d p_f + 0.25 p_f \frac{e}{r}\frac{Y}{H} \tag{5}$$

where r = radius of bin.

TABLE 10 Recommended Minimum Impact Factor C_i*

Ratio of volume dumped in one load to silo capacity	1:2	1:3	1:4	1:5	1:6 and less
Concrete bottom	1.4	1.3	1.2	1.1	1.0
Steel bottom	1.75	1.6	1.5	1.35	1.25

*Adapted from Ref. 9.

For circular silos it is conservative to compute the increased hoop forces by assuming the increased pressure to be distributed uniformly around the circumference. This will not enable the bending moments and shears caused by the actual distribution to be evaluated, but if the walls are reinforced at both faces the additional steel required by the increased hoop force should take care of the moments. However, for circular silos with large eccentricities of discharge, and for rectangular silos, it is advisable to consider more realistic pressure distributions such as those of Fig. 8.

5. Shock Effects from Collapse of Domes During emptying of a silo, domes of the stored material may form. They eventually collapse as emptying continues and their spans become too long, although for cohesive materials it is sometimes necessary to destroy them by external means to restore flow. Collapse of a dome is followed by free falling of the material above, and shocks are created when it strikes the material below. These shocks increase the emptying lateral pressure p_e and the vertical filling pressure q_f. The increased pressures p_d and q_d can be computed from[14]

$$p_d = p_e\left(1 + \frac{1.8S_d}{\sqrt{a}}\right) \qquad a \geq 3 \tag{6a}$$

$$q_d = q_f(1 + 2S_d) \tag{6b}$$

where S_d = coefficient which depends on material
a = horizontal distance, ft, between center of outlet and corresponding point on wall

The following values of S_d are from Ref. 14:

Material	S_d
Quartz gravel	0.10
Grain	0.15
Corn	0.30
Cement clinker	0.30–0.40

Shock effects may be disregarded if $1.8S_d/\sqrt{a}$ is less than 0.05.

Maximum shocks occur in silos with height-width ratios equal to or greater than 4. Experience has shown that shock effects are not significant if this ratio is less than 2. Linear interpolation can be used for intermediate height-width ratios.

Shock effects increase with decrease in the distance a from outlet to wall. In addition, the unevenly distributed shock pressure from eccentric outlets induces bending moments and shears in the silo walls. Therefore, it is recommended that only central outlets be used in silos for materials that are known to generate domes.

6. Pressures Induced by Dustlike Materials Dustlike materials such as cement, wheat flour, and limestone powder (talc) with an average particle diameter under 0.06 mm (0.0024 in.) become mixed with air during rapid filling of a silo. The mixture has been shown[14] to possess hydrostatic characteristics and is able to develop a maximum lateral pressure

$$p_{max} = 1.6\, \gamma_{min}\, v t_s k_f \tag{7}$$

where γ_{min} = minimum unit weight, Table 4
v = filling velocity measured as vertical rise of surface of material per hour
t_s = setting time of material, hr. Use 0.24 hr for talc, 0.19 hr for cement, and 0.14 hr for wheat flour
$k_f = p_f/q_f$ (Table 2)

The lateral pressure increases linearly from zero at the surface of the material to the maximum pressure, which occurs at a depth vt_s from the surface. The maximum pressure remains constant down to the bottom of the silo.

After the dustlike material has settled, and during emptying, pressures may be determined as for grainlike materials (Table 1).

It is necessary for design purposes to consider not only the effects caused by the rapid filling, but also the conventional pressures after the material has settled, and during emptying. Design of silos with small hydraulic radius may be controlled by p_{max} generated during rapid filling (Fig. 9*a*). Design of silos with larger hydraulic radius should be based on the envelope of maximum lateral pressures. This means that design will be governed at the top portion of the silo by lateral pressures generated during rapid filling and at the lower portion by lateral pressures generated during emptying (Fig. 9*b*).

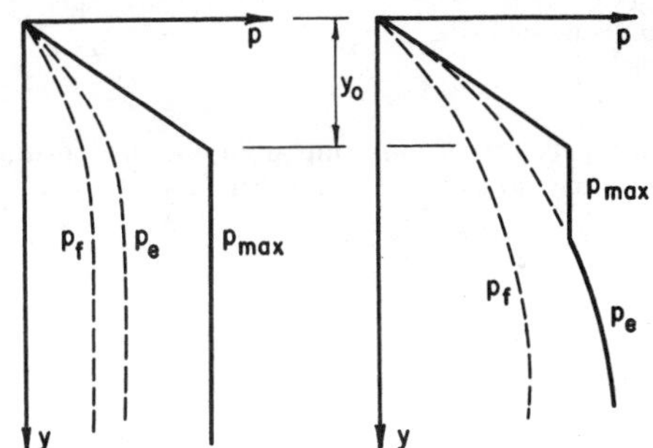

Fig. 9 Wall pressures from dust-like materials.

Pneumatic Emptying. Pneumatic devices to assist emptying (Fig. 10*a*) prevent a buildup of dustlike material and subsequent closing of the outlet. If the imposed air pressure p_u exceeds the emptying pressures p_e and q_e, it must be used for design of the lower portion of the silo (Fig. 10*b*). The range of height Δh above which this pressure must be accounted for is given by Pieper[14] as

$$\Delta h = 1.6 \frac{p_u}{\gamma_{min}} \tag{8}$$

The pressure variation for design is shown in Fig. 10*b*.

Homogenizing Silos. In these silos dustlike materials are thoroughly mixed with air. Design must satisfy the conventional pressures and loads of Table 1 as well as the hydrostatic pressures generated during mixing. For the latter

$$p = q = C\gamma_1 Y \tag{9}$$

where γ_1 is unit weight from Table 4. The coefficient C is given as 0.6 and 0.55 by ACI 313[13] and Pieper,[14] respectively.

7. Earthquake Forces The following minimum requirements have been proposed for design.[13] The total lateral seismic force H_e for shear at the base is given by

$$H_e = ZC_p(W_g + W_{\text{eff}}) \tag{10}$$

where Z = earthquake-zone factor = 3/16, 3/8, 3/4, and 1 for zones 1, 2, 3, and 4, respectively
W_g = weight of structure
W_{eff} = 80 percent of weight of stored material, applied at centroid of volume
C_p = 0.2 for silos with material stored on bottoms above ground and 0.1 when silo walls extend to ground and stored material rests directly on ground. For intermediate cases C_p may be obtained by linear interpolation

If the bin bottom-supporting system is independent of the walls, W_{eff} may be distributed between the two independent structures according to their relative stiffnesses.

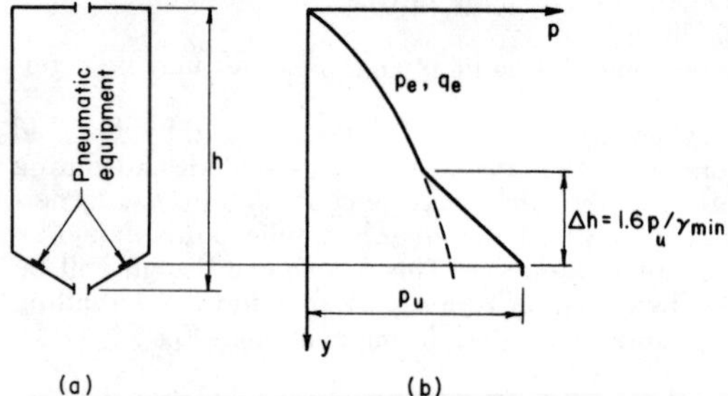

Fig. 10 Additional pressure generated by pneumatic equipment used during emptying of dust-like materials.

A dynamic analysis, using a design earthquake spectrum compatible with the seismic zone and with local foundation conditions, may be used instead of Eq. (10).

WALL FORCES

8. Circular Silos The hoop force F per unit height of wall due to the radial pressure p of the stored material is given by

$$F = \frac{pD}{2} \tag{11}$$

The increase ΔD in the diameter can be determined from

$$\Delta D \approx \frac{pD^2}{4tE_c} \tag{12}$$

where t = wall thickness.

Walls are also subjected to vertical compression from the roof, from their weight, and from wall friction of the stored material.

Eccentric Outlets. Forces due to the additional pressures caused by eccentric outlets (Fig. 7) are given in Figs. 11 and 12.

Grouped Circular Silos. Some groups frequently used are shown in Fig. 3. Initial design of individual circular cells of these groups is done as if they were isolated. The effects of interstices are determined by considering the adjacent circular cells to be empty (Fig. 13). Formulas for the case of an empty central circular cell with adjacent full circular cells are given in Fig. 14.

Analysis of other cross-section configurations and empty-full cases is possible by treating the whole horizontal cross section as a rigid frame. The curved members and variable wall thicknesses at wall intersections may make an "exact" analysis quite complex and laborious. Approximate methods based on local analysis of elements fixed at the sections where silo walls intersect are acceptable and may be the only practical thing to do in many cases.

Attention to local effects at wall intersections is important. Good design should provide strength for proper transfer of bending moments, shear, and direct axial force at these locations.

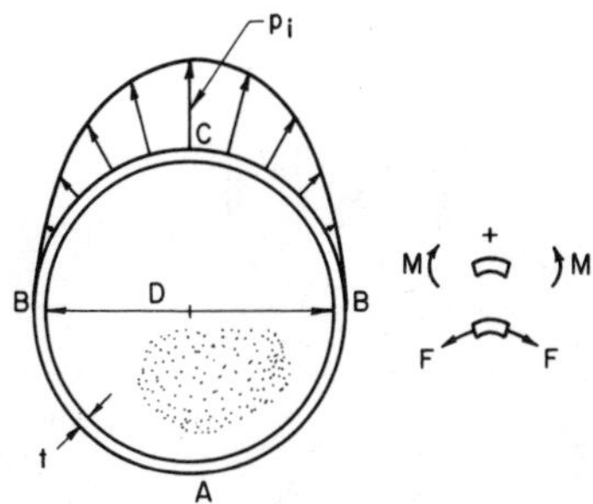

Bending moments

$M_A = -0.0183\ p_i (D+t)^2$

$M_B = 0.0208\ p_i (D+t)^2$

$M_C = 0.0234\ p_i (D+t)^2$

Hoop forces

$F_A = 0.1238\ p_i (D+t)$

$F_B = 0.1667\ p_i (D+t)$

$F_C = -0.2905\ p_i (D+t)$

Fig. 11 One-sided additional pressure on circular silo.

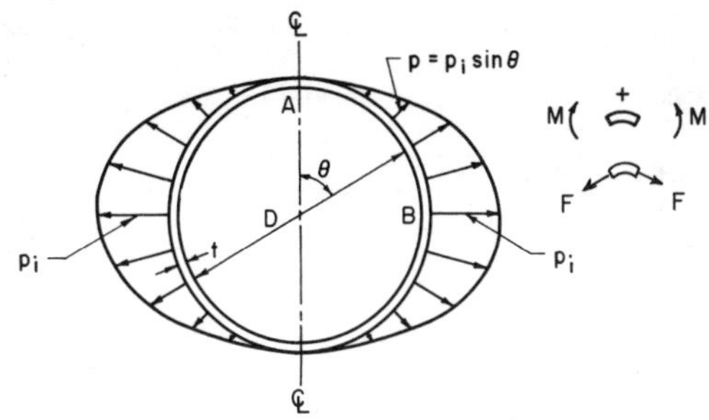

Bending moments

$$M = p_i \frac{(D+t)^2}{4} \left[\left(\frac{\pi}{4} - \frac{\theta}{2} \right) \cos\theta + \frac{\sin\theta}{2} - \frac{2}{\pi} \right]$$

$M_A = 0.0372\ p_i (D+t)^2$

$M_B = 0.0342\ p_i (D+t)^2$

Hoop forces

$$F = p_i \frac{(D+t)}{8} \left[(\pi - 2\theta)\cos\theta + 2\sin\theta \right]$$

$F_A = 0.393\ p_i (D+t)$

$F_B = 0.250\ p_i (D+t)$

Maximum Shear $V = 0.140\ p_i D$, at $\theta = 40.7°$

Fig. 12 Two-sided additional pressure on circular silo. *(After G. Ruzicka.)*

9. Rectangular and Polygonal Silos Walls in the pressure zones of square, rectangular, and polygonal silos are subjected to bending moment, horizontal shear, and horizontal tension due to the lateral pressure, and to vertical compression from the roof, from their weight, and from wall friction of the stored material.

Walls whose height is more than twice the width may be analyzed for one-way bending in the horizontal direction. Since adjoining walls are continuous at their junctures, moments may be determined as for a frame. Formulas for M, F, and V for a horizontal strip are given for rectangular walls in Fig. 15 and regular polygonal walls of N sides in Fig. 16.

Rectangular walls whose height is less than half the width may be analyzed for one-way bending in the vertical direction. The lower edge can usually be assumed fixed. The upper edge may be assumed fixed or simply supported, depending on the attached construction, or free if there is none.

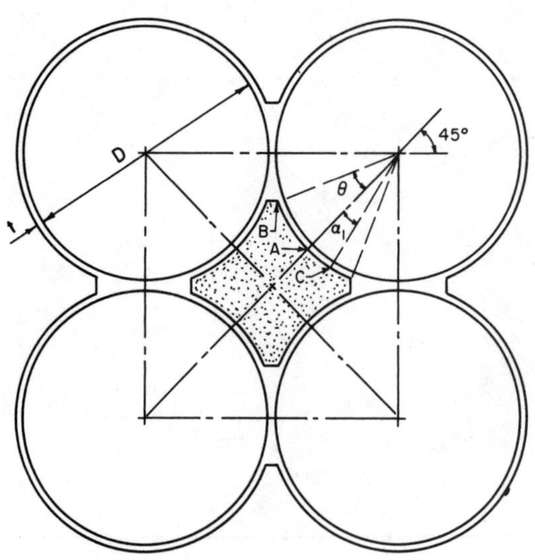

Bending moments

$$M_A = \frac{p}{4}(D+2t)(D+t)\sin\theta\left(1-\frac{\sin\theta}{\theta}\right)$$

$$M_B = \frac{p}{4}(D+2t)(D+t)\sin\theta\left(\cos\theta-\frac{\sin\theta}{\theta}\right)$$

$$M_C = \frac{p}{4}(D+2t)(D+t)\sin\theta\left(\cos\alpha_1-\frac{\sin\theta}{\theta}\right)$$

Compression forces

$$F_A = \frac{p}{2}(D+2t)(1-\sin\theta)$$

$$F_B = \frac{p}{2}(D+2t)(1-\sin\theta\cos\theta)$$

$$F_C = \frac{p}{2}(D+2t)(1-\sin\theta\cos\alpha_1)$$

Fig. 13 Full interstice and empty adjacent cells. *(Adapted from Ref. 18.)*

Moments in walls whose height is more than half the width but less than twice the width should be determined as for a plate supported on four edges or, if the upper edge is free, as a plate supported on three edges. Tables of moment coefficients for various cases are given in Refs. 8, 19, and 20.

Walls that are supported on columns are subjected to in-plane bending due to the load from an attached bottom. Analysis depends on the height of the wall relative to the spacing of the columns (Art. 17).

10. Thermal Effects A differential temperature ΔT between the interior and exterior faces of a silo creates a strain gradient $\alpha_t\Delta T/t$ in the wall section, where t is the thickness of the wall and α_t is the thermal coefficient of expansion of concrete. Because of the closed nature of silos, rotational restraint is imposed by continuity, and thus bending moments are generated in the presence of thermally induced strain gradients. For a section subject to bending moment M and axial force P caused by loading unrelated to thermal effects, Gurfinkel[21] has shown that the additional thermal moment M_t for a given strain gradient depends on the existing strain distribution, which in turn depends on M and P; iteration is required to determine M_t. ACI 313[13] calculates the thermal bending moments M_{xt} per unit of wall height and M_{yt} per unit of wall width as if generated in an uncracked section of wall subjected to a state of plane strain. Thus,

$$M_t = \frac{E_c t^2 \alpha_t \, \Delta T}{1 - \nu} \tag{13}$$

where E_c and ν are the concrete modulus of elasticity and Poisson's ratio, respectively.

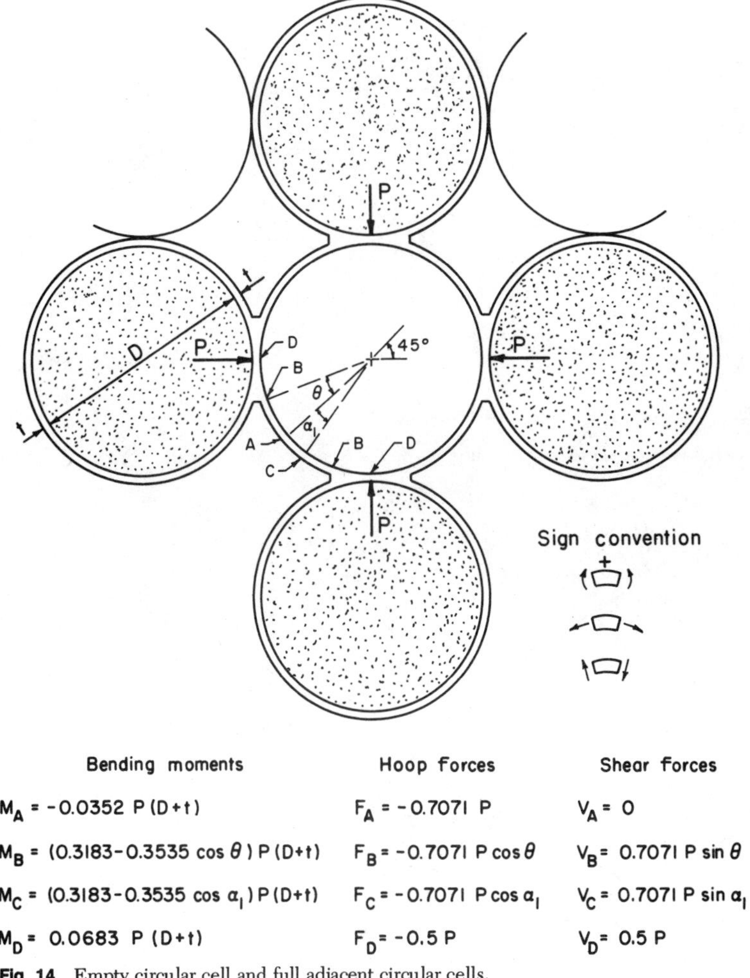

Fig. 14 Empty circular cell and full adjacent circular cells.

For normal-weight concrete $E_c = 57{,}000\sqrt{f'_c}$, $\alpha_t = 5.5 \times 10^6/°F$, and $\nu = 0.3$. This formula gives conservative values for the thermal moments, since any cracking of the wall section would reduce its stiffness and result in lower values.

Temperature Differential. The temperature differential between external and internal faces of a concrete silo wall containing hot stored material can be calculated from

$$\Delta T = (T_{i,\text{des}} - T_0)K_T \tag{14}$$

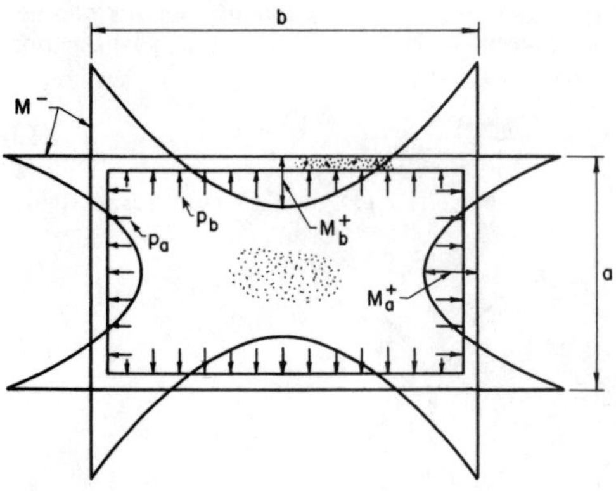

$$M^- = \frac{1}{12(1+n)}\left(p_a a^2 + n p_b b^2\right) \quad \text{where } n = \frac{b}{a}\left(\frac{t_a}{t_b}\right)^3$$

$$M_a^+ = \frac{p_a a^2}{8} - M^-$$

$$M_b^+ = \frac{p_b b^2}{8} - M^-$$

$$F_b = p\frac{a}{2} \qquad V_b = p\frac{b}{2}$$

$$F_a = \frac{pb}{2} \qquad V_a = \frac{pa}{2}$$

Fig. 15 Lateral pressure on rectangular silo.

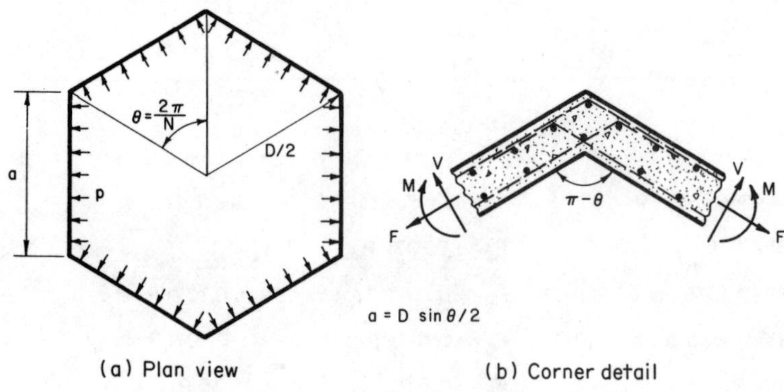

	At corner	At side midspan
F	$\frac{pD}{2}\cos\frac{\theta}{2}$	$\frac{pD}{2}\cos\frac{\theta}{2}$
V	$\frac{pD}{2}\sin\frac{\theta}{2}$	0
M	$-\frac{pD^2}{12}\sin^2\frac{\theta}{2}$	$\frac{pD^2}{24}\sin^2\frac{\theta}{2}$

Fig. 16 Lateral pressure on polygonal silo.

where $T_{i,des} = T_i - 80$, T_i = temperature of stored material, T_0 = design winter dry-bulb temperature, and K_T = ratio of thermal resistance of wall alone to that of the wall plus an outside surface film of air plus a thickness t_m of stored material acting as insulating material. If R_m and R_c = 0.08 represent the thermal resistances per unit thickness (resistivity*) of the stored material and the concrete wall, respectively, and R_a = 0.17 is the thermal resistance of the outer surface film of air, then

$$K_T = \frac{0.08t}{t_m R_m + 0.08t + 0.17} \tag{15a}$$

For silos storing hot cement ACI 313 suggests t_m = 8 in. and $t_m R_m$ = 3.92. This reduces K_t to

$$K_t = \frac{0.08t}{4.09 + 0.08t} \tag{15b}$$

The analysis is illustrated in Example 6.

DESIGN OF WALLS

Except as noted, design formulas in the following articles are in terms of strength design. Subscripts u denote ultimate values obtained by multiplying service-load forces by load factors. ACI 313 prescribes load factors of 1.7 for live load and 1.4 for dead load. ϕ is the capacity-reduction factor.

11. Minimum Thickness of Circular Walls To allow for noncalculable moments due to transient nonuniform pressure on the walls of circular silos, the following minimum thickness is recommended:[13]

$$t_{\min} = p\,\frac{D}{2}\,\frac{mE_s + f_s - nf_{ct}}{f_s f_{ct}} \tag{16}$$

in which m = shrinkage coefficient (may be taken as 0.0003), f_s = allowable steel stress (between 0.4 and $0.5f_y$), and f_{ct} = concrete stress in uncracked section under static lateral pressure (may be taken as $0.1\,f'_c$).

Final wall thickness is governed by practical considerations and by load requirements and permissible crack width. A minimum thickness of 6 in. should be used for cast-in-place silo walls.

12. Maximum Crack Width An important design consideration is the maximum crack width that may be tolerated. A limit of 0.008 in. is suggested[13] for the case of grain or cement storage silos and for other silos exposed to the weather. This helps to avoid penetration of water that causes corrosion of the reinforcement, and spoils the silo contents by inducing germination of the grain and hydration of the cement.

Lipnitski's method[9] allows a simple determination of crack width for walls which are subjected mainly to hoop tensions, and includes silo walls and hoppers and other membrane-type structures. The width w_{cr} of a vertical crack caused by the simultaneous action of short- and long-term loadings may be determined by

$$w_{cr} = w_1 - w_2 + w_3 \tag{17}$$

Values of w in Eq. (17) are given by

$$w_1 = \psi_1 \frac{A\beta}{\Sigma o}\frac{f_{s,\text{tot}}}{E_s} \qquad \text{where } \psi_1 = 1 - 0.7\frac{0.8Af'_t}{F_{\text{tot}}} \geq 0.3 \tag{18a}$$

*The unit of conductivity (Btu/ft²/hr/°F/in.) is the amount of heat in Btu that will flow in 1 hr through 1 ft² of a layer 1 in. thick of a homogeneous material per 1°F temperature difference between surfaces of the layer. Resistivity, which measures the insulating value of a material, is the reciprocal of conductivity.

$$w_2 = \psi_2 \frac{A\beta}{\Sigma o} \frac{f_{s,st}}{E_s} \qquad \text{where } \psi_2 = 1 - 0.7 \frac{0.8Af'_t}{F_{st}} \geqslant 0.3 \tag{18b}$$

$$w_3 = \psi_3 \frac{A\beta}{\Sigma o} \frac{f_{s,st}}{E_s} \qquad \text{where } \psi_3 = 1 - 0.35 \frac{0.8Af'_t}{F_{st}} \geqslant 0.65 \tag{18c}$$

where β = 0.7 and 1.0 for deformed and plain bars, respectively
A = cross-sectional area of wall per unit height
st = subscript indicating static load
tot = subscript indicating static load plus overpressure
$f'_t = 4.5\sqrt{f'_c}$ = tensile strength of concrete
f_s = steel stress
Σo = sum of perimeters of horizontal reinforcing bars per unit height of wall

Crack-width evaluation is illustrated in Example 2.

13. Walls in Tension The required reinforcement A_s per unit height of wall is given by

$$A_s = \frac{F_u}{\phi f_y} \tag{19}$$

ϕ in this equation may be taken as 0.9.

14. Walls in Tension and Flexure The following formulas are from Ref. 13. There are two cases.

Case I, $e = M_u/F_u \leqq t/2 - d''$ (Fig. 17*a*)

On the side nearer to F_u

$$A_s = \frac{F_u e'}{\phi f_y(d - d')} \tag{20}$$

and on the opposite side

$$A'_s = A_s \frac{e''}{e'} \tag{21}$$

Case II, $e = M_u/F_u = t/2 - d''$ (Fig. 17*b*)

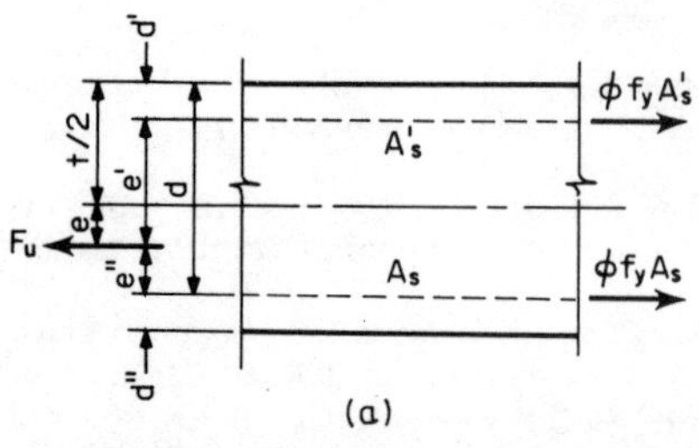

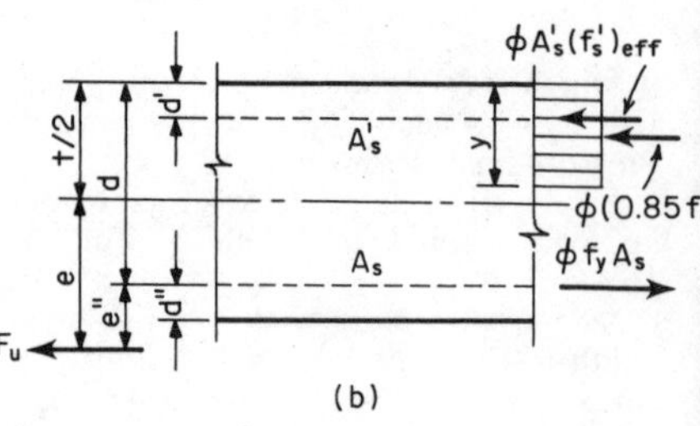

Fig. 17 Wall under tension and flexure.

1. Determine the depth y_L of a rectangular-section compression-stress block with 75 percent of balanced reinforcement (Sec. 1, Art. 7) from

$$\frac{y_L}{d} = 0.75\beta_1 \frac{87}{87 + f_y,\text{ksi}} \tag{22}$$

$$\beta_1 = 0.85 - 0.05 \frac{f'_c - 4000}{1000} \leqq 0.85 \tag{23}$$

Values of y_L/d for common values of f'_c and f_y are given in the following table.

	f_y, ksi		
f'_c, ksi	40	50	60
To 4	0.436	0.405	0.378
5	0.411	0.381	0.355
6	0.386	0.357	0.333

2. Determine the effective compressive-steel stress

$$(f'_s)_{\text{eff}} = 87\,\frac{y_L - \beta_1 d'}{y_L} - 0.85\,f'_c \ngtr f_y - 0.85\,f'_c \qquad \text{ksi} \tag{24}$$

3. If $(f'_s)_{\text{eff}}$ in step 2 is positive,

$$A'_s = \frac{F_u e''/\phi - 0.85 f'_c b y_L (d - y_L/2)}{(d - d')(f'_s)_{\text{eff}}} \tag{25}$$

4. If A'_s in step 3 is positive,

$$A_s = \frac{F_u/\phi + 0.85\,f'_c b y_L + A'_s (f'_s)_{\text{eff}}}{f_y} \tag{26}$$

5. If A'_s in step 3 is negative, no compressive steel is needed. In this case, whether or not steel is provided on the compression side, the wall is designed as singly reinforced according to

$$A_s = \frac{F_u/\phi + 0.85 f'_c b y}{f_y} \tag{27}$$

where $y \approx d - \sqrt{d^2 - \dfrac{2F_u e''}{0.85\phi f'_c b}}$

If $(f'_s)_{\text{eff}}$ in step 2 is negative, compression steel will be ineffective, and if a singly reinforced member would not be acceptable, either d must be increased or d' decreased.

Shear. The shear stress v_u is given by

$$v_u = \frac{V_u}{bd} \tag{28}$$

where v_u should not exceed v_c given by

$$v_c = 2\phi\left(1 - 0.002\,\frac{F_u}{bt}\right)\sqrt{f'_c}, \text{psi} \tag{29}$$

15. Walls in Compression ACI 318[22] allows walls for which the compressive force falls within the middle third to be considered as concentrically loaded. If buckling is not involved, the permissible compression is

$$f_c = 0.55\phi f'_c \tag{30}$$

where $\phi = 0.70$.

For rectangular walls, where slenderness may influence strength, buckling should be accounted for by using[13]

$$f_c = 0.55\phi f'_c\left[1 - \left(\frac{H_0}{40t}\right)^2\right] \tag{31a}$$

where $\phi = 0.70$ and H_0 = clear vertical distance between supports. If $H_0 > l_0$, where l_0 = clear horizontal distance between supports, use l_0 in place of H_0.

Circular walls in the pressure zone may be designed for the allowable stress of Eq. (30) if there are no openings. If there are unreinforced openings, Eq. (31*a*) should be used, with H_0 = height of opening.

Circular walls below the pressure zone, continuous throughout, should be designed for[13]

$$f_c = 0.55\phi f'_c\left[1 - \left(\frac{D}{120t}\right)^3\right] \tag{31b}$$

If there are unreinforced openings, use

$$f_c = 0.55\phi f'_c\left[1 - \left(\frac{H_0}{40t}\right)^2 - \left(\frac{D}{120t}\right)^3\right] \tag{31c}$$

16. Walls in Compression and Flexure These may be designed using the provisions of ACI 318, Chap. 10.[22] Combined compression and flexure is also discussed in Sec. 1, Art.

17. The interaction diagrams in that section for rectangular columns with reinforcement on opposite faces can be used for walls.

17. In-Plane Bending of Walls The in-plane bending behavior of a wall supported on columns depends on the height of the wall relative to the spacing of the columns.

Low Walls. The stiffness of a wall of a low bunker is of the same order as that of the hopper wall, and the two can be assumed to act together in transferring vertical load to the columns. For bin walls with $H/a \leq 0.5$ Ciesielski et al.[23] recommend that the wall and that portion of the hopper wall whose vertical projection is $0.4a$, where a is the length of the wall, be analyzed as a beam (Fig. 18a). The resulting bending stresses (Fig. 18b) are computed from $f = M/S$, where M is the moment due to the vertical loads. The stress in the part of the hopper wall not considered to be part of the beam is assumed to decrease linearly from the value of the bottom-fiber beam stress to zero at the vertex of the hopper wall. These stresses must be considered in combination with the moments and axial tensions due to lateral pressure (Fig. 15) in determining the wall reinforcement.

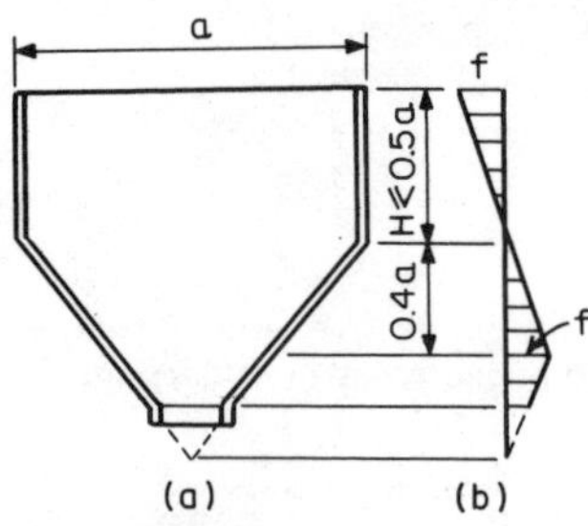

Fig. 18

A folded-plate analysis of the joint action of a low bunker wall and an adjoining hopper wall can also be made.[8]

High Walls. Experimental studies by a number of investigators of the beam behavior of reinforced-concrete walls are discussed in Ref. 24. Walls for which $H/l \leq 1$, where H = height and l = length, can be designed by the usual procedure for reinforced-concrete beams. For a single-panel wall with $H/l > 1$, simply supported on columns spaced l center-to-center and carrying a uniformly distributed load w, the tension T that must be furnished by tensile reinforcement is given by

$$T = \frac{0.14wl}{\sqrt{H/l}} \qquad 1 < \frac{H}{l} \leq 2 \tag{32}$$

The value of T for $H/l = 2$ is to be used for walls with $H/l > 2$. The reinforcement is to be distributed over a depth $0.1l$.

If one-half to two-thirds of the tensile reinforcement is bent up, the shear strength V of the panel is given by

$$V \leq 0.54 f'_c t^2 \sqrt{H/t} \tag{33}$$

If the shear $w_u l/2$ exceeds V by Eq. (33), or if no bars are bent up, web reinforcement must be provided to resist the tension T_s given by

$$T_s = \frac{w_u l/2}{\sqrt{2H/l}} \qquad 1 \leq \frac{H}{l} \leq 2 \tag{34}$$

It is assumed in this formula that the necessary web reinforcement is inclined about 60° to the horizontal. Of course, equivalent reinforcement in the form of stirrups can be used.

Equation (32) is based on tests in which the load was applied to the top of the panel. However, tests on panels loaded along the bottom edge showed that the formula can also be used for this case.

18. Walls Subjected to Thermal Stresses The required additional vertical and horizontal reinforcement per unit width or height is given by

$$A_{s,t} = \frac{1.4M_t}{f_y(d - d'')} \tag{35}$$

where M_t is given by Eq. (13). This steel should be placed near the cooler (usually outer) face of the wall. In singly reinforced walls it should be added to the main hoop steel, which should be near the outer face. In doubly reinforced walls it should be added to the outer layer, but for simplicity an equal amount is often added to the inner layer to avoid having bar sizes or spacings differ from one layer to the other.

Vertical tensile thermal stress is usually offset by vertical dead-load compressive stress so that additional temperature steel is often not needed.

19. Vertical Reinforcement This is required not only in outside walls of silo groups but also on all inside walls. Vertical steel distributes lateral overpressures to adjacent horizontal reinforcement. Gurfinkel reported a silo where failure was averted when vertical steel redistributed lateral pressures that could not be resisted by hoop reinforcement that was in an advanced stage of corrosion.[25] Vertical steel also resists tension caused by bending moments due to restraint against circular elongation, eccentric loads from hopper edges or attached auxiliary structures, and temperature differentials between inside and outside wall surfaces or between silos.

20. Details and Placement of Reinforcement Table 11 summarizes the requirements of ACI 313.[13] Bar splices, both horizontal and vertical, are staggered. Adjacent hoop-reinforcing splices in the pressure zone are staggered horizontally by not less than one lap length in 3 ft and do not coincide in vertical array more frequently than every third bar.

TABLE 11 Minimum Reinforcement Requirements

Region	Horizontal steel	Vertical steel (No. 4 or larger)
Pressure zone	As required by calculations	Exterior walls: $0.0020t$; max spacing $4t$ or 18 in. Interior walls: $0.0015t$; max spacing $4t$ or 24 in.
Below pressure zone	Continue A_s from above for a distance equal to six times wall thickness. Below this provide $0.0025t$ per unit height of wall	$0.0020t$
Bottom of walls and columns		Dowels as needed to prevent uplift and shifting by earthquake or wind loading
Wall intersections subjected to moment	Provide as required	Provide as required
Adjoining silos	Provide as required to prevent separation	
Circular walls, single-reinforced	Place nearest to the outer faces	

Slipforming should not be considered an excuse for not tying reinforcement together. Haeger considers the normal tying of the ends of hoop reinforcement, with additional ties every 4 to 5 ft between, to be acceptable.[26] Vertical steel should not be omitted to provide access for concrete buggies in slipform construction; instead it may be spaced farther apart at specified access locations. The total amount of vertical steel is unchanged; only the spacing is affected. The conventional practice of leaving the slipforming jack rods embedded in the concrete is fine, but widely spaced jack rods should not be construed as the equivalent of vertical reinforcement.

Typical reinforcing patterns at wall intersections are shown in Fig. 19.

Wall Openings. Table 12 summarizes the requirements of ACI 313.[13] Figure 20 shows a typical detail of the reinforcement of a narrow silo wall between openings.

DESIGN OF BOTTOMS

21. Bottom Pressure Static unit pressure q_α normal to a surface inclined at an angle α to the horizontal is given by

$$q_\alpha = p \sin^2 \alpha + q \cos^2 \alpha \tag{36}$$

Silo bottoms are designed to resist q_α. In seismic zone 4, q in Eq. (36) should be computed for the effective weight (80 percent) of the stored material because of the loss of friction against the silo walls due to seismically induced lateral vibrations, that is, $q = 0.8\gamma Y$. In other seismic zones q should be increased by the following fractions of the increased pressure $(0.8\gamma Y - q)$ for zone 4: ¾ for zone 3, ⅜ for zone 2, and 3/16 for zone 1. ACI 318 load factors are suggested for ultimate-strength design under seismic load.

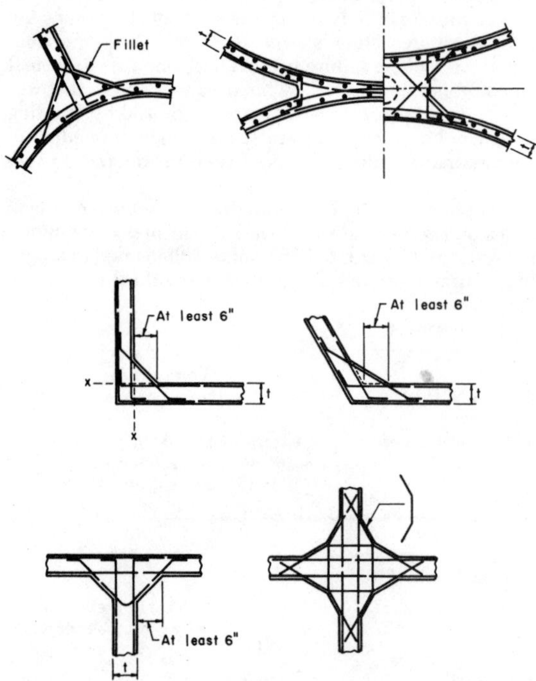

Fig. 19 Reinforcement at intersecting walls. (*Adapted from Ref. 13.)*

TABLE 12 Reinforcement at Wall Openings

Openings	Horizontal steel	Vertical steel
In pressure zone	Add at least 1.2 times area of interrupted reinforcement, ½ above the opening and ½ below	Provide by assuming narrow strip of wall, $3t$ in width on each side of opening, to act as column within opening height subjected to its own vertical load plus that from ½ wall span above opening. Add steel at least equal to that eliminated by opening
Outside pressure zone	Add no less than the normal reinforcement interrupted by opening, distributed as above	As above
Closely spaced		See Fig. 20

22. Plane Bottoms Design loads for horizontal slab bottoms are dead load, vertical pressure q, and thermal load from the stored material. For inclined slab bottoms q_a should be used. Allowance for earthquake forces should be made as described in Art. 21.

Formulas for bending moments and deflections of circular slabs, with and without a central hole, are given in Ref. 19. Tables of coefficients are given in Ref. 8. Moments and deflections in rectangular slabs can be computed by the ACI 318 procedure for two-way slabs.

23. Conical Hoppers Walls of these structures are subjected to meridional and circumferential tensile membrane forces F_m and F_t (Fig. 21). Values of F_m and F_t per unit width at any horizontal section are given by

$$F_m = \frac{qD}{4 \sin \alpha} + \frac{W}{\pi D \sin \alpha} \tag{37}$$

$$F_t = \frac{q_\alpha D}{2 \sin \alpha} \tag{38}$$

where q, q_α = pressures computed at the section
D = diameter at the section
W = weight of that portion of hopper and hopper contents below the section

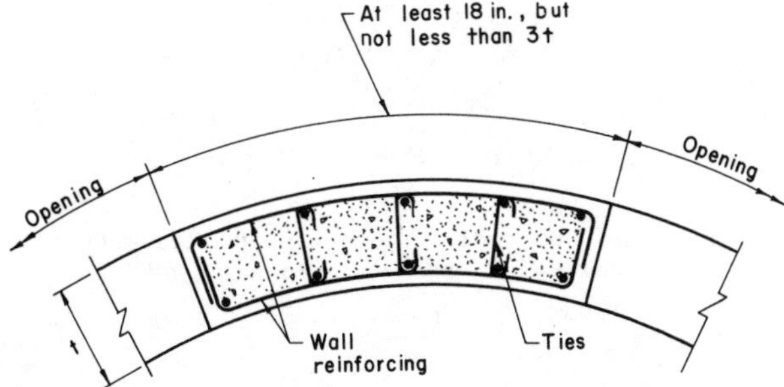

Fig. 20 Reinforcement for narrow silo wall between openings. *(From Ref. 13.)*

F_{mu} and F_{tu} for strength design are obtained by multiplying q and q_α by the load factor 1.7 and computing W by

$$W = 1.4W_h + 1.7W_m \tag{39}$$

where W_h = weight of hopper below the section and W_m = weight of hopper contents below the section. These weights are given by

$$W_h = \frac{\pi(D^2 - d^2)\gamma_h}{4 \cos \alpha} \tag{40}$$

$$W_m = \frac{\pi(D^3 - d^3)\gamma_m \tan \alpha}{24} \tag{41}$$

where γ_h = weight of hopper wall per unit area and γ_m = weight of material per unit volume.

The required reinforcement is given by Eq. (19). A minimum wall thickness of 5 in. is recommended, and the crack width should not exceed an acceptable value.

Design of a conical hopper is illustrated in Example 4.

24. Pyramidal Hoppers Walls of these structures are subjected to tensile membrane forces F_m and F_t (Fig. 22) and plate-type bending. There will also be in-plane bending if the hopper is not supported continuously along its upper edge.

If the vertical components of the meridional forces are assumed to be distributed uniformly on the perimeter of a horizontal section of a symmetrical hopper of rectangular cross section, F_m is given by

$$F_m = \frac{W + (q_a + q_b)\, ab/2}{2(a + b) \sin \alpha} \tag{42}$$

where a is the length and b the width of the section, q_a and q_b are the vertical pressures corresponding to sides a and b, respectively, W is the weight of that portion of the hopper and hopper contents below the section, and α is the angle with the horizontal of the hopper wall.

If $a > b$, $q_a < q_b$ (because the hydraulic radius is smaller for the longer side). This suggests that the vertical component of the meridional force on wall b may be larger than

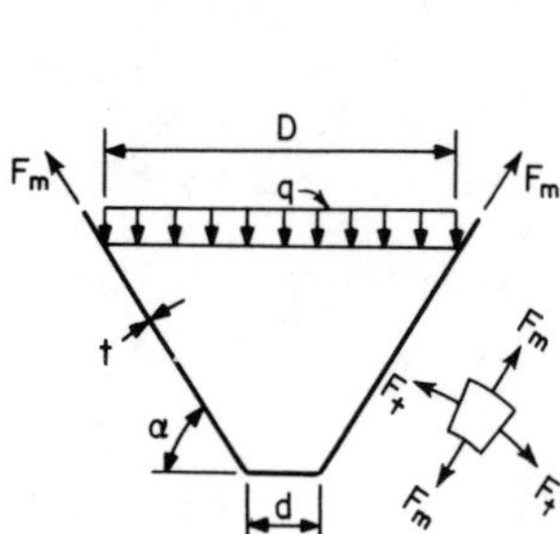

Fig. 21 Forces in conical hopper.

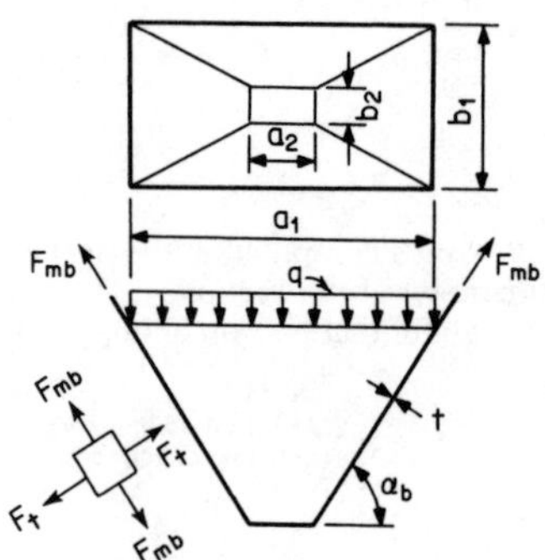

Fig. 22 Forces in pyramidal hopper.

on wall a. Assuming $W/4$ and the resultant vertical pressure on the triangular area adjacent to each wall to be carried by that wall, the following formulas result:

$$F_{ma} = \left(\frac{W}{a} + q_a b\right)\frac{1}{4 \sin \alpha_a} \tag{43a}$$

$$F_{mb} = \left(\frac{W}{b} + q_b a\right)\frac{1}{4 \sin \alpha_b} \tag{43b}$$

The horizontal membrane force F_t is given by

$$F_{ta} = \frac{1}{2}(q_{\alpha b} + \gamma_h \cos \alpha_b)\ b \sin \alpha_a \tag{44a}$$

$$F_{tb} = \frac{1}{2}(q_{\alpha a} + \gamma_h \cos \alpha_a)\ a \sin \alpha_b \tag{44b}$$

where γ_h = weight of hopper wall per unit of area.

F_{mu} and F_{tu} for strength design are obtained by multiplying q by the load factor 1.7 and computing W by Eq. (39). The weights W_h and W_m are given by

$$W_h = h_h\gamma_h\left(\frac{a_1 + a_2}{\sin \alpha_a} + \frac{b_1 + b_2}{\sin \alpha_b}\right) \tag{45}$$

$$W_m = \frac{h_h\gamma_m}{6}\left[(2a_1 + a_2)\ b_1 + (2a_2 + a_1)b_2\right] \tag{46}$$

where a_1 is the length and b_1 the width of the section at which F is being computed, and a_2 and b_2 are the corresponding dimensions of the hopper opening.

Plate Bending under Normal Pressures. Bending moments in triangular walls may be approximated by the bending moments in the equivalent rectangular plate shown in Fig. 23a.[8] Tables of coefficients for the analysis of triangular walls with various types of edge support are available.[8]

Bending moments in trapezoidal walls for which $a_2/a_1 \geqq 4$ can be approximated by the moments in the triangular wall formed by extending the sloping sides to their intersection[8] (Fig. 23b). Therefore, trapezoidal walls of these proportions can also be solved by using the equivalent rectangle of Fig. 23a.

Bending moments in trapezoidal walls for which $a_2/a_1 < 4$ can be approximated by the moments in an equivalent rectangular wall (Fig. 23c) with the dimensions[8]

$$a_{eq} = \frac{2a_2(2a_1 + a_2)}{3(a_1 + a_2)} \tag{47a}$$

$$b_{eq} = h - \frac{a_2(a_2 - a_1)}{6(a_1 + a_2)} \tag{47b}$$

Tables of coefficients for the analysis of trapezoidal plates with $a_1 = \frac{3}{8}a_2$ and $a_1 = \frac{1}{2}a_2$ with various types of edge support are available.[8]

Edge conditions (fixed, simply supported, etc.) of hopper walls depend on the adjoining construction. They should be considered fixed at their junctures with adjoining walls. In adjoining walls of unequal lengths, the average of the unequal end moments may be used, or they may be distributed in the ratios given by the negative-moment formula in Fig. 15; moments in the central regions of the plates should be adjusted to correspond. Upper edges which are continuous with silo or bunker walls, with or without an intervening edge beam, may be assumed to be fixed. On the other hand, the upper edge of a pyramidal bunker which has no roof must be considered to be free.

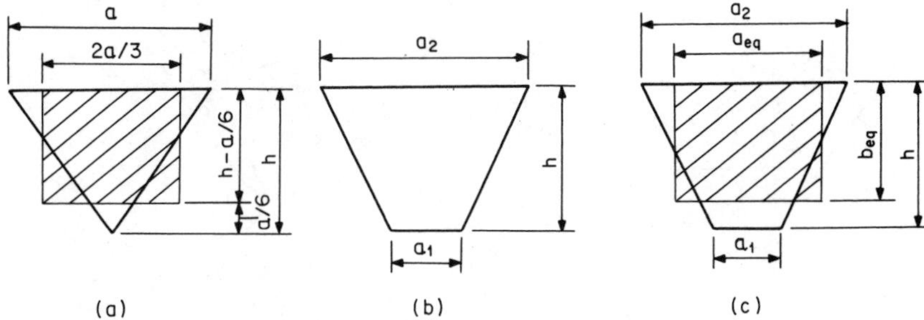

Fig. 23

Bending in Plane of Wall. In-plane bending of a hopper wall acting in conjunction with a low vertical wall is discussed in Art. 17 (Fig. 18). A similar analysis can be made for the bunker without vertical walls, using an effective beam depth of $0.5a$ at midspan[8] (Fig. 24). The stress in the portion below the effective-beam depth is assumed to vary linearly to zero at the vertex, as shown. The bending moment can be computed from $M = F_m a^2/8$, where F_m is the meridional tension computed by Eq. (42) or Eqs. (43). These bending stresses must be combined with the horizontal membrane forces to determine wall reinforcement.

Concentrated Forces at Pyramidal Bunker Supports. Pyramidal bunkers and pyramidal hoppers supported independently on columns at the four corners are subjected to concentrated forces at the supports (Fig. 25). The tensile force T along the edge of adjoining walls is given by

$$T = \frac{P}{\sin \alpha} \tag{48}$$

where α is the angle of the edge with the horizontal. The compressive forces are given by

$$C_a = P \cos \alpha \cos \beta_a \tag{49a}$$

$$C_b = P \cos \alpha \cos \beta_b \tag{49b}$$

where β_a and β_b are the angles between the edges a and b and the diagonal of the horizontal cross section of the hopper.

Since these forces are localized, they need be provided for only in the vicinity of the column.

25. Hopper-Supporting Beams Concrete hoppers are usually supported by edge beams cast integrally with the hopper wall (Fig. 26*a*,*b*). A conical steel hopper supported by a ring beam is shown in Fig. 26*c*. The beams may be supported continuously by a wall, or by pilasters or columns.

If the hopper wall does not intersect the centroid of the supporting-beam cross section, a twisting moment acts on the beam. In ring beams this produces a bending moment which is uniform around the circumference. Both moments can be neglected if the beam is supported on a wall, because the deformation that would be generated is prevented by the silo wall and the bearing wall. Therefore, such a ring beam can be designed for only the horizontal component of the hopper meridional tension, which produces a uniform compressive force, equal to $(F_m \cos \alpha)(D/2)$, in the ring (Example 4).

Ring beams supported on columns or pilasters, as in Fig. 26*a*, must be designed for the moments, shears, and torsion due to F_m, as well as the ring compression. A design procedure for this case is discussed by Safarian.[27]

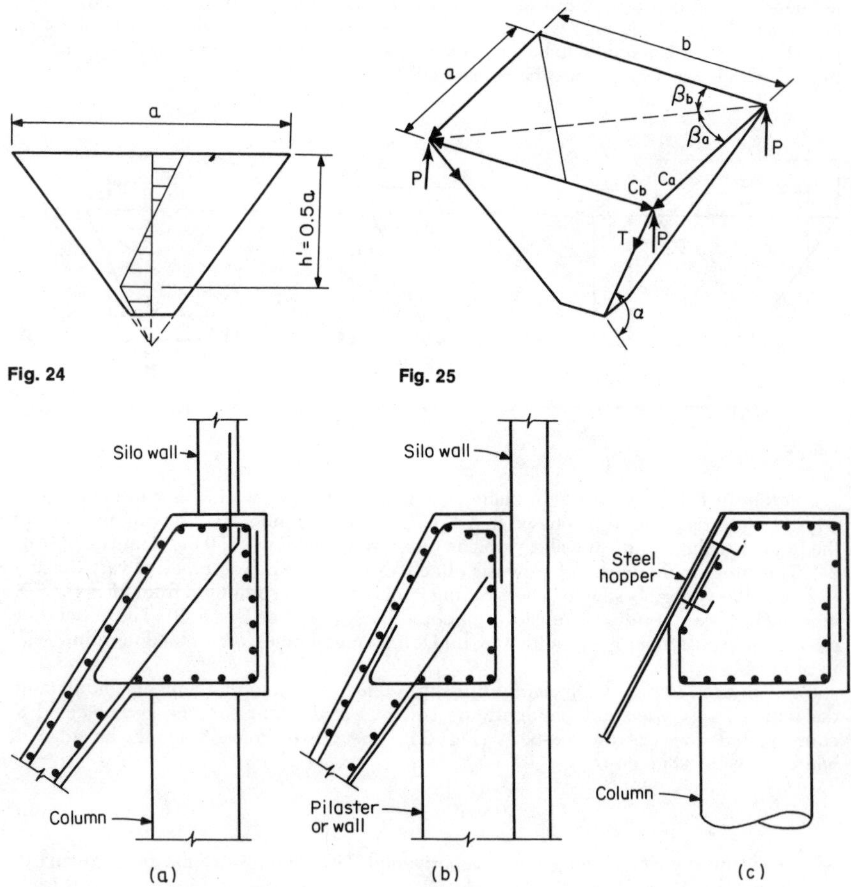

Fig. 24

Fig. 25

Fig. 26 Typical hopper-supporting beams. *(From Ref. 13.)*

No bending moments are produced by twisting of an eccentric edge beam of a rectangular hopper. Torsion can be neglected in edge beams on bearing walls. Edge beams on columns should be doweled to the silo walls to resist torsion.

Where supporting beams are not used and the hopper is keyed to the silo wall, reinforcement must be provided for the hopper-wall bending moments at the juncture.

26. Columns Columns supporting silo shells, and particularly silo bottoms, will be subjected to a live load due to the stored material that is substantially larger than the dead load of the structure. Long sustained periods of material storage cause reinforced-concrete columns to creep. As a result, concrete stresses decrease and the load carried by the steel reinforcement increases. Subsequent emptying of the silo may place the concrete in tension as the reinforcement recovers elastically. Once the tensile stress in the concrete exceeds its tensile strength, the column develops severe transverse cracking. The situation can be dangerous if transverse cracking is accompanied by longitudinal cracking, as could occur with high bond stresses during unloading. To prevent this condition, it is recommended that the reinforcement ratio ρ not exceed 0.02 and that the total amount of reinforcement not exceed L/f_y, where L is the live load on the column. If lateral loads must be resisted, larger columns should be used to keep the steel ratio low. All other provisions of ACI 318, Chap. 10 for the design of columns should be followed.

The analysis described here is illustrated in Example 5.

27. Roofs Designers are divided on the subject of attachment of concrete roof slabs to silo walls. Some believe that the slab should be supported only vertically at the walls (on elastomeric material or heavy tar paper) so as to be free to contract or expand with temperature changes and to move slightly during earthquakes. To prevent total freedom of horizontal displacement, the slab may be attached at one central location, usually the elevator tower. On the other hand, attaching the roof slab to the walls stiffens tall silos against wind and earthquake loads and reduces lateral deformations. Continuity with the silo walls also makes the roof slab stiffer and reduces vertical deflections under live load.

In long installations expansion joints are provided to reduce cracking of the slab in winter and undue longitudinal forces on the silos in summer. Good design calls for an expansion joint to cut across the silo group by extending down to the foundation, especially if the roof slab is attached to the silo walls.

The steel beams which support the wood platform during slipforming of silo walls are used later to support the roof slab, thus reducing its span considerably. Ample bearing on the concrete wall should be provided at the ends of the steel beams, and the concrete below and to each side of the beam should be reinforced to prevent undue cracking or even a concrete fallout after some years of service.

28. Failures Three major reasons for failures of reinforced-concrete silos are foundation failure, incorrect determination of loads, and improper detailing and faulty workmanship.

Concrete silos are supported on pile foundations, or on extended rafts in the case of a stiff subsoil. Theimer[2] cites various reasons for failures: (1) the weight is usually great and may shift considerably with unbalanced filling and emptying of the numerous cells, causing major overstress in the foundation, (2) dredging in an adjoining river may weaken the pile foundation, (3) batter piles may fail after having been damaged by a ship collision, (4) piles which have been eroded by aggressive groundwater may buckle, (5) underlying soft soil may shift, causing tilting of the raft and elevator. Prevention of foundation failure requires thorough investigation of subsoil conditions including a number of test-pile loadings. Thorough control and inspection of pile-driving operations and cast-in-place piles (especially those without steel shells to prevent mud intrusions) are necessary. Batter piles should always be used to support high silos against wind action and seismic motions. Raft foundations should be designed as continuously reinforced concrete mats with two grids to resist bending moments caused by unbalanced loading of the silos.

Incorrect determination of silo pressures has resulted invariably in underreinforced structures. Application of the conventional Janssen theory without any allowance for overpressures generated during emptying and by eccentric outlets, combined with the use of higher allowable stresses for concrete and steel, has caused unacceptable major cracking and even total failure in numerous instances. ACI 313,[13] DIN 1055,[12] or Pieper's recommendations[14] will give a more accurate evaluation of load effects.

Major damage and even collapse of silos have resulted from improper detailing and faulty construction. Irregular and excessive spacing of hoop reinforcement, particularly in slipformed silos, may seriously reduce strength. In addition, radial displacement of hoop bars is frequent, and by reducing concrete cover, the capacity of the lap splices is limited. If hoop bars are placed without tying, circumferential shifts may occur, thus making some lap lengths shorter and some longer than the specified length. Absence of vertical steel in combination with excessive spacing of hoop reinforcement may leave large portions of concrete unreinforced and cracked, a situation which may eventually result in concrete fallouts.[28] Vertical reinforcement of walls and columns should be anchored to the foundation by dowels to prevent uplift and shifting under earthquake or wind loading. Insufficient cover for the hoop reinforcement causes it to corrode in a rather short period of time. Sloppy workmanship may cause all splices of hoop reinforcement to be at the same locations without staggering, thereby increasing the possibility of bond failure (with splitting) and thus weakening the wall. Improper detailing at wall intersections of interstices and pocket bins may omit wall fillets and double layers of reinforcement that are necessary to rest local bending moments and shears. Proper attention to detail, followed by inspection at the jobsite and proper control and organization of construction are required to prevent mistakes that may lead to collapse.

29. Dust Explosions in Grain Elevators and Flour Mills Major destruction of these installations occurs when dust from grain products ignites and releases great amounts of energy. An extremely rapid pressure rise, of the order of 2000 tons/ft²/sec, originates a pressure wave of such high intensity that normal vents for the release of explosion pressures are insufficient to prevent the installation from blowing up. Theimer[29] cites three principal causes for these explosions: a dust cloud, a source of ignition, and the presence of oxygen.

It is necessary to have a minimum concentration of 0.02 oz/ft³ of grain or flour dust (resembling a dense fog) before it can become ignited. If the concentration is greater than 2 oz/ft³, incomplete combustion of the particles retards ignition and prevents the explosion. Dry dust that accumulates on floors, walls, ledges of doors and windows, steel beams, overhead ducts, etc., is highly oxygenated and quite dangerous. Good housekeeping calls for constant removal of such dust. Suction is the principal method to control dust clouds by inducing air currents supplied through dust-collecting systems. Venting is also recommended for bins, heads of bucket elevators, and scale hoppers.

Ignition temperatures vary between 750 and 930°F when the air relative humidity is between 30 and 90 percent. Sources of heat that can ignite dust clouds are: (1) open flames (lights, matches, burning cigarettes); (2) heat generated on pulleys of bucket elevators due to belt slip; (3) hot surfaces of radiators, bearings, light bulbs; (4) sparks caused by metal parts in rotating machinery, electric equipment, and friction; (5) static electricity; and (6) welding, cutting, and soldering. In addition, Theimer[29] cites a case of spontaneous ignition due to constant increase in the temperature of the material caused by inadequate heat dispersion. Obviously, all prevention measures should be taken to avoid ignition heat sources.

Designers should consider the layout of various buildings in an installation to decrease its vulnerability to explosions as a whole. Theimer suggests leaving as large a space as possible between the various buildings. Explosion reliefs such as vents, light brick walls, and light roof construction should be provided.

EXAMPLES

Example 1 Determine the hoop and meridional reinforcement for a 30-ft-diameter silo 100 ft high with a 6-in. wall, and a conical hopper 20 ft deep, to contain wheat (Fig. 2*a*). Use ACI 313 (Art. 4) with f'_c = 4000 psi and A615 Grade 60 deformed bars.

From Table 5, γ = 50 lb/ft³, ρ = 28°, and μ' = 0.44. The basic pressures and forces are obtained from Janssen's formulas (Table 1) except that $V = (\gamma Y - 0.8q)\,R$ as required by ACI 313.

$$R = \frac{D}{4} = \frac{30}{4} = 7.5 \text{ ft} \qquad k = \frac{1 - \sin\rho}{1 + \sin\rho} = \frac{1 - \sin 28°}{1 + \sin 28°} = 0.361$$

$$Y_0 = \frac{R}{\mu' k} = \frac{7.5}{0.44 \times 0.361} = 47.22 \text{ ft}$$

$$q = \gamma Y_0 (1 - e^{-Y/Y_0}) \qquad p = qk$$

Basic pressures are multiplied by the overpressure factors C_d from Table 9. The results are given in Table 13.

Hoop reinforcement is determined in Table 14, using the design pressures of Table 13. The hoop force $T = p_{des}D/2$ and $F_u = 1.7F$. The required steel area is given by $A_s = T_u/(0.9 \times 60)$. Spacing of No. 5 bars can be computed from $s = 0.31 \times 12/A_s$.

Spacing is given by zones starting from a minimum No. 5 at 12 in. for the top 40 ft of wall and ending with No. 5 at 5½ in. for the lower 20 ft above the hopper level. The latter spacing is continued for a distance of six times the wall thickness below the pressure zone (Table 11). With t = 6 in. this calls for No. 5 at 5½ in. to be continued another 3 ft below hopper level. From there down to the foundation, Table 11 requires a minimum $A_s = 0.0025t$. Thus $A_s = 0.0025 \times 6 = 0.015$ in.²/in. = 0.18 in.²/ft, i.e., No. 5 at 20.7 in.; use No. 5 at 12 in. (0.31 in.²/ft).

Meridional reinforcement is usually determined to satisfy minimum requirements (Table 11). For this purpose $A_s = 0.0020t = 0.0020 \times 6 = 0.012$ in.²/in. = 0.144 in.²/ft, i.e., No. 4 at 16.3 in.; use No. 4 at 16 in. (0.147 in.²/ft).

WALL COMPRESSION. The maximum meridional force on the walls above the hopper level is at Y = 100 ft. The factored loads are

$$\begin{aligned} \text{Weight of wall} &= 1.4 \times 100 \times 0.15 \times 6/12 &= 10.5 \\ \text{Weight of 6-in. roof} &= 1.4 \times (30/4) \times 0.15 \times 6/12 &= 0.8 \\ 1.7\, C_d V \text{ from Table 13} &= 1.7 \times 47.56 &= 80.9 \\ & & 92.2 \text{ kips/ft} \end{aligned}$$

$$f_c = \frac{92.2}{12 \times 6} = 1.28 \text{ ksi}$$

The allowable value from Eq. (30) is $f_c = 0.7 \times 0.55 \times 4 = 1.54$ ksi.

TABLE 13 Values of q, p, and V for Silo of Example 1

			Basic pressures and forces			Design pressures and forces			
Y, ft	Y/Y_0	$1 - e^{-Y/Y_0}$	q, lb/ft²	p, lb/ft²	V, kips/ft	C_d	C_dq, lb/ft²	C_dp, lb/ft²	C_dV, kips/ft
0	0	0	0	0	0	0	0	0	0
10	0.212	0.191	451	163	1.04	1.6	722	261	1.66
20	0.424	0.346	817	295	2.60	1.7	1389	502	4.42
30	0.635	0.470	1110	401	4.59	1.7	1887	682	7.80
40	0.847	0.571	1348	487	6.91	1.8	2426	877	12.44
50	1.059	0.653	1542	557	9.50	1.8	2776	1003	17.10
60	1.271	0.719	1698	613	12.31	1.9	3226	1165	23.39
70	1.483	0.773	1825	659	15.30	1.9	3468	1252	29.07
80	1.694	0.816	1927	695	18.44	1.9	3661	1321	35.04
90	1.906	0.851	2009	725	21.70	1.9	3817	1378	41.23
100	2.118	0.880	2078	750	25.03	1.9	3948*	1425*	47.56
110	2.330	0.903	2132	770			3061	1105	
120	2.541	0.921	2174	785			2174	785	

*Design pressures are reduced linearly from this level to basic pressures at bottom of hopper.

TABLE 14 Hoop Reinforcement for Silo of Example 1

	Design pressure	Hoop forces		Steel required	Steel provided	
Depth Y, ft	p_{des}, lb/ft²	F, kips/ft	F_u, kips/ft	A_s, in.²/ft	A_s, in.²/ft	Spacing
0	0	0	0	0	0.31	No. 5 at 12 in.
10	261	3.92	6.66	0.12		
20	502	7.53	12.80	0.24		
30	682	10.23	17.39	0.32		
40	877	13.16	22.37	0.41	0.47	No. 5 at 8 in.
50	1003	15.05	25.59	0.47		
60	1165	17.48	29.72	0.55	0.62	No. 5 at 6 in.
70	1252	18.78	31.93	0.59		
80	1321	19.82	33.69	0.62	0.68	No. 5 at 5½ in.
90	1378	20.67	35.14	0.65		
100	1425	21.38	36.35	0.67		

ECCENTRIC DISCHARGE. If the silo of this example is provided with an inclined hopper slab (Fig. 2f) instead of a central conical hopper, the design pressures should be determined by Eq. (5). Since the outlet is at the wall, $e = r$ and $p_{des} = C_dp + 0.25pY/H$, where p and C_dp are given in Table 13. These pressures are compared in Table 15 with the values from Table 13 for a central outlet.

Example 2 Determine the maximum crack width in the walls of the silo of Example 1. Maximum effects can be expected just above the top of the hopper (Y = 100 ft). From the data of Table 13 the hoop tensions for Eqs. (18) are

$$F_{st} = pD/2 = 750 \times 30/2 = 11{,}250 \text{ lb/ft}$$
$$F_{tot} = p_{des}D/2 = 1425 \times 30/2 = 21{,}375 \text{ lb/ft}$$

From Table 14, A_s = 0.68 in.²/ft. Then

$$f_{s,st} = 11{,}250/0.68 = 16{,}640 \text{ psi}$$
$$f_{s,tot} = 21{,}375/0.68 = 31{,}620 \text{ psi}$$

From Eqs. (18), with $f'_t = 4.5\sqrt{f'_c} = 4.5\sqrt{4000} = 285$ lb/in., $\beta = 0.7$ for deformed bars, and $A = 6 \times 12 = 72$ in.²

$$\psi_1 = 1 - \frac{0.7 \times 0.8 \times 72 \times 285}{21{,}375} = 0.462$$
$$\psi_2 = 1 - \frac{0.7 \times 0.8 \times 72 \times 285}{11{,}250} = -0.02 < 0.3;\ \text{use } 0.3.$$
$$\psi_3 = 1 - \frac{0.35 \times 0.8 \times 72 \times 285}{11{,}250} = 0.489 < 0.65;\ \text{use } 0.65.$$

For No. 5 bars at 5½ in. Σo = 1.96 × 12/5.5 = 4.28 in./ft

$$w_1 = \frac{0.462 \times 72 \times 0.7 \times 31{,}620}{4.28 \times 29 \times 10^6} = 0.00593 \text{ in.}$$
$$w_2 = \frac{0.3 \times 72 \times 0.7 \times 16{,}640}{4.28 \times 29 \times 10^6} = 0.00203 \text{ in.}$$
$$w_3 = \frac{0.65 \times 72 \times 0.7 \times 16{,}640}{4.28 \times 29 \times 10^6} = 0.00439 \text{ in.}$$

Crack width at Y = 100 ft is given by Eq. (17)

$$w_{cr} = w_1 - w_2 + w_3 = 0.00829 \text{ in.}$$

This crack width is 3.6 percent larger than the suggested limit 0.008 in., but may be acceptable. Reducing the spacing to 5 in. would reduce the crack width to 0.00829 × 5/5.5 = 0.0075 in.

TABLE 15 Pressures with Concentric Outlet and Eccentric Outlet for Silo of Example 1

	p_{des}	
Y, ft	Concentric	Eccentric
0	0	0
10	261	265
20	502	517
30	682	712
40	877	926
50	1003	1073
60	1165	1257
70	1252	1367
80	1321	1460
90	1378	1541
100	1425	1613

Example 3 Determine the pressures and forces for a flat-bottomed silo 30 ft in diameter and 120 ft high to contain wheat, using Pieper's recommendations (Art. 3).

Assume the outlet diameter d = 2 ft. From Table 3, c = 0.85 for unfinished concrete walls. From Fig. 6,

$$\alpha_2 = 32(\sqrt[4]{\rho_f c}) = 32\sqrt[4]{31 \times 0.85} = 72.5°$$
$$h_{F2} = \frac{D - d}{2}\tan\alpha_2 = \frac{30 - 2}{2}\tan 72.5° = 44.4 \text{ ft}$$

Since H = 120 > $2h_{F2}$, the silo is classified as tall. Therefore, the design is governed by emptying pressures (Art. 3).

The data and formulas are shown in Fig. 27 and the calculated pressures and forces in Table 16. Comparison of values in this table shows that the largest lateral pressures are generated during emptying and the largest vertical pressures during filling.

Over the height h_{F2} = 44.4 ft above the outlet, emptying pressures vary linearly from the value of p_e at 44.4 ft (Y = 75.6 ft) to p_f at the level of the outlet, while increases occur in the region 1.4D = 1.4 × 30

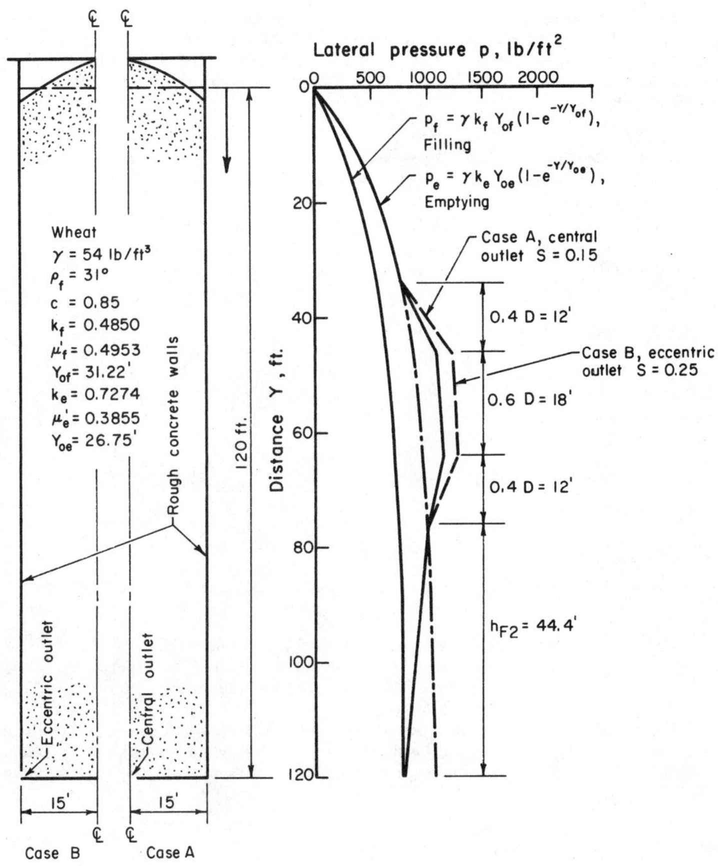

Fig. 27 Silos of Example 3.

TABLE 16 Pressures and Forces for Silo of Example 3

	Filling					Emptying			
Y, ft	Y/Y_{0f}	$1 - e^{-Y/Y_{0f}}$	q_f, lb/ft²	p_f, lb/ft²	V, kips/ft	Y/Y_{0e}	$1 - e^{-Y/Y_{0e}}$	q_e, lb/ft²	p_e, lb/ft²
0	0	0	0	0	0	0	0	0	0
10	0.320	0.274	463	224	0.58	0.374	0.312	451	328
20	0.641	0.473	797	387	2.12	0.748	0.527	761	553
30	0.961	0.618	1041	505	4.34	1.122	0.674	974	708
40	1.281	0.722	1218	591	7.07	1.495	0.776	1121	815
50	1.602	0.798	1346	653	10.16	1.869	0.846	1222	889
60	1.922	0.854	1439	698	13.51	2.243	0.894	1291	939
70	2.422	0.894	1507	731	17.05	2.617	0.927	1339	974
80	2.563	0.923	1556	755	20.73	2.991	0.950	1372	998
90	2.883	0.944	1592	772	24.51	3.365	0.965	1395	1014
100	3.203	0.959	1618	785	28.37	3.738	0.976	1410	1026
110	3.523	0.971	1636	794	32.28	4.112	0.984	1421	1034
120	3.844	0.979	1650	800	36.23	4.486	0.989	1428	1039

= 42 ft above the 44.4-ft level (Fig. 7*c*). From Art. 3, values of *S* are taken at 0.15 for a central outlet (Case *A* of Fig. 27) and 0.25 for an outlet at the wall (Case *B*). Final lateral design pressures are given in Table 17. The horizontal distribution of these additional pressures is shown in Fig. 7*d*. The pressures in Case A produce additional hoop forces, while Case B produces bending moments as well as additional hoop forces (Fig. 11).

TABLE 17 Final Lateral Design Pressure, Example 3

Depth Y, ft	Basic pressure p_e, lb/ft²	Central outlet p_e, lb/ft²	Eccentric outlet p_e, lb/ft²
0	0	0	0
10	328	328	328
20	553	553	553
30	708	708	708
33.6	751	751	751
45.6	860	989	1075
63.6	953	1096	1191
75.6	988	988	988
80	998	969	969
90	1014	927	927
100	1026	885	885
110	1034	842	842
120	1039	800	800

Example 4 Design a conical hopper for the silo of Example 1 according to ACI 313 (Art. 4). The hopper is 8 in. thick, 20 ft high, and has top and bottom diameters of 30 and 3 ft, respectively.

The tangential and meridional forces are given by Eqs. (37), (38), and (39), with W_h and W_m by Eqs. (40) and (41). Values of q_α are computed by Eq. (36).

Results are given in Table 18. Areas of reinforcement per unit length required for strength are given in the last two columns. Some adjustment upward was necessary to satisfy a crack width $w < 0.008$ in. This was easily accomplished by increasing the length of the meridional bars and reducing the spacing of the hoop bars.

TABLE 18 Design of Conical Hopper for Silo of Example 1

Depth Y, ft	Basic pressures q, lb/ft²	Basic pressures p, lb/ft²	Design pressures q_{des}, lb/ft²	Design pressures p_{des}, lb/ft²	Design pressures $q_{\alpha,des}$, lb/ft²	Tangential force F_{tu}, lb/ft	Meridional force F_{mu}, lb/ft	Tangential steel $A_{s,t}$, in.²/ft	Meridional steel $A_{s,m}$, in.²/ft
100	2078	750	3117	1125	1725	52,600	55,400	0.97	1.03
105	2105	760	3158	1140	1748	41,300	42,500	0.76	0.79
110	2132	770	3198	1155	1771	29,700	29,800	0.55	0.55
115	2154	778	3231	1167	1789	17,700	17,400	0.33	0.32
120	2174	785	3261	1178	1806	5,500	5,000	0.10	0.09

RING BEAM (Art. 25). From Table 18, F_{mu} = 55.4 kips at the top of the hopper. From Fig. 28, the slope of the hopper wall is 56°.7. Therefore, the ring compression is

$$P = (F_{mu} \cos \alpha)(D/2) = (55.4 \cos 56°.7)(30/2) = 456 \text{ kips}$$

The ring shown in Fig. 28 is 15 × 20 in. with 8 No. 6 bars. Then

$$P_u = \phi[0.85f'_c(A_g - A_s) + f_yA_s]$$
$$= 0.7[0.85 \times 4(300 - 3.52) + 60 \times 3.52] = 853 \text{ kips}$$

Although P_u is considerably larger than P, use of a smaller ring is questionable. The 15-in. width gives projections to facilitate forming, and the depth gives bending strength to bridge openings that might later be cut into the bearing wall.

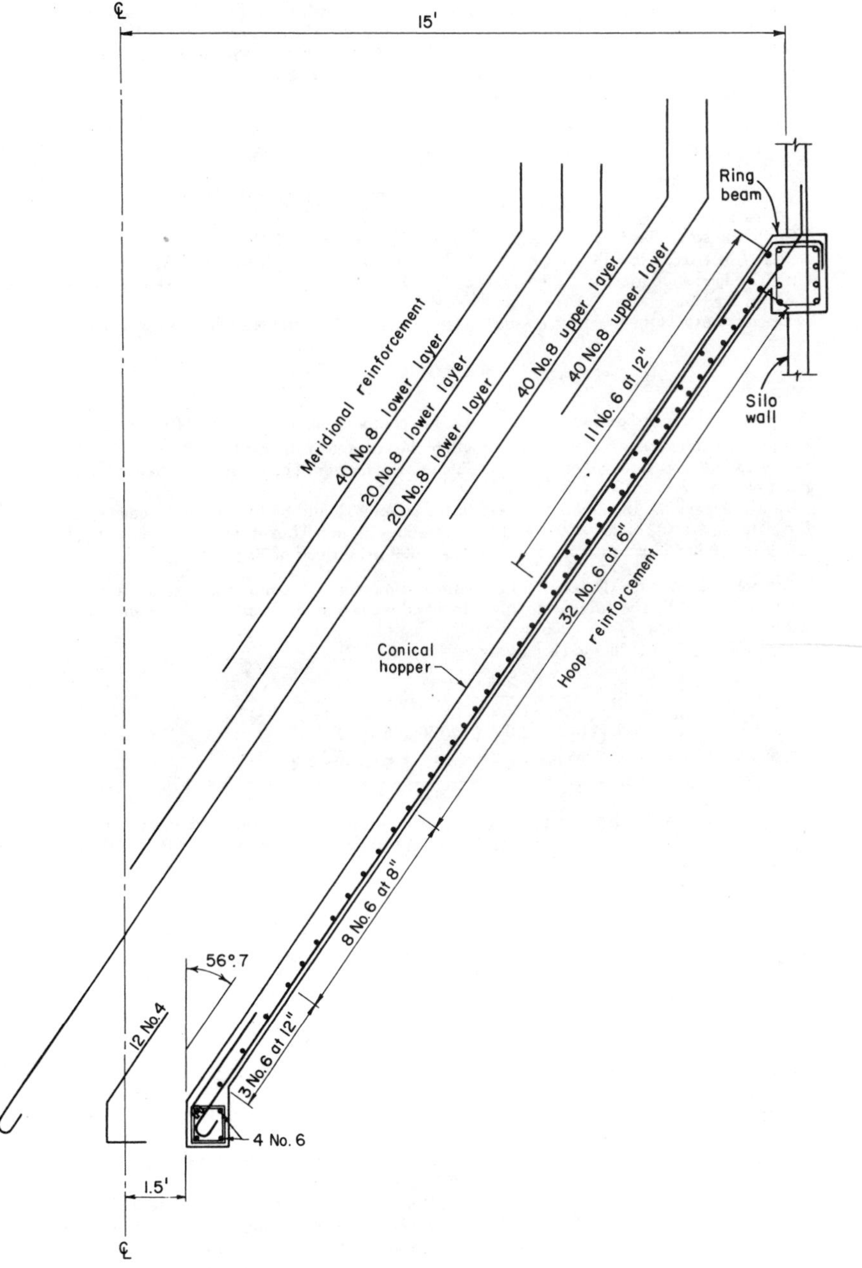

Fig. 28 Conical hopper and ring beam, Example 4.

Example 5 A 20-in.-square reinforced-concrete column with 4 No. 18 bars and No. 4 ties at 12 in. is one of a group supporting a silo hopper as shown in Fig. 2*e*. The column loads are $D = 150$ kips and $L = 600$ kips. $f'_c = 3$ ksi, $f_y = 60$ ksi, $E_c = 3000$ ksi, and $E_s = 29{,}000$ ksi. Unloading of the silo may occur a long time after filling. Check the suitability of the design (Art. 26).

The initial strain ϵ_i in the concrete is

$$\epsilon_i = \frac{D + L}{[A_g + (E_s/E_c - 1)A_s]\,E_c}$$
$$= \frac{150 + 600}{[20 \times 20 + (29{,}000/3000 - 1)16]3000} = 0.000464$$

Assume the strain has trebled because of creep. Thus $\epsilon_t = 3 \times 0.000464 = 0.0014$. The stress in the steel is $f_s = \epsilon_t E_s = 0.0014 \times 29{,}000 = 40.6$ ksi and can support a load $P_s = f_s A_s = 40.6 \times 16 = 649.6$ kips. The load supported by the concrete is $P_c = D + L - P_s = 150 + 600 - 649.6 = 100.4$ kips. The stress in the concrete is $P_c/(A_g - A_s) = 100.4/(20 \times 20 - 16) = 0.261$ ksi.

Upon removal of the live load, elastic unloading occurs. The strain $\Delta\epsilon$ which is recovered is given by

$$\Delta\epsilon = \epsilon_i \frac{L}{D + L} = 0.000464 \times \frac{600}{150 + 600} = 0.000371$$

The stress in the concrete is reduced from 0.261 in compression to $0.261 - 0.000371 \times 3000 = 0.852$ ksi in tension. This exceeds the cracking strength of the concrete, estimated at $4.5\sqrt{f'_c} = 4.5\sqrt{3000} = 246$ psi. As a result, concrete will crack during unloading of the silo. The stress in the steel after unloading is given by $f_s = D/A_s = 150/16 = 9.38$ ksi.

To avoid cracking, the amount of steel in the column should be limited to the smaller of $0.02A_c$ or L/f_y. This gives $0.02 \times 20 \times 20 = 8$ in.^{2} and $600/60 = 10$ in.^{2}. Thus A_s for this column should be 8 in.^{2} instead of 16 in.^{2} This reduction requires $f'_c = 4000$ psi instead of 3000 psi.

Example 6 (See Art 10) Determine the thermal reinforcement required for a concrete silo in a region where $T_0 = -20$°F. The silo has doubly reinforced walls 8 in. thick and stores cement for which $T_i = 400$°F. $f'_c = 4000$ psi, $f_y = 60{,}000$ psi.

Using Eqs. (15*b*), (14), and (13) gives

$$K_t = \frac{0.08 \times 8}{4.09 + 0.08 \times 8} = 0.135$$
$$\Delta T = [400 - 80 - (-20)]0.135 = 46°$$
$$M_t = \frac{57{,}000\sqrt{4000} \times 8^2 \times 5.5 \times 10^{-6} \times 46}{1 - 0.3} = 83{,}200 \text{ ft-lb}$$

Assume net covers of 1.5 and 0.75 in., respectively, for the exterior and interior layers of hoop reinforcement. If No. 5 bars are used, $d = 8 - 1.5 - 0.62/2 = 6.2$ in. and $d' = 0.75 + 0.62/2 = 1.1$ in. Then from Eq. (35)

$$A_{s,t} = \frac{1.4 \times 83{,}200}{0.9 \times 60{,}000(6.2 - 1.1)} = 0.42 \text{ in.}^2/\text{ft}$$

Place this amount in the outer layer and, for the sake of equal spacing, the same amount in the inner layer.

REFERENCES

1. Janssen, H. A.: Versuch über Getreide-druck in Silozellen, *VDI Z.* (Düsseldorf), vol. 39, Aug. 31, 1895.
2. Theimer, O. F.: Failures of Reinforced Concrete Grain Silos, *ASME Publ.* 68-MH-36, New York, 1968.
3. Kovtun, A. P., and P. N. Platonov: The Pressure of Grain on Silo Walls, *Mukomol'no Elevat. Promst.*, Moscow, USSR, vol. 25, no. 12, December 1959.
4. Pieper, K., G. Mittelman, and F. Wenzel: Messungen des Horizontalen Getreidedruckes in einer 65 m hohen Silozelles, *Beton und Stahlbetonbau*, Berlin, vol. 59, no. 1, November 1964.
5. Pieper, K.: Investigation of Silo Loads on Measuring Models, *J. Eng. Ind. Trans. ASME*, May 1969.
6. Reimbert, Marcel, and André Reimbert: "Silos—Traité Theoretique et Practique," Editions Eyrolles, Paris, 1961.
7. Jenike, A. W., and J. R. Johanson: Bin Loads, *J. Struct. Div. ASCE*, April 1968.
8. Fischer, W.: "Silos und Bunker in Stahlbeton," VEB Verlag für Bauwesen Berlin, DDR, 1966.
9. Lipnitski, M. E., and S. P. Abramovitsch: "Reinforced Concrete Bunkers and Silos" (in Russian), Izdatelstvo Literaturi Po Stroitelstvu, Leningrad, 1967.
10. Turitzin, A. M.: Dynamic Pressure of Granular Material in Deep Bins, *J. Struct. Div. ASCE*, April 1963.

11. Homes, A. G.: Lateral Pressures of Granular Materials in Silos, *ASME Publ.* 72-MH-30, New York, 1972.
12. "Lastannahmen für Bauten. Lasten in Silozellen," DIN 1055 Sheet 6, November 1964. Also supplementary provisions, May 1977.
13. ACI 313–77: Recommended Practice for Design and Construction of Concrete Bins, Silos, and Bunkers for Storing Granular Materials, *J. ACI,* October 1975.
14. Pieper, K., Technische Universität Braunschweig, unpublished work in private communication to E. H. Gaylord, 1977.
15. Gurfinkel, G.: Collapse Investigation of Inclined Hopper of Reinforced Concrete Silo in Shellburn, Indiana, Report for Sullivan County Farm Bureau Co-op, Sullivan, Ind., March 1976.
16. Ukazania P_0 Proectirovaniu Silosov Dlia Siputschich Materialov (Instructions for Design of Silos for Granular Materials), Soviet Code CH-302-65, Gosstroy, USSR, Moscow, 1965.
17. Safarian, S. S.: Design Pressures of Granular Material in Silos, *J. ACI,* August 1969.
18. Kellner, M.: Silos à Cellules de Grande Profondeur, *Travaux,* October 1960.
19. Timoshenko, S., and S. Woinowski-Krieger: "Theory of Plates and Shells," 2d ed., McGraw-Hill Book Company, New York, 1959.
20. Rectangular Concrete Tanks, Portland Cement Association, IS003.020, Chicago, 1969.
21. Gurfinkel, G.: Thermal Effects in Walls of Nuclear Containments, Elastic and Inelastic Behavior, *Proc. 1st Int. Conf. Structural Mechanics in Reactor Technology,* vol. 5, part J, Berlin, Germany, September 1972.
22. Building Code Requirements for Reinforced Concrete, ACI 318-77, American Concrete Institute, Detroit.
23. Ciesielski, R., et al.: Behalter, Bunker, Silos, Schornsteine, Fernsehturme und Freileitungsmaste, Wilhelm Ernst & Sohn KG, Berlin, 1970.
24. Schütt, H.: Über das Tragvermögen wandartiger Stahlbetonon und Stahlbetonbau, Beton and Stahlbetonbau, October 1956.
25. Gurfinkel, G.: Investigation of Silos at Seneca, Illinois, Report for Continental Grain Co., Regional Office in Chicago, Ill., October 1974.
26. Discussion of ACI 313–77, *J. ACI,* June 1976.
27. Safarian, S. S.: Design of a Circular Concrete Ring-Beam and Column System Supporting a Silo Hopper, *J. ACI,* February 1969.
28. Gurfinkel, G.: Structural Adequacy of Reinforced Concrete Silo Complex in Leverett, Illinois, Report for Thomasboro Grain Co., Thomasboro, Ill., December 1976.
29. Theimer, O. F.: Cause and Prevention of Dust Explosions in Grain Elevators and Flour Mills, *ASME Publ.* 72-MH-25, New York, 1972.

Appendix

TABLE A1 Torsional Properties of Solid Cross Sections*

Cross section	Torsional stiffness J	Shear stress
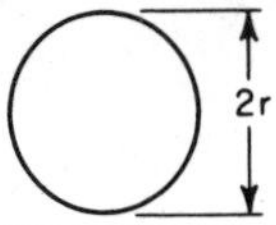 Circle, diameter 2r	$\frac{1}{2}\pi r^4$	$\frac{2T}{\pi r^3}$
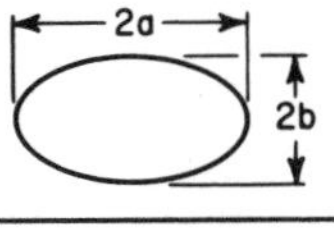 Ellipse, axes 2a, 2b	$\frac{\pi a^3 b^3}{a^2+b^2}$	$\frac{2T}{\pi ab^2}$ at ends of minor axis
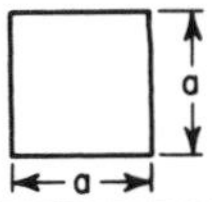 Square, side a	$0.141a^4$	$\frac{T}{0.208a^3}$ at midpoint each side
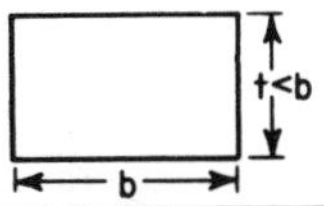 Rectangle, b × t, t<b	$\frac{bt^3}{3}\left[1-0.63\frac{t}{b}+0.052\left(\frac{t}{b}\right)^2\right]$	$\frac{3T}{bt^2}\left(1+0.6\frac{t}{b}\right)$ at midpoint each long side
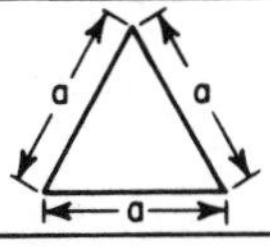 Equilateral triangle, side a	$\frac{a^4\sqrt{3}}{80}$	$\frac{20T}{a^3}$ at midpoint each side

* $T = GJ\theta$, where T = torque, G = shearing modulus of elasticity, J = torsional stiffness, θ = angle of twist, radians per unit length.

TABLE A2 Torsional Properties of Closed Thin-walled Sections*

Cross section	Torsional stiffness J	Shear stress
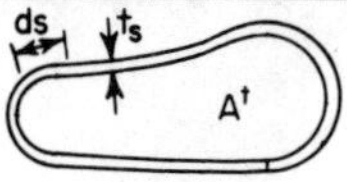 ds, t_s, A†	$\frac{4A^2}{\int ds/t_s}$	$\frac{T}{2At}$
 r	$2\pi r^3 t$	$\frac{T}{2\pi r^2 t}$
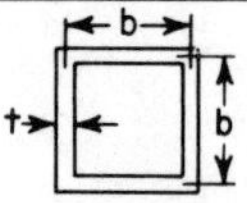 b, t, b	$b^3 t$	$\frac{T}{2b^2 t}$
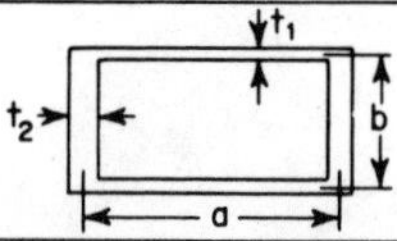 t_1, t_2, b, a	$\frac{2a^2b^2}{\frac{a}{t_1}+\frac{b}{t_2}}$	$\frac{T}{2abt}$

* $T = GJ\theta$, where T = torque, G = shearing modulus of elasticity, J = torsional stiffness, θ = angle of twist, radians per unit length.

† A = area bounded by midline of wall.

NOTE: Warping constant C is usually negligible for closed thin-walled cross sections.

TABLE A3 Torsional Properties of Open Cross Sections

Cross section	Location e of shear center S	Warping constant C
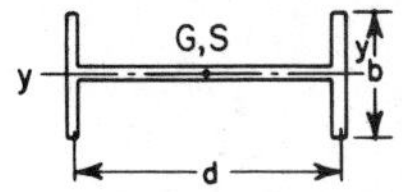 y, G,S, b, d		$\frac{d^2 I_y}{4}$
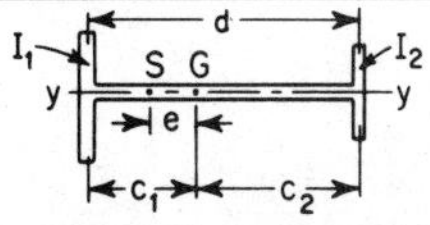 I_1, I_2, S G, y, e, d, c_1, c_2	$\frac{c_1 I_1 - c_2 I_2}{I_y}$	$\frac{d^2 I_1 I_2}{I_y}$
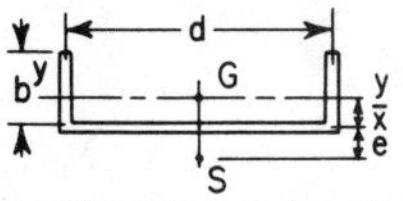 d, b, y, G, $\bar{x}$, e, S	$\frac{\bar{x}}{4}\left(\frac{d}{r_x}\right)^2$	$\frac{d^2 I_y}{4}\left[1 - \frac{\bar{x}(e-\bar{x})}{r_y^2}\right]$
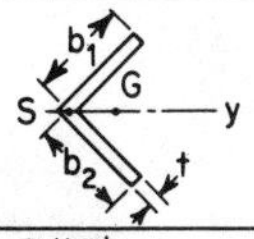 b_1, b_2, G, S, y, t		$(b_1^3 + b_2^3)\frac{t^3}{36}$
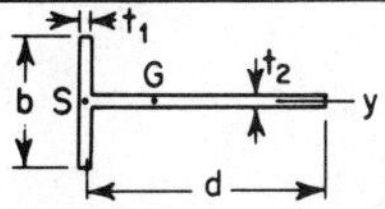 t_1, t_2, b, S, G, y, d		$\frac{t_1^3 b^3}{144} + \frac{t_2^3 d^3}{36}$
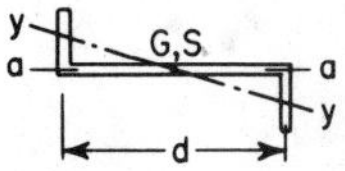 y, a, G,S, d		$\frac{d^2}{4} I_a$

NOTE: The torsional stiffness J for cross sections in this table can be determined closely enough for most applications by $J = \Sigma bt^3/3$. The warping constant C is usually negligible for the angle and the T.

TABLE A4 Effective Length Coefficients for Columns

Case	Ends	K
	er–fr	$\left(\frac{10}{\beta_1}+4\right)^{1/2}$
	er–fx	$\frac{4+\beta_1}{2+\beta_1}$
	er–p	$\frac{3+0.7\beta_1}{3+\beta_1}$
	er–er	$\frac{2+0.5\beta_1}{2+\beta_1}$
	er–fx	$\frac{2.1+0.5\beta_1}{3+\beta_1}$
Central elastic support 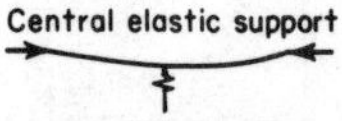	p–et–p	$\frac{1}{(1+\beta_2)}$ if $\beta_2<3$
	p–et–p	0.5 if $\beta_2>3$
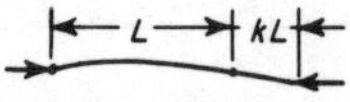 L, kL	p–p–fr	1+2k
	fx–p–fr	0.7+2k
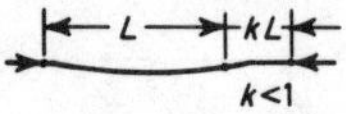 L, kL, k<1	p–p–p	0.7+0.3k
	fx–p–fx	0.5+0.2k
 P, L, L, P_o	p–p–p	$0.9+0.1\frac{P_o}{P}$
Central load, 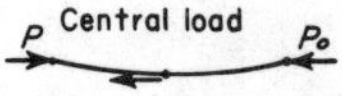 P, P_o	p–p	$0.75+\frac{P_o}{4P}$
Intermediate loads, 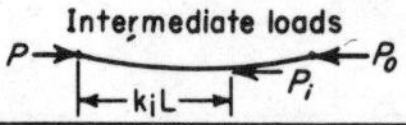 P, P_o, P_i, k_iL	p–p	$\left(\frac{\Sigma P_i k_i}{\Sigma P_i}\right)^{1/2}$

NOTE: p = pinned, fx = fixed, fr = free, er = elastic rotational restraint, et = elastic translational restraint.

$\beta_1 = \dfrac{\alpha_1 L}{EI}$ where α_1 = moment to produce 1 radian rotation

$\beta_2 = \dfrac{\alpha_2 L^3}{53EI}$ where α_2 = load to produce unit deflection of the central support

* From "The Strength of Aluminum," Aluminum Company of Canada, Ltd., 1965.

TABLE A5 Buckling of Plate under Edge Stress, Four Edges Simply Supported*

Case	Range of application		Buckling coefficient $k^{\dagger}$
1	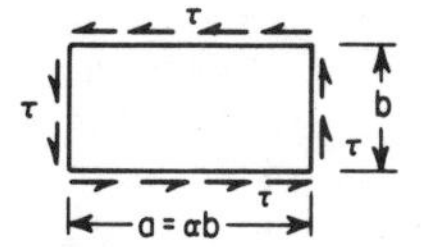	$\alpha \geq 1$	$5.34 + \frac{4}{\alpha^2}$
		$\alpha \leq 1$	$4.00 + \frac{5.34}{\alpha^2}$
2	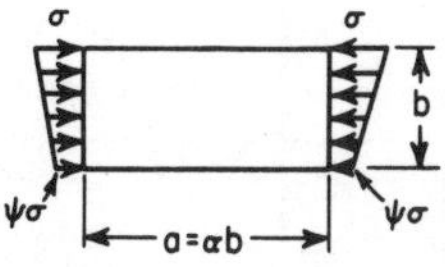 $0 \leq \psi \leq 1$	$\alpha \geq 1$	$\frac{8.4}{1.1+\psi}$
		$\alpha < 1$	$\frac{2.1}{1.1+\psi}\left(\alpha + \frac{1}{\alpha}\right)^2$
3	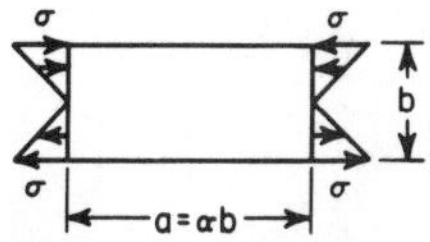	$\alpha \geq \frac{2}{3}$	23.9
		$\alpha < \frac{2}{3}$	$15.87 + \frac{1.87}{\alpha^2} + 8.6\alpha^2$
4	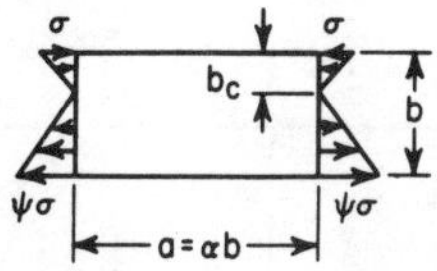 $\psi \geq 1$	Same as Case 3 except use $b = 2b_c$ to compute α and b/t	
5	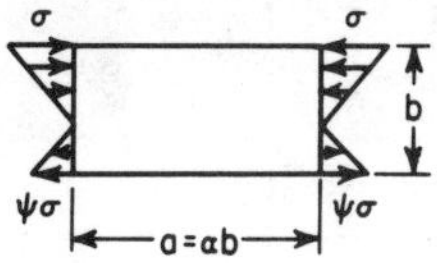 $0 \leq \psi \leq 1$	$k = (1-\psi)k_2 + \psi k_3 - 10\psi(1-\psi)$ where $k_2 = k$ for Case 2 with $\psi = 0$ $k_3 = k$ for Case 3	

* From German Specification DIN4114.

† Critical (buckling) stress σ (or τ) $= \dfrac{k\pi^2 E}{12(1-\nu^2)(b/t)^2}$.

TABLE A6 Stiffened Beam Webs

Case	λ	Buckling coefficient		Stiffener	
		k	Range	I_s/bt^3	Range
1	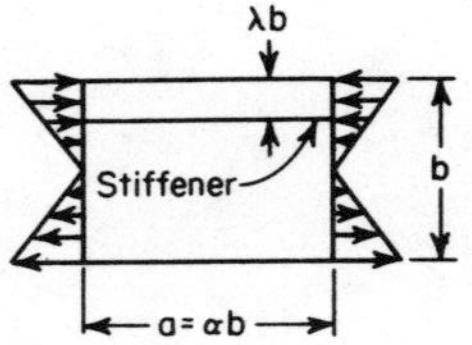 1/2	35.6	$\alpha \geq \frac{2}{3}$	0.12	
	1/4	101	$\alpha \geq 0.4$	$0.31\left[3.7\alpha^2\left(1+4\frac{A_s}{A_w}\right)-\alpha^3\right]$	$\alpha \leq 1.6$
				$0.22\left(1+7.7\frac{A_s}{A_w}\right)$	$\alpha \leq 0.5$
				$1.1\left(1+7.7\frac{A_s}{A_w}\right)(\alpha-0.3)$ but need not exceed $1.47\left(1+12.5\frac{A_s}{A_w}\right)$	$\alpha \geq 0.5$
	1/5	129		$1.15\left[0.4+\left(1+6\frac{A_s}{A_w}\right)\alpha^2\right]$	$0.5 \leq \alpha \leq 1.5$
				$0.355+0.47\alpha+0.81\alpha^2\left(1+8.8\frac{A_s}{A_w}\right)$	$0.5 \leq \alpha \leq 1.5$
2	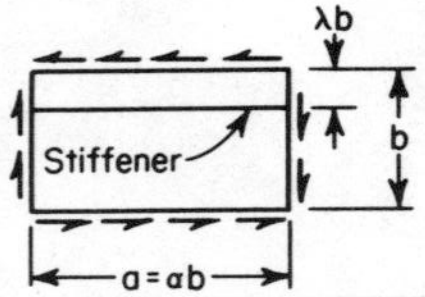 1/2	Table A6, Case 1		$0.5\alpha^2(-1+2\alpha+2.5\alpha^2-\alpha^3)$	$0.5 \leq \alpha \leq 2$
	1/4	Table A6, Case 1		$0.66\alpha^2(1-3.3\alpha+3.9\alpha^2-1.1\alpha^3)$	$0.5 \leq \alpha \leq 2$
3	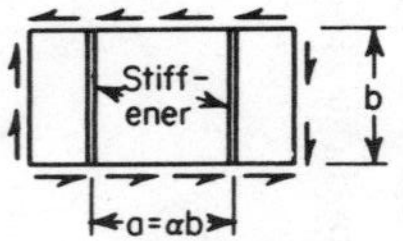	Table A6, Case 1		$0.37\left(\frac{7}{\alpha}-5\alpha\right)$	

NOTE: t = web thickness; I_s = moment of inertia of stiffener (for one-sided stiffeners usually taken at face of web); A_s = cross-sectional area of stiffener; A_w = area of web = bt.

On the basis of tests, Massonnet suggests that, to keep longitudinal stiffeners practically straight to collapse of girder, the theoretical values of I_s be multiplied by n, where n = 3 for $\lambda = \frac{1}{2}$, 4 for $\lambda = \frac{1}{3}$, 6 for

Index